D1301107

Geophysical Monograph Series

American Geophysical Union

Geophysical Monograph Series

A. F. SPILHAUS, JR., *managing editor*

The Structure and Physical
Properties of the Earth's Crust

geophysical monograph 14

The Structure and Physical

Properties of the Earth's Crust

JOHN G. HEACOCK editor

American Geophysical Union
Washington, D. C.
1971

Published with the aid of a grant
from the Charles F. Kettering Foundation

International Standard Book Number 87590-014-3

Copyright © 1971 by the American Geophysical Union
2100 Pennsylvania Avenue, N. W.
Washington, D. C. 20037

Library of Congress Catalog Card Number 75-182370

WILLIAM BYRD PRESS, RICHMOND, VIRGINIA

Preface

This monograph is based on the proceedings of a Symposium on the Structure and Physical Properties of the Earth's Crust, held July 27 through 31, 1970, at the University of Colorado. This meeting was jointly sponsored by the Office of Naval Research (ONR) and the Cooperative Institute for Research in the Environmental Sciences (CIRES); the latter was formed by the University of Colorado and the Environmental Research Laboratories of the National Oceanographic and Atmospheric Administration, Department of Commerce. Thirty-one papers were presented, the majority of which are published in this volume.

The Symposium was initiated by the editor with a special view toward answering fundamental environmental questions about the nature of the earth's crust, particularly at depths below a few kilometers, and in adjacent portions of the upper mantle. Conditions at these depths have rarely been the subject of multidisciplinary study in the past.

Participants were invited from a broad range of specialists in appropriate areas of geophysics, physics, geochemistry, and geology. The number of participants was intentionally kept small (some fifty scientists) to ensure effective, informal interaction. A practical question—the possible existence of a lithospheric electromagnetic wave guide—provided a useful sense of direction to the symposium presentations and discussions.

The editor's work was greatly assisted by the editorial board and readers, whose services are sincerely appreciated. Special acknowledgment is extended to J. C. Harrison, Director of CIRES, who was responsible for the excellent meeting arrangements. In addition, he undertook the major task of compiling the taped discussions.

The work done in the past through the Upper Mantle Project has led us to recognize the importance of the lithosphere in any solution of the mysteries of plate tectonics and processes occurring within the mantle. The origin and evolution of the lithosphere has been identified as a major focal theme in connection with the United States program for the Geodynamics Project.

This monograph constitutes a current review of various important aspects of the upper portion of the lithosphere. It is hoped that the volume will serve as a stimulus to further efforts to understand this important part of the solid-earth environment.

JOHN G. HEACOCK

EDITOR

vii

Contents

1. INTRODUCTION

Intermediate and Deep Properties of the Earth's Crust, a Possible Electromagnetic Wave Guide

JOHN G. HEACOCK

Earth Physics Program
Office of Naval Research, Arlington, Virginia   22217

Abstract. The idea that an electromagnetic wave guide may exist in the earth was postulated some 10 years ago on the assumption that conductive, wet surface layers and conductive, hot mantle layers enclose a resistive zone of dry 'basement' rocks. Little is known in detail about the precise nature of the materials and physical properties of the deeper parts of the earth's crust, where the resistive 'basement' rocks were presumed to exist. If a zone of sufficiently resistive rocks exists in the deeper continental crust, the case is argued on the basis of evidence from seismic data that continuous zones of resistive rocks may exist and may continue beneath tectonic belts and geologic province boundaries relatively undisturbed. Recent studies of crustal resistivities suggest that, at least in stable continental areas, magnitudes of 10^6 to 10^7 ohm-m may exist; the latter value is high enough to give attenuations of less than 0.1 db/km in wave guide propagation. The need for conducting laboratory studies in coordination with geophysical field programs in order to interpret the materials, processes, and physical properties in the deeper parts of the earth's crust is emphasized. Major benefits to mankind can result for a variety of reasons from an improved understanding of this part of our environment.

Continuous zones may exist in the deeper parts of the earth's continental crust that could form an electromagnetic wave guide and provide new pathways for communication. To form such a wave guide, these earth zones must not only be continuous but also have a resistive interior bounded by conducting walls. A high interior resistivity of 10^6 or 10^7 Ω-m is necessary if attenuations appreciably less than 1 db/km and significant ranges are to be achieved (Wait, this volume; Wait and Spies, this volume).

Neither the magnitude of deep earth resistivities nor their continuity over extended regions has been established, since the extensive data required to supply this information are not available. The purpose of this paper, however, is to determine from existing data whether indications are favorable for the possible existence of an electromagnetic wave guide in the earth.

LITHOSPHERIC WAVE GUIDE

Wait [1954] first noted that electromagnetic energy might propagate through earth layers with low attenuation provided a wave guide configuration could be found. As postulated by *Wheeler* [1961], approximate wave guide conditions may exist in the earth because surface layers are made conductive by the presence of ground water; the dielectric resistive interior of the wave guide is formed by dry basement rocks, and the lower conductive wall is formed by heated rocks in the lower crust or upper mantle. Later reviews emphasizing assumed resistivity-depth profiles in the earth are based on

1

reasonable estimates from laboratory and field data. Papers on this subject were published by *Levin* [1962, 1966], *Ames et al.* [1963], *Keller* [1963], and others.

Electromagnetic propagation through an earth wave guide has been studied by *Ghose* [1960], *Wait* [1963a, b, 1966a, b, c, d], *Schwering et al.* [1968], *Tsao and de Bettencourt* [1968], and others. Wait (this volume) reviews the basic mathematics of the major aspects of these treatments; see also *Wait* [1963b]. *Ryazantsev and Shabel'nikov* [1965] wrote an extensive review of papers by United States authors on this subject. *Dolukhanov* [1970] discusses several types of underground communication and speculates that underground radio propagation will find many applications as technical development proceeds.

The essential element lacking in this field is a clear knowledge of the physical properties of the earth at the depths of interest; in particular, we must determine: first, the distribution of resistivity and permittivity with depth; second, whether the wave guide conditions postulated by Wheeler exist; and third, where these conditions are continuous and the ranges over which energy will propagate effectively. The other articles in this volume review the current state of our knowledge of the crust on a broad multidisciplinary basis. They provide the building blocks that will ultimately enable us to evaluate not only the validity of the wave guide concept but also the nature of the earth's crustal environment.

CONTINUITY REQUIREMENT FOR WAVE GUIDE

Earlier analyses for crustal wave guides have assumed an average resistivity-depth profile. Yet, most geologists and geophysicists strongly question the possibility that zones of continuous physical properties extend over significant ranges in the earth's crust. Attempts to trace seismic reflection horizons have rarely succeeded over distances of more than a few kilometers [*James and Steinhart*, 1966]. Surface geology and geological province boundaries are clear evidence for strong lateral discontinuities. Even subsurface structures and stratigraphic pinchouts from shallow crustal data support the concept that the crust is broken and disturbed laterally.

Rather recently, however, a sequence of papers has been published that describe a widespread seismic low-velocity zone in the continental crust that occurs in a fairly characteristic depth range at approximately 10-km depth. These results tend to provide a new picture of crustal structure at 10 km and below and suggest that there is an increasing tendency toward continuity of various physical properties at these depths in the crust.

CRUSTAL SEISMIC LOW-VELOCITY ZONES

The concept of a low-velocity zone in the continental crust was first discussed critically by *Gutenberg* [1950, 1951, 1954]. It was not widely accepted then nor is it yet. The goal of this study is not to argue the existence of the crustal low-velocity zone. Instead, to simplify the argument, the existence of these zones is assumed as described elsewhere in the literature and as summarized briefly below.

The question of a crustal seismic low-velocity zone was recently revived by *Mueller and Landisman* [1966], who identify the characteristics of an arrival they call P_c. This arrival is associated with wide-angle reflections and refractions from the bottom of a shallow seismic low-velocity zone [*Mueller and Landisman*, 1966, Figure 2b, p. 531; or see Figure 2 of Landisman et al., p. 14, this volume]. This shallow or sialic low-velocity zone [*Mueller and Landisman*, 1966] is typically found at depths of 10 ± 3 km, is usually several kilometers thick, and is assigned by these authors a probable velocity of 5.5 km/sec on the basis of earthquake data. It has been observed at roughly the same depths beneath a variety of geological provinces, including shield, sedimentary, and active tectonic regions.

Based solely on the characteristics of P_c, *Mueller and Landisman* [1966] identify the shallow low-velocity zone at approximately twelve locations in the crust beneath all the continents except Antarctica, where they have not yet studied the matter. Although the results have not been published, M. Landisman (personal communication, 1970) also identifies P_c, on records taken from the Argentine continental shelf.

Data have been used for the identification of P_c as just described, and data have also been inverted to a velocity-depth function in the following areas: the Black Forest southeast to the German Alpine foreland [*Landisman and*

Mueller, 1966]; the south German triangle in central Western Germany [*Fuchs and Landisman*, 1966]; the Alps [*Choudhury et al.*, 1967]; the Ukrainian shield [*Pavlenkova*, 1969]; South Africa [*Bloch et al.*, 1969]; the Basin and Range province [*Mueller and Landisman*, 1971]; southeastern Oklahoma, where a deep low-velocity zone is noted but where a shallow zone is barely developed [*Mitchell and Landisman*, 1970; Figure 6, p. 83, this volume]; the Rhine graben and vicinity, in Germany [*Ansorge et al.*, 1970; see Figure 7 of Landisman et al., p. 85, this volume]; the Precambrian Grenville-Superior boundary in northeastern Canada (M. J. Berry and K. Fuchs, personal communication, 1971); and the Great Slave Lake region of west-central Canada (K. G. Barr, personal communication, 1971).

From the apparently widespread occurrence of the sialic low-velocity zone, the impression is created that at least an appreciable degree of lateral continuity of physical properties may exist at depths below roughly 10 km in the continental crust. It is still true, however, that more data are needed to establish the characteristics of the shallow seismic low-velocity zone in relation to various geological provinces. If it is valid, the sialic low-velocity zone seems to signal the shallowest depth at which an appreciable degree of lateral continuity can be found.

Prodehl [1970] indicates that the sialic low-velocity zone may be more limited in extent than is suggested by the earlier results. Using data acquired earlier from 64 profiles of the United States Geological Survey, he finds a low-velocity zone at 10 km in both the Basin and Range province and in the southern Cascade Mountains of northern California. Prodehl does not find a low-velocity zone beneath the Colorado Plateau, the Sierra Nevada Mountains, the Great Valley, the Coast Ranges, or the Transverse Ranges of California.

It is conceivable that, if additional studies are made in these regions of the western United States specifically in search of seismic low-velocity zones, detection of energy from phases associated with low-velocity zones can be increased by using different filter settings or closer station spacings. It is also conceivable that the low-velocity zones may be more extensive than Prodehl's data indicate. It should be noted, for example, that through the use of different interpretational techniques, *Mueller and Landisman* [1971], in studying the Basin and Range data, obtain different results in that they find an additional low-velocity zone at 20 km, whereas none is reported by Prodehl. Nevertheless, the Prodehl study indicates that the sialic low-velocity zone is discontinuous in some areas.

FACTORS CONTROLLING THE DISTRIBUTION OF PHYSICAL PROPERTIES IN THE CRUST

For any attempt to gain insight into the relation between the distribution of various physical properties in the crust, it should be noted that seismic velocities in a given section depend on the distribution of the materials and their physical properties, as determined by the development of the crust in geological time or by processes that are currently active in the crust. More basically, seismic velocities depend on elastic parameters and densities and may be affected by certain dissipative processes. These factors are determined by the lithology, porosity, mechanical strength properties, and stress distributions as they affect fracture patterns, the presence or absence of pore fluids, and temperatures and pressures as they affect the mineralogical phases present and their elastic constants. In addition, of course, intrusives and tectonic forces can disrupt established velocity patterns.

Internal frictional processes and their influence on seismic velocity are discussed by Gordon and Rader in this volume. Melting conditions will probably occur no shallower than 20 km in the crust, as reviewed by Wyllie in this volume, or perhaps below 30 km, as suggested by Boettcher in this volume. The effect of melting on velocity is discussed by *Vaisnys* [1968] and by *Walsh* [1969]. Lithologic control by the intrusion of granitic laccoliths is proposed for the sialic low-velocity zone by Landisman et al. in this volume.

Clearly, factors that control the seismic velocities in earth materials are different from those that determine the distribution of other physical properties, e.g. density, resistivity, and magnetic susceptibility. Resistivities, for example, are controlled primarily by porosity and the presence of fluids or, in dry rocks, by the resistivities of the mineral components themselves.

This raises the question of whether the distribution of various additional physical properties of the crust can be deduced from the apparently widespread occurrence of the sialic low-velocity zone at typical depths. This question relates directly to the possible existence of an electromagnetic wave guide, since, to be effective, the resistivity distribution defining such a wave guide would have to be continuous, or subject only to minor discontinuities.

PRIMARY CONTROL OF THE SIALIC LOW-VELOCITY ZONE—TEMPERATURE-PRESSURE

Despite the need for much additional data to aid in our adequate understanding of the deep crustal environment, it may nevertheless be useful as a guide to future research to attempt to deduce possible additional crustal relationships from the evidence we now have. It is clear that, if we knew the causes of the sialic low-velocity zone, we would be in a better position to infer the distribution of physical properties other than seismic velocity at similar and greater crustal depths. Yet, on considering the possibilities described in the preceding section for the control of seismic velocity, I constantly find myself logically returning to the idea that, in some way, temperature and pressure must be primarily responsible for the occurrence of the zone at characteristic depths.

Initially, the characteristic-depth result seems complex if it is truly independent of geological province, since it implies that the low-velocity zone persists at or about the same depths despite tectonic uplifts and heat flow rates in various geological provinces. As Blackwell (this volume) points out, uniform mantle heat-flow values are characteristic of certain broad geothermal provinces across the continental United States, whereas surface heat-flow values differ from region to region and depend on the variations in near-surface radioactivity content. Hence, temperatures at the depth of the sialic low-velocity zone are only indirectly related to surface heat-flow values. Further details on the depth of the sialic low-velocity zone in various regions may demonstrate slight variations that can be correlated with mantle heat flow or with regional tectonic uplift, for example. If this is true, it could greatly simplify the problem of identifying the cause of the sialic low-velocity zone. It opens the possibility that a direct in situ temperature-pressure dependence may control the location of the sialic low-velocity zone.

As noted by Boettcher (this volume) and Wyllie (this volume), the material must not be molten at the relatively shallow depth of 10 km. If it is intruded (Landisman et al., this volume), a mechanism for depth control is needed. Therefore, although in situ temperature-pressure control seems reasonable, further study is needed to determine the precise nature of the controlling mechanism.

RELATION BETWEEN THE SEISMIC LOW-VELOCITY ZONE AND CRUSTAL RESISTIVITY

If the effect of temperature and pressure is primary in controlling the location of the sialic low-velocity zone, as postulated, their effect may produce additional zones of continuous physical properties at greater depths. If we consider the possible complexities in the control of seismic velocity, such a simple result suggests that it would not be surprising if other physical properties exhibit an equally simple dependence on in situ temperature and pressure. Resistivity, for example, the property most important in determining the existence of a crustal wave guide, has a relatively simple dependence on external factors in comparison to those which control seismic velocity. Since, as noted by Brace (this volume), rocks whose pore spaces are saturated with water of a salinity appropriate to the crustal environment will have resistivities no higher than about 10^5 ohm-m, it will be necessary to find a zone of dry rocks if sufficiently high resistivities are to be found for low-attenuation wave guide propagation.

If zones of dry rock can be found at depth in the earth, if we take into account the apparent tendency for the sialic low-velocity zone to occur at fairly characteristic depths, and if zones of dry rock are also controlled by in situ temperature and pressure (which seems likely), then it seems reasonable that these zones of high resistivity will also tend to be continuous. Furthermore, since the sialic low-velocity zone seems to be the shallowest continuous zone for which we have any evidence at all in the continental crust, it is probable that continuous high-resistivity zones, if they exist, will not be found much shallower than 10 km. At greater depths in the crust, there should be an increased tendency for layers to form and a greater likelihood that they will be more nearly continuous than shallower ones because of the greater geochemical reaction rates caused by higher temperatures, provided that any vertical motions occurring at a given depth are slower than the rates at which the physically identifiable zones are formed there.

Although few data exist on the effect of pressure on reaction rates and although it is not usually possible to predict which way pressure will drive a reaction rate, temperature is the predominant factor, as noted by *Turner and Verhoogen* [1960, p. 48], who state 'Reaction rates increase exponentially with increasing temperature; the effect of pressure is, in comparison, very small.' On this basis, provided we are dealing with a geochemical rate process, it is probable that, if a highly resistive zone can be found at a depth below the sialic low-velocity zone, it will tend to be more extensive and more nearly continuous than the shallower seismic zone.

PHYSICALLY IDENTIFIABLE ZONES BELOW 10 KILOMETERS

Having reasoned that, owing to increased geochemical reaction rates, deeper zones can be anticipated that are more extensive and nearly continuous than the sialic low-velocity zone, let us look at the evidence. Strangely, although mantle properties have been extensively studied, little effort has been concentrated on a study of the deeper crust. There are, however, several references reporting a seismic low-velocity zone at about 20 km beneath the continents.

Mitchell and Landisman [1970; see also their work in this volume] describe a deeper seismic low-velocity zone at a depth of 17 to 18 km that continues beneath the Wichita Mountains of Oklahoma. A shallower seismic low-velocity zone is not very strongly formed, but it seems to occur near 10 km in the depth range associated with the sialic low-velocity zone, and it bottoms at 12 to 13 km. Unlike the deeper zone, the shallower low-velocity zone is not continuous through the Wichita Mountain belt. *Ansorge et al.* [1970; see also their work in Figure 7, Landisman et al., this volume] demonstrate both a sialic low-velocity zone near 10 km and a deeper low-velocity zone near 20 km, both of which continue beneath the Rhine graben. M. J. Berry and K. Fuchs (personal communication, 1970), on studying the boundary between the Grenville and Superior provinces in northeastern Canada, find a possible low-velocity zone at 25 km whose presence is allowed by the data but that is not necessarily required by their interpretational technique. *Mueller and Landisman* [1971] find a low-velocity zone at both 10 and 20 km in the Basin and Range province.

There are not enough examples of the deeper low-velocity zone to establish its properties under various geological conditions or to determine how extensively it occurs. The fact that the deeper zone exists, however, may support the idea that increased geochemical reaction rates tend to produce additional layers at greater depths in the crust. The one example cited above contrasting the continuity of the shallow and deep low-velocity zones tends to support the idea that the deeper zones are less disrupted by vertical motions than the sialic low-velocity zone. Applied to the wave guide concept, this line of reasoning suggests that, if a deep highly resistive zone can be found, it may have a reasonable chance for continuity beneath various tectonic zones or between different geological provinces.

NEED FOR MULTIDISCIPLINARY APPROACH TO A STUDY OF THE CRUSTAL ENVIRONMENT

Even if they are controlled in situ by temperature and pressure, as postulated, various physical rock properties behave in different ways in response to these factors. Thus, phase changes, resistivity distributions, and seismic velocities, et cetera, are not necessarily subparallel to each other. However, to speculate on the detailed nature of these interrelationships at this time, when few data are available, would be wasteful. By obtaining data on the structure and physical properties of the deeper parts of the crust, where a degree of continuity is suggested, and by noting the interrelationships of physical properties here, the exciting possibility exists that we may be able to infer both the characteristics of this environment and the processes that actively or historically have produced it.

Not only must seismic reflection and refraction studies be performed, but measurements of the electrical properties of the crust must also be conducted in the same areas in a variety of geological provinces. These studies and measurements, combined with gravity, magnetic, electromagnetic and geothermal measurements, must be coupled with laboratory studies of the physical properties of rocks made at the high temperatures and pressures corresponding to those found at the depths of interest in the earth. Such studies will be required for the identification of materials at depth, for the gradual elimination of ambiguities that currently beset the clear identification of various physical properties in the crustal environment.

OBSERVED DISTRIBUTION OF RESISTIVITY IN THE EARTH'S CRUST

Having postulated an ultimate dependence of various physical properties, including resistivity, on the in situ temperature and pressure in the crust and shallow upper mantle, and having concluded that a certain degree of continuity is a reasonable expectation, it is now appropriate to consider the actual depth dependence of crustal resistivity according to data and interpretations that are currently available.

Keller [1971 and in this volume] concludes that certain generalizations can be drawn regarding the variation of electrical resistivity with depth from one geological province to another. Keller's summary is illustrated by his Figure 12 (this volume). The maximum resistivity is given as just under 10^6 ohm-m for stable continental areas. A slightly lower maximum is found for mobile crustal areas, and a very low resistivity of between 1 and 10 ohm-m is found in volcanic rift areas and subduction zones at rather shallow depths. Thus, it appears from present data that marked lateral-resistivity changes occur from one geologic province to another. This result means that, if highly resistive layers exist in the regions of high average conductivity, either they have not yet been detected in the more conductive background or zones of high resistivity may be confined to certain geologic provinces.

There are, however, two considerations that seem to encourage the possibility that 10^6 and 10^7 ohm-m resistivities can be found in the resistive sections of the crust. First is the general inability of measuring methods, i.e., of the magnetotelluric technique, to measure the magnitude of highly resistive sections on an absolute scale or of present electrode measurements to

resolve the problem, since they give only a resistivity-thickness product, as discussed by Madden (this volume). Furthermore, lateral discontinuities tend to distort measured resistivity values as discussed by Porath, in this volume. Second, *Keller* [1971] finds that, on using existing resistivity data, if in stable continental regions he allows for a transition zone between a highly resistive zone followed by a resistivity that gradually decreases at mantle depths, then resistivities of 10^6 and 10^7 are quite possible in the resistive zone. The values that Keller infers approach the maximum of 10^8 to 10^9 ohm-m observed in the laboratory for dry rocks [*Parkhomenko and Bondarenko*, 1963].

At temperatures below those causing significant solid-state conduction, the existence of high resistivities in deep rocks depends greatly on whether they are dry or wet with pore fluids. As indicated by the studies of Brace (this volume), a finding of resistivities of 10^6 ohm-m or greater will indicate that dry rocks do exist at depth in the earth.

RESISTIVITY OF SEISMIC LOW-VELOCITY ZONES

It is interesting to note the results of Word et al. (Figure 13, p. 163, this volume), which show a high conductivity layer at a depth of roughly 10 km in the Texas Gulf coast, seaward of their station 3. No correlations between the sialic low-velocity zone and resistivity have been reported in the literature, but Mitchell and Landisman (this volume) note a possible correlation between the 20-km seismic low-velocity zone in eastern New Mexico and a low-resistivity zone observed in a nearby area. Caution is needed on this point, however, until combined experiments are conducted in the same areas to establish this relationship clearly.

OCEANIC CRUSTAL AND UPPER-MANTLE RESISTIVITIES

Only limited studies of earth resistivities beneath the ocean have been conducted. Cox (this volume) reports some of the pioneering work on this problem. Based on temperature, earth materials in the upper mantle beneath the oceans may be highly resistive, provided they are dry. Whether these rocks are dry and what absolute resistivities they have remain to be determined. Cox, however, finds that appreciably conductive rocks do not occur until depths of roughly 30 km beneath the ocean. Clearly, the impact of major geological structures in the ocean crust on the distribution of resistivities is an open question.

CONCLUSIONS

Although sections of sufficiently high resistivity for long-range wave guide propagation have not yet been reported in the published literature, recent calculations by *Keller* [1971] indicate that, if transition zones are included in the calculation, existing data in tectonically stable areas yield zones of 10^6 and 10^7 ohm-m resistivity. If these zones exist, wave guide attenuations as low as 0.1 db/km (Wait and Spies, this volume) are possible. Laboratory measurements on rocks indicate 10^8 and 10^9 ohm-m resistivities for dry rocks, which makes it possible that highly resistive zones may exist in the earth's crust. Specialists in high-pressure high temperature mineralogy indicate that free water in the earth's crust is possible but not necessary and that the geophysicist will have to resolve this question through experimentation (Wyllie, this volume, and Boettcher, this volume). If the rocks are wet, Brace (this volume) shows that 10^5 ohm-m resistivities will be maximum. The Keller results indicate that dry rocks may exist at depth.

Present resistivity measuring techniques are subject to ambiguities (Madden, this volume, and Porath, this volume) and have not yet answered the question of whether sufficiently resistive zones exist in the earth to make long-range electromagnetic propagation feasible. The question of continuity in resistive earth zones remains unanswered on the basis of available field data. However, the argument is made intuitively that in situ temperature and pressure must control the presence of an apparently widespread seismic low-velocity zone reported elsewhere as occurring in a characteristic depth range at about 10 km in the continental crust. Assuming an in situ temperature and pressure control for resistivity, it appears reasonable to inquire into the possibility that, should a highly resistive zone in dry rocks be found at depths below 10 km, such a zone could extend across geological province boundaries, perhaps relatively undisturbed by tectonic activities.

If a zone of sufficiently resistive rocks can be found, bounded top and bottom by conducting material, it could form an electromagnetic wave guide whose range would depend not only on its electrical properties (Wait, this volume, and Wait and Spies, this volume) but also on its extent and degree of lateral continuity.

Because of the difficulties involved in measuring the electrical resistivity distribution in the earth's crust, it seems probable that a multidisciplinary approach must be made to this problem. Laboratory studies of physical rock properties, combined with multidisciplinary field data, should help us to identify the materials that exist at depth. Through such studies, it should be possible to eliminate some of the ambiguity that now makes it difficult to determine the distribution of electrical resistivity with depth in the earth.

An improved knowledge of the crustal environment should contribute major potential benefits to mankind, which could be used in the development of new techniques for communication, in the identification of deep mechanical properties in order to reduce earthquake hazard in stressed regions, and in the solution of environmental problems related both to waste disposal and to the acquisition of new natural resources. The need for geochemical experiments combined with laboratory studies of physical properties and with carefully designed multidisciplinary studies in given areas of various geological provinces cannot be overemphasized. The development of techniques to identify precisely the nature of the earth's internal environment is a major need today.

Acknowledgment. Professors Mark Landisman and Stephen Mueller have been most generous in discussing their findings over a number of years. Professors M. Landisman, G. V. Keller, T. R. Madden, and W. F. Brace and Dr. P. C. Badgley have critically read the manuscript and offered helpful suggestions. I sincerely appreciate their help.

This study was supported by the Office of Naval Research as part of its effort to understand the environment more thoroughly.

REFERENCES

Ames, L. A., J. W. Frazier, and A. S. Orange, Geological and geophysical considerations in radio propagation through the earth's crust, *Trans. IEEE Antennas Propagat., 11*(3), 369–371, 1963.

Ansorge, J., D. Emter, K. Fuchs, J. P. Lauer, S. Mueller, and E. Peterschmitt, Structure of the crust and upper mantle in the rift system around the Rhinegraben, in *Graben Problems,* edited by H. Illies and S. Mueller, pp. 190–197, Schweizerbart, Stuttgart, Germany, 1970.

Bloch, S., A. Hales, and M. Landisman, Velocities in the crust and upper mantle, of southern Africa from multi-mode surface wave dispersion, *Bull. Seismol. Soc. Amer., 59*(4), 1599–1629, 1969.

Choudhury, M., P. Giese, and G. de Visintini, Crustal structure of the Alps—Some general features from explosion seismology, 15 pp., paper given at Int. Union of Geodesy and Geophysics General Assembly, Switzerland, 1967.

Dolukhanov, M., Underground propagation of radio waves, *Radio* (in Russian), *1,* 42–43, 1970.

Fuchs, K., and M. Landisman, Detailed crustal investigation along a north-south section through the central part of Western Germany, in *The Earth Beneath the Continents, Geophys. Monogr. Ser.,* vol. 10, edited by J. S. Steinhart and T. J. Smith, pp. 433–452, AGU, Washington, D. C., 1966.

Ghose, R., The long range sub-surface communication system, in *Communications, the Key to an Expanding Universe, IRE Sixth Nat. Commun. Symp.,* pp. 110–120, Institute of Radio Engineers, New York, 1960.

Gutenberg, B., Structure of the earth's crust in continents, *Science, 111*(2872), 29–30, 1950.

Gutenberg, B., Crustal layers of the continents and oceans, *Geol. Soc. Amer. Bull., 62*(5), 427–440, 1951.

Gutenberg, B., Effects of low-velocity layers, *Geofis. Pura Appl., 29,* 1–10, 1954.

James, D. E., and J. S. Steinhart, Structure beneath continents: A critical review of explosion studies 1960–1965, in *The Earth Beneath the Continents, Geophys. Monogr. Ser.,* vol. 10, edited by J. S. Steinhart and T. J. Smith, pp. 293–333, AGU, Washington, D. C., 1966.

Keller, G. V., Electrical properties in the deep crust, *Trans. IEEE Antennas Propagat., 11*(3), 344–357, 1963.

Keller, G. V., Electrical properties of the earth's crust—A survey of the literature, *Tech. Rep. March.,* Office of Naval Research, Arlington, Virginia, 1971.

Landisman, M., and S. Mueller, Seismic studies of the earth's crust in continents, 2, Analysis of wave propagation in continents and adjacent shelf areas, *Geophys. J., 10*(5), 539–548, 1966.

Levin, S. B., Geophysical factors in electromagnetic propagation through the lithosphere, *IRE/NEREM Record,* 3.5, 1962.

Levin, S. B., Lithospheric radio propagation—A review, Trans. *NATO/AGARD Symp. Subsurface Commun.* reprinted by the Institute for Exploratory Research, U.S. Army Electronics Command, Fort Monmouth, New Jersey, 1966.

Mitchell, B. J., and M. Landisman, Interpretation of a crustal section across Oklahoma, *Geol. Soc. Amer. Bull., 81*(9), 2647–2656, 1970.

Mueller, S., and M. Landisman, Seismic studies of the earth's crust in continents, 1, Evidence for a low-velocity zone in the upper part of the lithosphere, *Geophys. J., 10*(5), 525–538, 1966.

Mueller, S., and M. Landisman, An example of the unified method of interpretation for crustal seismic data, submitted to *Geophys. J.,* 1971.

Parkhomenko, E. I., and A. T. Bondarenko, An investigation of the electrical resistivity of rocks at pressures up to 40,000 kg/cm² and temperatures up to 400°C, *Akad. Nauk SSSR, Izv., Geophys. Ser., 12,* 1106–1111, 1963.

Pavlenkova, N. I., On low velocity layers within the Earth's crust of the Ukrainian shield, *Akad. Nauk Ukr. SSSR, Geophys. Commun., 30,* 36–43, 1969.

Prodehl, C., Seismic refraction study of crustal structure in the western United States, *Geol. Soc. Amer. Bull., 81*(9), 2629–2645, 1970.

Ryazantsev, A. M., and A. V. Shabel'nikov, Propagation of radio waves through the earth's crust (a review), *Radio Eng. Electron. Phys., 11,* 1643–1657, 1965.

Schwering, F. K., D. W. Peterson, and S. B. Levin, A model for electromagnetic propagation in the lithosphere, *Proc. IEEE, 56*(5), 799–804, 1968.

Tsao, C. K. H., and J. T. deBettencourt, Subsurface radio propagation experiments, *Radio Sci., 3*(11), 1039–1044, 1968.

Turner, F. J., and J. Verhoogen, *Igneous and Metamorphic Petrology,* 2nd ed., 694 pp., McGraw-Hill, New York, 1960.

Vaisnys, J. R., Propagation of acoustic waves through a system undergoing phase transformation, *J. Geophys. Res., 73*(24), 7675–7683, 1968.

Wait, J. R., An anomalous propagation of radio waves in earth strata, *Geophysics, 19*(2), 342–343, 1954.

Wait, J. R., The possibility of guided electromagnetic waves in the earth's crust, *Trans. IEEE Antennas Propagat., 11*(3), 330–335, 1963a.

Wait, J. R., (Ed.), *Electromagnetic Waves in the Earth, IEEE Trans. Antennas Propagat., 11,* 206–371, 1963b.

Wait, J. R., Some factors concerning electromagnetic wave propagation in the earth's crust, *Proc. IEEE, 54*(8), 1020–1025, 1966a.

Wait, J. R., Electromagnetic propagation in an idealized earth crust waveguide, *Radio Sci., 1*(8), 913–924, 1966b.

Wait, J. R., Some highlights of a symposium on subsurface propagation of electromagnetic waves, *Radio Sci., 1*(9), 1115–1118, 1966c.

Wait, J. R., Influence of a sub-surface insulating layer on e/m ground wave propagation, *Trans. IEEE, Antennas Propagat., 14*(6), 755–759, 1966d.

Walsh, J. B., New analysis of attenuation in partially melted rock, *J. Geophys. Res., 74*(17), 4333–4337, 1969.

Wheeler, H. A., Radio wave propagation in the earth's crust, *Radio Sci., 65D*(2), 189–191, 1961.

DISCUSSION

Alldredge: What do you expect to learn from the seismic crustal low-velocity layer and how can you relate the seismic data to electrical properties?

Heacock: I am using the layer as a marker. If it can be shown to be continuous across tectonic provinces then we have reason to believe that other properties may also be continuous.

2. PHYSICAL PROPERTIES OF THE CONTINENTAL CRUST

Review of Evidence for Velocity Inversions in the Continental Crust[1]

M. LANDISMAN, S. MUELLER[2], AND B. J. MITCHELL

Geosciences Division, University of Texas at Dallas, Dallas, Texas   75230

Abstract. Observations of sub-basement echoes at near-normal incidence and in the under-critical distance range, in combination with recordings of strong later arrivals in seismic refraction experiments, furnish evidence for velocity reversals in the continental crust. These low-velocity layers may significantly alter the location and strength of wave amplitude maximums near the critical points associated with higher-velocity refractors lying just below the proposed inversions. The velocity contrast at the fairly abrupt intervening interfaces can be cited in support of reported observations of sub-basement reflections. The sialic low-velocity zone, postulated to have a lower boundary at depths between approximately 8 and 15 km, has been most often interpreted in areas that have been the site of major tectonic adjustments at some time in their geologic history. This low-velocity region might be associated with semi-continuous acidic intrusions having lower velocities and densities than the surrounding country rocks. Although not universal in extent, many of the world's continents have produced data compatible with this feature. It is significant that most of the shallow earthquakes in continental regions have depths similar to those postulated for the sialic low-velocity zone. A second low-velocity region, which may be continuous over large portions of North America and Eurasia, has been proposed for depths of the order of 20 km. This feature has been associated with a low-resistivity zone in several areas for which suitable data are available. More basic layers at greater depths were found to be continuous in Oklahoma, even though faults with large surface displacements extend through the shallower layers. The degree to which this deeper inversion is developed may depend on the temperature and pressure at corresponding depths in the crust, and it is possible that enhanced crustal seismicity may be associated with a well developed relatively deep crustal low-velocity zone.

Strong later arrivals observed at several tens of kilometers from the shot point in crustal refraction experiments, in combination with near-vertical and wide-angle reflections from intra crustal depths, may be cited as evidence in favor of velocity reversals in the continental crust. These reversals can be related to significant increases in the amplitudes and decreases in the distances observed for the wave maximums as-sociated with the critical distance ranges for the higher-velocity materials that lie immediately below the inversions. Further, if a velocity inversion is sufficiently well developed, relatively high-frequency normal-incidence and wide-angle reflections from intracrustal depths may be correlated with the boundaries of the inversion.

Evidence for velocity reversals in the crust can be cited for most of the world's continents, although not for all regions within each continent. There are some regions in which velocity reversals are not likely to occur, others where they might be difficult to delineate, and still

---

[1] Contribution 183, Geosciences Division, University of Texas at Dallas.

[2] On leave from Universität Fridericiana, Karlsruhe, Germany.

others in which they probably occur but have not yet been identified. Demonstration of the existence of low-velocity zones requires travel times from high quality refraction profiles in combination with reflection times and-or amplitude data. Relevant evidence from other geophysical disciplines should be considered wherever possible. These measurements should preferably be made in regions which are undisturbed by faults that might disrupt the continuity of otherwise observable refraction branches. The requisite compilations are being made for a number of regions, and although the results are better defined only for those areas where intensive work has proceeded over a span of several years, a number of important features may be noted at this time.

The sialic low-velocity zone has been proposed to have a lower boundary at depths of 8 to 15 km. Evidence collected from most continents indicates that this feature is more likely to occur in areas where the crust has undergone intensive deformation at some time in its geologic history. It is possible that this low-velocity zone can be correlated with gravimetrically derived, semi-continuous, low-density intrusions that have been postulated for shallower parts of the crust in several tectonic regions, including fold-belt areas. We consider it significant that accurate epicentral studies of shallow earthquakes have located an increasingly large percentage of these events at depths lying within the range to be expected for the sialic low-velocity zone.

The observed velocities for the first sub-basement refraction signal P_g appear to have been relatively consistent and are usually reported in the vicinity of 6.0 km/sec for explosion measurements. Earthquake studies, by contrast, have often reported P_g values of the order of 5.5 km/sec [see *Gutenberg*, 1954, 1955]. We take this difference, which greatly exceeds the error of measurement for either type of experiment, as evidence that the velocity in the sialic low-velocity zone is of the order of 10% less than that in the overlying crystalline basement just beneath the sediments.

Another crustal velocity inversion, at depths of the order of 20 km, may be associated with a zone of increased electrical conductivity (low resistivity), if available results from the few areas for which suitable data exist are indicative

of more widespread conditions. High-velocity layers below this inversion have been found to be continuous in Oklahoma, even though faults with surficial displacements of several kilometers disturb the shallower layers. A few of the implications of these proposals will be discussed in the text.

THE REFRACTION ARRIVAL P_c

An earlier report by *Mueller and Landisman* [1966] showed that there are widespread observations of the refracted phase P_c, which is a previously unexplained strong later arrival that is typically observed with a nearly constant delay of one second after the P_g onset at distances from several tens of kilometers to well beyond 100 km from the shot point. The presence of this strong signal at small distances was proposed as evidence for the shallow crustal velocity inversion [*Mueller and Landisman*, 1966]. This positive type of evidence for a velocity inversion seems to be more widespread than the stronger condition of the disappearance of the basement refraction P_g, cited earlier by *Gutenberg* [1950, 1951, 1954, 1955]. Local geological conditions, as might be expected, strongly influence the signal quality and greatly affect the character of this and all other refracted arrivals [*Steinhart and Meyer*, 1961; *Giese*, 1963; *Pakiser*, 1963; *Prodehl*, 1964].

Recent work in various stages of completion has yielded several observations of P_c that supplement the observations reported by *Mueller and Landisman* [1966] and *Landisman and Mueller* [1966]. The earlier set of data was compiled from recordings made in Germany, southern Africa, Pennsylvania, eastern Montana, southeastern Australia, Japan, and the Baltic shield; more recent examples come from the eastern United States and Canada, the western United States, and Germany.

The present discussion of evidence for the sialic low-velocity zone may be brought into sharper focus by presenting just two of these many examples of the refraction arrival P_c, extracted from observations recorded in different parts of the world. The first example, in the upper part of Figure 1, comes from western Utah. It was recorded about 60 km west of the shot point near the town of Delta, which lies approximately 130 km south-southwest of Salt Lake City. The P_c arrival follows the P_g onset in this example by about one second and has an amplitude that is at least five times as large as that of the first arrival. Another example, shown in the lower part of Figure 1, comes from a recording site approximately 90 km northeast of Mannheim in West Germany. The seismic source in this case was a quarry near Hilders, about 65 km to the northeast. The delay between P_g and P_c and the ratio of the amplitudes for these two signals are very similar to those for the example from the western United States. Calculations for models without the sialic low-velocity region predict that the large critical reflections that form the secondary arrivals would be 5 or 10 km further than those observed for these two profiles.

The nearly constant delay of the P_c arrivals with respect to the basement refraction P_g and the large amplitudes observed for P_c at small distances from the shot point may both be interpreted as direct results of the sialic low-velocity zone. It is instructive to discuss an example of this crustal velocity inversion taken from a region having as few other complications within the crust as possible. Such an example comes from an early set of seismic measurements that were made in southern Germany. The refraction profile starts at a quarry near Böhmischbruck in the Bohemian Massif and runs southwest through this area into the Bavarian Molasse basin in the foreland north of the Alps. Readings of first arrivals and strong later arrivals recorded in this early experiment have been plotted as white circles in the travel time diagram in the upper part of Figure 2 [*Mueller and Fuchs*, 1966]. The distances (Δ, km) of the recording stations from the shot point are measured along the abscissa, and the reduced arrival times of the stronger signals are displayed along the ordinate for each station. Reduced times, which are travel times minus station distances divided by 6 km/sec, have been used in this and following figures. The reduction of the time scale increases the time resolution and emphasizes small differences between velocities close to the value 6.0 km/sec, which are often observed for the basement refractions.

The continuous lines in the upper part of Figure 2 are reduced travel times calculated for the simple crustal model interpreted for this area of basement outcrop (lower left corner). A thin weathered section at the top of the model grades

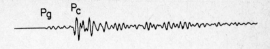

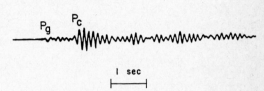

Fig. 1. The upper seismogram was recorded about 60 km from the Delta-West shot point in Utah (adapted from *Mueller and Landisman* [1971]). The lower seismogram was recorded at a distance of about 65 km on the profile Hilders-West in Germany (adapted from *Fuchs and Landisman* [1966]). The large (P_c) signals that follow the P_g onsets by about one second are interpreted to be near-critical reflections from the bottom of the sialic low-velocity layer.

into a layer with a velocity of about 5.9 km/sec. These shallow layers are associated with the P_g refraction branch, which is represented by an upward sloping line that runs through the first arrivals recorded from the shot point to distances slightly greater than 110 km. It was difficult for the observers to find this arrival at greater distances.

This disappearance of the P_g branch is a severe condition that is greatly dependent on the local geologic setting and the gradients and attenuation in the uppermost portions of the crust. This phenomenon will not be considered in the present discussion of evidence for the sialic low-velocity layer because there is a high degree of variability associated with it.

The P_c arrivals associated with the sialic low-velocity channel are best represented by the critical reflections observed about one second after the P_g waves, at distances of about 50 km from the shot point. The associated P_c refractions, with apparent velocities of slightly greater than 6 km/sec, can be followed to distances beyond 200 km. The calculated times for the model in Figure 2 predict that, at distances beyond the p_g-P_n cross-over, the P_c refractions will be followed by a sequence of arrivals. The first of these are wide-angle reflections from the base of the crust (not shown in Figure 2); they are followed by wide-angle reflections from the base of the sialic low-velocity region (curved line with arrow

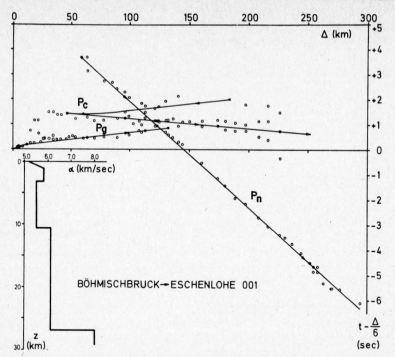

Fig. 2. Travel-time plot and model for the refraction profile Böhmischbruck-Eschenlohe in southern Germany. The refraction branches P_g, P_c, and P_n, which arrive sequentially at distances of less than 100 km, contain information about the top, middle, and base of the crust, respectively. The delay between the P_g and P_c arrivals is about one second near the critical distance for the P_c branch (after *Mueller and Fuchs* [1966]).

head pointing left). At large distances, all these arrivals are preceded by the rays that penetrate the entire crust and return from the uppermost part of the mantle. These first arrivals correspond to the P_n observations that extend from distances of about 60 km to nearly 300 km, with a velocity slightly greater than 8 km/sec.

The delay between the branches P_g and P_c can be ascribed to the extra time required for the seismic disturbance to penetrate the sialic low-velocity region, which of course produces no refraction branch of its own. The P_c signals, which may be observed after the P_g arrivals at distances of several tens of kilometers from the shot point, can be attributed to strong constructive interference between the energetic critical reflections and the refractions from the large velocity contrast at the base of the velocity inversion. An abrupt lower boundary and a moderately large velocity difference (>0.2 km/sec) between the values for the layers below and above the inversion can combine to produce energetic P_c arrivals at small distances.

This model of the crust, introduced as an example, represents an over simplification of the lower portions of the crust for most regions. The reverse for this profile runs northeast from Eschenlohe, in the Bavarian Molasse basin, in the northern Alpine foreland toward Böhmischbruck, in the Bohemian massif. A seismogram montage or record section for this set of data is presented in Figure 3 [after *Giese*, 1968]. A large P_c arrival can be seen at a reduced time of about 2.5 sec approximately 45 km from the shot point. Some of the complications that seem to be typical of the lower crust in many areas can be found in this record section. A number of refractions arrive between the branches P_c and P_n at distances of less than 100 km; they can be taken as indications of additional refracting horizons in the lower crust. This record section also contains a conspicuous set of large-amplitude wide-angle reflections from the lower crust, at distances beyond those for which the P_n branch crosses the shallower crustal refractions.

The P_c refraction arrival, associated with the sialic low-velocity zone, has been interpreted most often in data from folded and tectonically mobilized areas. The α-β quartz transition, which was suggested by *Gutenberg* [1951] as a causal mechanism for his proposed shallow crustal velocity reversal, requires temperatures far greater than the normal geothermal values of slightly more than 300°C at depths near 10 or 15 km.

As a possible alternative, we suggest that an intrusion of relatively low-density, acidic (i.e., granitic) rock might be considered as a possible explanation for this shallow zone of low velocity in tectonic areas. A density contrast of nearly 0.1 or 0.2 g/cm³ could be caused by the juxtaposition of acidic intrusions having densities of about 2.67 g/cm³ next to country rocks having densities ranging from 2.75 to 2.85 g/cm³. These density values and the resulting contrast might reasonably be expected, according to the empirical velocity-density relation derived by Nafe and Drake [in *Talwani et al.*, 1959], if the intrusive rocks had compressional velocities of 5.5 km/sec and if the surrounding country rocks were associated with values of about 6.0 to 6.2 km/sec. These velocities are in accord with those proposed for the sialic low-velocity zone and those measured for adjacent refracting horizons. The authors suggest that gravity studies provide important support for the intrusion hypothesis. Density contrasts of about 0.1 to 0.2 g/cm³ and maximum depths similar to the 10+ km proposed for the sialic low-velocity region have been inferred from several independent gravity interpretations of relatively light, buried intrusive bodies such as stocks and plutons. These studies were based on gravity data collected in areas where the crystalline basement has been folded and deformed [*Bott and Smith*, 1958; *Bott et al.*, 1958; *Thompson and Talwani*, 1964; *Smithson*, 1965; *Bott and Smithson*, 1967; *Smithson and Ramberg*, 1970]. The buried intrusive bodies implied by these gravity interpretations are directly related to continuations of the injection features that extend to the surface in some areas. Field studies in older, deeply eroded regions [*Simonen*, 1969] and uplifted active areas [*Smithson*, 1965; *Hamilton*, 1969] provide direct visual evidence for the tops of granitoid intrusions, which have been termed 'roofless batholiths' [*Hamilton*, 1969]. We sug-

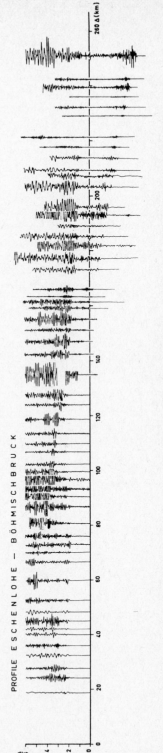

Fig. 3.   Record section from the profile Eschenlohe-Böhmischbruck in southern Germany (adapted from *Giese* [1968]).

gest that a semi-continuous laccolithic zone, of the type just proposed for the sialic low-velocity region, would appear fairly continuous for seismic refraction signals having wavelengths of several kilometers. In accord with numerous field observations, a zone of this type probably would not be observed as a continuous reflecting horizon when probed with high-frequency normal-incidence and wide-angle reflected waves [*Mueller*, 1968].

NORMAL-INCIDENCE SUB-BASEMENT ECHOES

Deep crustal reflections constitute another type of evidence that may be invoked in favor of velocity inversions within the continental crust and of the abrupt velocity increases below. These signals have been observed in a number of locations in North America and Europe over the past few decades. Perhaps the most extensive program of sub-basement normal-incidence

reflection seismology has been carried out in West Germany, where there has been active cooperation on this problem between industrial and university scientists since 1952. The data to be analyzed were most often obtained with standard prospecting equipment which was permitted to record for several seconds after the returns from horizons of commercial interest had been received.

Several examples of deep crustal reflections are shown in Figure 4. The signal, with a vertical reflection time of about 4.0 sec, has been widely observed with only very slight differences in its time of arrival [*Liebscher*, 1962, 1964]. The same studies have also shown that the late-arriving signals cannot be attributed to multiple shallow reflections. The 4-sec echo had earlier been attributed to the 'Foertsch' (F) discontinuity, postulated for a depth of about 10 km by *Schulz* [1957] and *Dohr* [1957, 1959], but

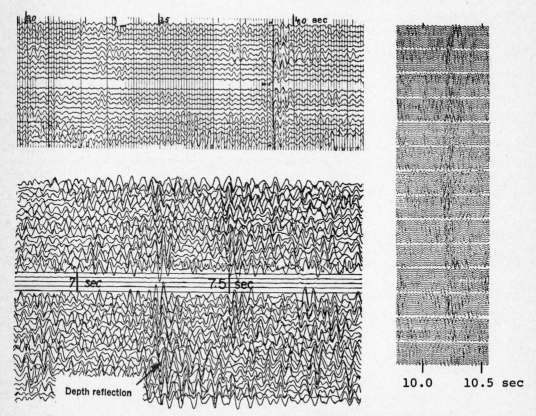

Fig. 4. Normal incidence crustal reflections in West Germany. The signal near 4-sec echo time is from the abrupt velocity increase below the sialic low-velocity zone. The later strong signals at about 7-sec and 10+-sec echo time arise from the Conrad and Mohorovicic discontinuities, respectively (after *Schulz* [1957] and *Liebscher* [1964]).

there were great difficulties in resolving these reflection data with the refraction interpretations then being reported, since only monotonic increases of velocity were being considered. A class of models having a sialic low-velocity zone at a depth of about 10 km was proposed by *Mueller and Landisman* [1966] and *Landisman and Mueller* [1966] as a means for resolving this appearent discrepancy and also explaining the presence of the P_e arrivals discussed above. The 4-sec echo was then attributed to a return from the high-contrast interface at the bottom of the sialic low-velocity zone.

Reflected signals from the Conrad discontinuity, for arrival times near 7.2 sec, and a correlated set of echoes at about 10.3 sec from the Mohorovicic boundary roughly 30 km beneath the surface of the Bavarian Molasse basin are also shown in Figure 4 [after *Liebscher*, 1964]. The remarkable clarity of these observations may perhaps serve to further the search for deep crustal reflections in other areas.

Figure 5 shows histograms which represent a statistical description of deep crustal reflection data from two areas, one in the western and one in the eastern part of the Bavarian Molasse basin [*Liebscher*, 1962], separated by about 130 km. The numbers of arrivals during each tenth of a second are read from a large body of data for a given area and are plotted as a function of echo time. The largest peaks at about 2.7, 4.2, 7.3, and 10.5 sec represent consistently observed signals from intra crustal interfaces characterized by large velocity contrasts. Although those familiar with seismic results in Bavaria did not seriously question the correlation of the peaks at 7.3 and 10.5 sec with the 'Conrad' and Mohorovicic discontinuities established earlier from refraction studies, there was some question when Liebscher reported in 1962 that the 4.2-sec echo corresponded to a major discontinuity at 10 km, since no corresponding arrivals had been found in refraction experiments. We suggest that this problem and a number of other outstanding

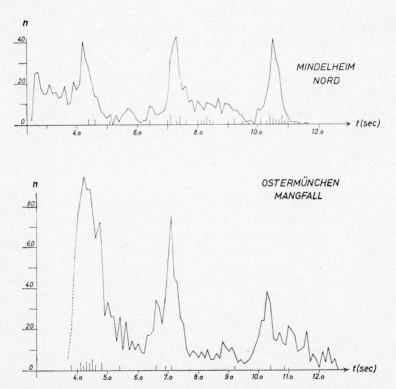

Fig. 5. Histograms of reflection times observed in the western Molasse basin (upper) and eastern Molasse basin (lower) in the northern foreland of the Alps. The peaks near 4, 7, and 10.5 sec correspond to reflections from the base of the low-velocity channel, the Conrad, and the Mohorovicic discontinuities, respectively (after *Liebscher* [1962]).

problems in crustal seismology can be resolved by the introduction of the sialic low-velocity zone in the upper part of the continental crust in tectonic regions. According to the interpretation proposed in this study and in earlier works by us, the echoes at 2.7 and 4.2 sec are explained as signals returning from the top and bottom of the sialic low-velocity region.

A few North American deep crustal reflection experiments [*Junger*, 1951; *Shor*, 1955; *Narans et al.*, 1961; *Dix*, 1965; *Perkins*, 1970] have indicated reflected horizons at depths similar to those found in southern Germany. Other investigators [*Tuve and Tatel*, 1948; *Mead*, 1950] may not have observed deep crustal reflections because of local geological conditions and differences in field techniques.

FEATURES AT GREATER DEPTHS WITHIN THE CRUST

Recent detailed studies of the crust have revealed that complexities in crustal properties are not confined solely to the shallower parts of the crust but appear to extend to greater depths. Early reconnaissance surveys, having station separations of tens of kilometers and employing whatever collection of recording equipment could be assembled, tended to produce data of such variable quality as to prevent any detailed interpretation. More closely spaced recording sites and modernization of equipment and reduction techniques have improved the quality of the data sufficiently to reveal some of the detailed properties of the deeper portions of the crust in a few areas. Improved amplification systems combined with magnetic tape recording in the field have permitted the study of much of the signal character that was lost in the early visual seismic recordings. Digital processing of the data has also introduced a degree of flexibility that was previously unattainable with analog equipment. Another important improvement was the introduction of the digitally plotted record section or seismogram montage, which permits the correlation of signal character across an entire data suite. This and other digital processing techniques are only now beginning to be applied, and their implications for seismic crustal studies are still not fully apparent.

Mid-continental United States. Some of these improvements in the data and the reduc-

tion techniques are used in two reinterpretations of seismic measurements in the southern Great Plains of the United States by *Mitchell and Landisman* [1970, 1971a]. A summary of the most recent results for this region can be found in another paper in this volume [*Mitchell and Landisman*, 1971b]. Figure 7, on page 85 of this volume, is one of the record sections produced during these studies; it illustrates several of the points under discussion.

Figure 7, on page 85 of this volume, was prepared from a set of data produced by the Gnome explosion in eastern New Mexico. Several of the stations were occupied by workers in the Crustal Studies Branch of the United States Geological Survey [*Stewart and Pakiser*, 1962], who generously permitted us to study the FM field tapes. Additional data were recorded by volunteer teams under program Plowshare [*Westhusing*, 1962]. Digitization of these data was followed in some cases by digital filtering or deconvolution [*Mitchell and Landisman*, 1970]. These operations often led to significant clarification of the seismic signals. Some of the wide-angle reflection arrivals associated with the curve having an intercept slightly greater than 6 sec are good examples of signals that showed significant improvements in quality after digital processing. These phases have been interpreted as reflections from the upper boundary of a velocity inversion at a depth of about 20 km. Note that these signals do not merge into a refraction branch at greater distances and may be interpreted as reflections from the upper surface of a velocity inversion.

The geophysical studies by *Mitchell and Landisman* [1970, 1971a, b] appear to indicate that this deeper velocity inversion may well extend across the Great Plains for at least several hundred kilometers. The cited papers also correlate this seismic inversion with a zone of low electrical resistivity located at similar depths in west Texas. The base of the seismic inversion apparently forms a continuous horizon beneath a set of major normal faults bounding the Wichita uplift in Oklahoma. Geological evidence indicates that the surface layers have undergone a relative vertical displacement of about 10 km across the frontal Wichita fault system [*Ham et al.*, 1964]. Refraction branches for the seismic horizons above the deeper crustal velocity inversion are interrupted by the set of faults;

by contrast, corresponding refractions from the base of this inversion and from three deeper horizons have the same apparent velocities in both directions along a reversed profile running normal to the strike of the fault system. *Mitchell and Landisman* [1970, 1971a] suggest that the deeper velocity inversion may be close to the lower boundary of the major tectonic movements that have occurred in this part of the Great Plains. It is possible to speculate that this anomalous zone in the crust beneath the southern Great Plains might be the site of mobile materials that accommodate large relative displacements at the surface.

Western United States. The tectonically active region in the western United States has been the site of a major program of seismic crustal studies by the United States Geological Survey. Record sections prepared from FM tape recordings have been presented in a number of the reports by that group. It is difficult to choose a single example for the present review from this wealth of material distributed over such a wide range of tectonic features.

The seismic investigations of southern Germany discussed in an earlier section of this paper show the importance of having normal-incidence reflection data available for a combined interpretation with refraction observations. Accordingly, an area in Utah (Figure 6, lower right corner) was chosen for which both reflection and refraction observations have been reported. The refraction studies consist of the U.S.G.S. profile Delta-West, which lies within the Basin and Range province, running westward from a shot point (*D* on the map in Figure 6) near the town of Delta in central Utah [*Eaton et al.*, 1964a, b]. Additional refraction measurements in a southerly direction (from Promontory shot point, shown by *P* on the map in Figure 6), roughly following the strike of the major structural trends, were reported earlier by *Berg et al.* [1957, 1960]. *Berg et al.* [1960] also reported normal-incidence reflection measurements for areas near Rozel hills (*R* on map) and Promontory (*P*), Utah [*Berg et al.*, 1959; *Narans et al.*, 1961]. The normal-incidence reflections at these two sites can be correlated, and the patterns of arrival are similar. Furthermore, the complexity of these signals suggests that there is greater complexity in the crust than might be inferred from a simple interpretation of the refraction data.

A record section of the U.S.G.S. profile Delta-West was reinterpreted by *Mueller and Landisman* [1971], as shown at the top of Figure 6. The continuous lines drawn across the seismogram montage show the correlations in the cited reinterpretation for the three well-recorded refraction branches having apparent velocities of 5.90, 6.42, and 7.43 km/sec. These velocities differ slightly from the values of 5.73 and 6.33 km/sec, which were derived from scattered readings by *Berg et al.* [1960] for the pattern of receivers south of Promontory. *Berg et al.* [1960] also reported a velocity of 7.59 km/sec. This velocity was inferred from first arrivals having smaller scatter, using the Promontory source and a line westward to recording sites at Gold Hill, in Utah, and Wells, Elko, and Eureka, in Nevada, at distances of 173 to 355 km from the shot point. *Berg et al.* [1960] concluded that there is no significant warping of the corresponding refracting horizon as it continues beneath the mountains to the west. *Berg et al.* [1960] also measured an apparent velocity of 7.44 km/sec, using first arrivals running eastward from the nuclear explosion Blanca, detonated at the Nevada test site and recorded by stations across Utah, at distances between 380 and 631 km from the source.

A somewhat higher apparent velocity of about 7.8 km/sec was reported by *Eaton et al.* [1964b] for the profile Delta-West, for stations near Eureka, Nevada, at distances approximately 150 to 250 km from the shot point. The corresponding interface was reported to be dipping approximately 1° to the west, based on apparent velocities of 7.6 km/sec measured across the same part of the Utah-Nevada border area for waves from the Shoal nuclear event, located about 500 km to the west. These higher velocity signals may also correspond to the 7.81-km/sec arrivals reported by *Diment et al.* [1961] for distances betwen 158 and 426 km along a line southeast from the Nevada test site toward Kingman, Arizona. These refractions may possibly be associated with a slightly deeper horizon than the one indicated by the 7.43-km/sec line in Figure 6. *Prodehl* [1970], in advance of publication, kindly furnished us with working copies of the recordings for the Delta-West profile at greater distances than those shown in

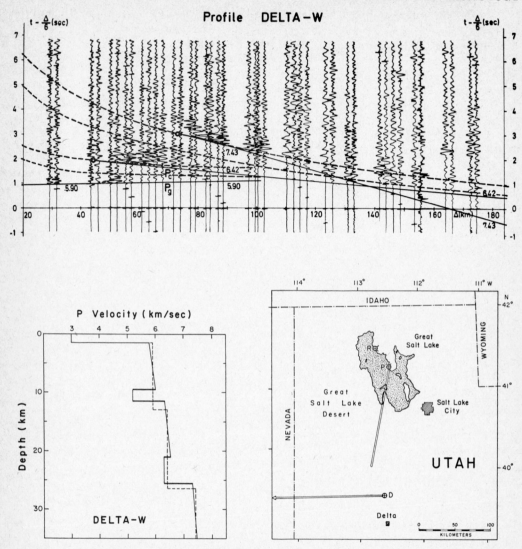

Fig. 6. A record section from the U.S.G.S. refraction profile Delta-West is shown in the upper half of the figure. The heavy solid lines indicate refraction times, and the small circles denote ray-optical critical points. The velocity distributions from a simplified interpretation of apparent velocities and intercept times (dotted line) and the combined interpretation of apparent velocities, intercept times, and normal-incidence reflection times (continuous line) appear at the lower left. Lower right is a map showing locations of the Delta-West profile, *D*, and sites of the reflection experiments at Promontory, *P*, and Rozel Hills, *R* (after *Mueller and Landisman* [1971], reproduced with permission from the *Geophysical Journal*).

Figure 6. It appears possible to us that these data may permit the distinction of two separate refraction branches of 7.4+ and 7.8+ km/sec.

Among the important observations reported by *Berg et al.* [1960] is a set of strong, critically reflected, later arrivals for stations about 75 to 105 km south of the shot point at Promontory [see *Berg et al.*, 1960, Fig. 5]. This is almost identical to the distance range for which similar arrivals are seen in the Delta-West recordings of Figure 6 and indicates a continuity of the refraction observations for this deeper horizon. This point will be discussed more fully after consideration of the reflection data, which are also indicative of continuous reflectors at depth in the crust.

The reflection times for Rozel Hills and Promontory, reproduced in columns 1 and 2 of Table 1, display nearly constant differences of only 0.3 sec for corresponding reflections, even though the two sites are 32 km apart. These nearly constant differences imply that the crust at these two sites is quite similar below some extremely shallow depth. The differences of 0.3 sec must then be generated at small depths corresponding to the sedimentary veneer plus, perhaps, a thin weathered layer at the top of the basement. Since little is known of this surficial cover for the Delta-West line (the recordings begin almost 30 km from the shot), we followed the ordinary convention of using an assumed sedimentary velocity, here taken to be 3 km/sec.

The separate internal consistencies of the reflection and refraction data then can be used to transfer the reflection observations along strike to the Delta shot point, if proper corrections are made for the delays in the shallow layers. For a model based on the assumed sedimentary velocity and the P_g refraction in Figure 6, the echo time to the basement for the Delta shot point is 1.0 sec. Berg et al. [1960] report that the corresponding value is 0.1 sec at Promontory. The differential shallow delay of 0.3 sec gives the equivalent value of 0.4 sec for Rozel hills (see first row of Table 1). Subtracting the values 0.1 sec at Promontory and 0.4 sec at Rozel hills from the 1.0-sec value at Delta yields differential delays of 0.6 and 0.9 sec. All the observed reflections for Rozel hills and Promontory, in columns 1 and 2, respectively, were then corrected to the values that should be expected for the area near the Delta shot point. The results are given in columns 3 and 4. The consistency of the observations and the method of adjustment are confirmed by the closely comparable times in columns 3 and 4.

These reflection data were then used to check and to supplement the refraction observations. In the simplest type of interpretation, in which no velocity reversal is assumed, the velocities and intercepts of the three refraction branches in the upper part of Figure 6 can be used to derive layer velocities and thicknesses. The results of this simplified interpretation are shown as a set of broken lines in the lower left corner of Figure 6. These velocities and thicknesses

are similar to those reported by Berg et al. [1960]. However, as concluded by Narans et al. [1961], the number of reflections encountered may indicate the existence of a complex sub-basement structure. Their conclusion is supported by the calculated echo times for this simplified interpretation, given in column 5 of Table 1. The discrepancies between the calculated times are serious; however, even more serious is the fact that many, if not most, of the echoes have no corresponding explanation.

Many models were then tried and discarded in an attempt to find interfaces that might cause the observed reflections. This was not an easy task, since the simplified model produces only two sub-basement reflections and the data appear to indicate that there are perhaps five or more discontinuities in the crust. Difficulties arise if these extra reflections are to be associated with increases of velocity, because, for example, no comparable refraction arrives between the onset of the basement line and the well defined 6.42-km/sec refraction branch indicated by the first continuous line in the record section. In addition, the introduction of intermediate velocities compounds the problem of explaining the position of the large critical point arrivals, since the reduction in contrast serves only to move the calculated critical point of the 6.42-km/sec refraction branch toward greater distances. As a result, the shallow echo at 3.2 sec in Table 1 is still unexplained.

There was greater success, however, with four of the later echoes. Fuchs and Landisman [1966] present a simple calculation that locates the boundaries of a low-velocity channel after specifying the normal-incidence reflection time to the bottom of the inversion; the intercept time corresponds to the refraction from the lower boundary of the channel and the velocities for the layers above, in, and below the inversion.

This simple formula was used with values taken from the Delta-West profile, and an attempt was made to satisfy the reflections in columns 3 and 4 of Table 1 as closely as possible. Both the thickness and velocity of the low-velocity channel can be determined if reflection times from both the top and bottom of the channel are available. The echoes at 3.8 and 4.3 sec in columns 3 and 4 were thus interpreted as returns from the top and bottom of the sialic low-velocity region, with a lower boundary not

TABLE 1.   Normal-Incidence Reflection Times

Rozel Hills	Promontory	Rozel Hills*	Promontory*	Simplified†	Combined‡
0.4	0.1	1.0	1.0	1.0	1.0
2.6	2.3	3.2	3.2	?	?
3.2	2.9	3.8	3.8	· · ·	3.7
3.7	3.4	4.3	4.3	4.9	4.5
6.7	6.8	7.3	7.7	· · ·	7.4
8.1		8.7			
	8.4		9.3	9.1	8.9
8.9		9.5			

* Adjusted for differential shallow delays at Delta.
† Calculated for simple apparent velocity-intercept model for profile Delta-West.
‡ Calculated for interpretation using adjusted reflection times and observed refraction arrivals for shot point near Delta.

very far from the depth found earlier for the top of the intermediate layer.

The problem just discussed for the 6.42-km/sec refraction also arises for the 7.43-km/sec line. The calculated critical point for the 7.43-km/sec branch is displaced toward greater distances than those observed on either the Delta-West or the Promontory profiles when intermediate-velocity layers are inserted. For this reason, the echoes in the third and fourth columns at 7.3 and 7.7 sec are assigned to the top of a second velocity inversion below the intermediate layer, at depths slightly greater than 20 km. Similarly, the echoes at 8.7 and 9.3 sec were correlated with the bottom of this inversion. The discrepancies between the calculated times in column 6 and the 'observed' times in columns 3 and 4 are less than 0.2 sec for the upper pair and 0.4 sec for the lower pair of reflections.

A further demonstration of the consistency of all the seismic observations and of the time corrections just discussed can be derived by comparing the Promontory refraction observations for the 7+-km/sec branch with the Delta-West seismograms. This demonstration provides strong support for the correction scheme adopted and for the lateral homogeneity of the sub sedimentary crustal layers, since the Promontory refractions were not used in the derivation.

In Table 1, 0.9 sec was added to reflections observed at Promontory in order to correct them to the Delta shot point. This same time can be added to readings for the 7+-km/sec line of the Promontory refraction profile [Berg

et al., 1960], and the modified results can then be compared to the Delta-West seismograms. Berg et al. [1960] recorded a later arrival near the critical distance at a station 76.5 km from the shot point. The travel time was about 14.7 sec. Subtracting the 6-km/sec reduction time of 12.75 sec and adding the 0.9-sec differential correction yield a reduced time of +2.8(5) sec. A first arrival at 27.64 sec was recorded at 173.3 km. Subtracting the 28.88-sec reduction time and adding the 0.9-sec differential correction yields a reduced time of −0.34 sec. Both these values are concordant with the Delta-West arrivals for the 7.43-km/sec branch shown in Figure 6, in support of the correction scheme used in Table 1.

The distances from the source to the large critical reflections observed on the profiles Delta-West and Promontory can also be used as an independent test of the proposed solution. The critical distances for the model with velocity inversions are shown by small circles on the 6.42- and 7.43-km/sec lines on the record section, using a ray-optical calculation that includes the effects of gradients and sphericity [Landisman et al., 1966]. The critical distances for the simpler model without velocity inversions are greater; the critical distance corresponding to the intermediate layer is moved more than 10 km farther from the shot point. Even after allowance for the effects of topography, which might introduce dips of as much as 1°, a model without the sialic low-velocity zone would have a critical point at least 5 km farther from the shot point than the preferred model. Červený [1966] has shown that the largest amplitudes for finite wavelengths can be about 20% farther from the source than the zero-wavelength critical point. It is therefore possible that the simpler interpretation is at some variance with the refraction observations themselves.

After the derivation of the preferred model was completed, we calculated wide-angle reflection times in order to compare them with the arrivals on the record section. The broken lines in the seismogram montage at the top of Figure 6 correspond to calculated reflections from the top and bottom of the two velocity inversions. The earliest reflections from shallow interfaces are situated nearer the bottom of the figure; those from deeper interfaces that came at later times are higher up in the figure. The curves for the lower boundaries are tangent to the appropriate refraction branches at their critical points.

Studies by *Fuchs and Kappelmeyer* [1962] and *Fuchs* [1969] have shown that wide-angle reflections tend to be weak except near the critical distance range. The results cited above provide a possible explanation for the weakness of the wide-angle reflected signals in the record section in Figure 6. The poor response of the recording equipment at high frequencies may also be partly responsible. These effects will be considered in relation to the signals recorded along the Delta-West profile.

The reflections along the earliest two-reflection hyperbolas, corresponding to the top and bottom of the sialic low-velocity zone, are extremely weak, except perhaps for distances between about 50 and 80 km, in the critical distance range. The high-frequency response of the recording equipment for this profile was not as great as that of the equipment used by some of the volunteer teams that recorded good wide-angle reflections for the Gnome profile (which is shown in Figure 7, p. 85, in this volume) or of the prospecting equipment used to record normal-incidence intra crustal reflections in Germany. This lack of high-frequency response may possibly contribute to the restriction of the observed reflections to high-signal areas near the critical points and to lower frequency components. Examples of the latter may be seen in the vicinity of the two slowest hyperbolas at greater distances.

The time interval between the second and third reflection hyperbolas in Figure 6 is char-

acterized by well developed ringing for seismograms between about 50 and 80 km. This time-distance interval corresponds to arrivals from the intermediate layer between the two velocity inversions. *Fuchs* [1968, 1969] has shown that ringing of this nature may be produced by a sequence of very thin layers having contrasting velocities. The appearance of the ringing at times corresponding to arrivals from between the two velocity inversions may be taken as an indication that the laminations thus inferred are not to be found within the velocity inversions. This fine structure may well be present, however, in the intermediate portion of the crust between the two velocity inversions.

REVIEW OF RESULTS FOR THE RHINE RIFT AREA

An intensive program of geophysical and geological investigation in the neighborhood of the Rhine rift system, including the Vosges and Black Forest uplifts, the south German triangle and the Molasse basin, has been in progress for almost twenty years, with active contributions by the German Research Group for Explosion Seismology, the Deutsche Forschungsgemeinschaft (German Research Association), and equivalent government, industry, and university groups in France and Germany. Many of the results of this effort have been presented in *Illies and Mueller* [1970]; a few a these results will be summarized here.

Figure 7 shows an ESE-WNW cross section across the Rhine rift area in the vicinity of Karlsruhe. Heavy bars in the diagram indicate places where individual seismic refraction and reflection experiments have been inserted into the interlocking network of geophysical measurements used to derive this interpretation. The compressional velocity distribution B-B corresponds to the center of the graben, where commercial reflection equipment was used to obtain echoes from interfaces down through the crust to the Mohorovicic discontinuity, which overlies 8.2-km/sec upper mantle material. The layer of 7.6–7.7-km/sec material that appears to be particularly associated with the rift system and that is known as the 'rift cushion' has a velocity similar to that found in the western United States at roughly equivalent depths. The velocity distribution above this layer shows general features that resemble those interpreted for the Utah-Nevada border area (Figure 6). The

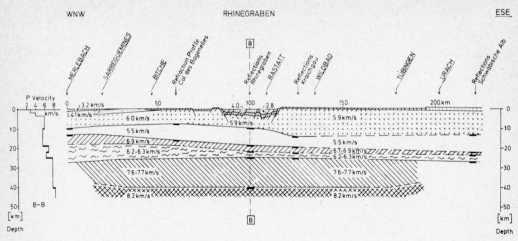

Fig. 7. Crustal model of the Rhine graben rift system (reproduced from *Ansorge et al.* [1970], with permission of Schweizerbart'sche Verlag).

thickening of the sialic low-velocity zone in the center of the structure and its greater depth toward the eastern side are reflected in the patterns found from studies of local seismicity, which apparently tend to follow the sialic low-velocity zone [*Hiller et al.*, 1967]. Refraction data for this area show clear evidence of the P_e phase, as shown in Figure 1. Studies of refraction data, in combination with appropriate reflection data when available, have contributed greatly to the study of the sialic low-velocity region in the upper part of the crust and of a second, deeper, inversion in the lower crust [*Mueller and Landisman*, 1965; *Landisman and Mueller*, 1966; *Mueller et al.*, 1967, 1969; *Ansorge et al.*, 1970].

The velocity distributions found from explosion seismic experiments have been verified by studies of the frequency-dependent interaction of the crustal layering with various types of seismic energy. Rayleigh wave phase velocities measured for several paths across the Rhine rift system have been shown to be closely comparable to calculations for the velocity distribution given in Figure 7 [*Seidl et al.*, 1970].

Studies of the Rhine graben by *Bonjer and Fuchs* [1970], using the crustal transfer function [*Phinney*, 1964], provide additional confirmation for the model in Figure 7. In this experiment, teleseismic long-period P waves were recorded at Stuttgart, which lies on the eastern flank of the Rhine graben at a site corresponding to the area between Wildbad and

Tübingen in the cross section of Figure 7. The ratios formed by the spectrum of the vertical component of ground motion divided by the spectrum of the longitudinal component for several teleseismic recordings were compared to calculations for a few models, as shown in Figure 8. Among the test cases was the velocity distribution designated STU K10, taken from the appropriate part of the cross section in Figure 7, with two velocity inversions and a rift cushion (arrow). The earlier models in Figure 8 predict minimums in their theoretical transfer ratios at just those frequencies for which maximums are observed and vice versa, whereas the model with detailed features including the rift cushion, as derived from the combined interpretation of the network of refraction and reflection measurements, is concordant with the observations. This agreement is all the more surprising since it was obtained without any adjustment of the model during the process of comparison.

ROLE OF WATER OF HYDRATION IN THE CRUST: HYPOTHESIS AND IMPLICATIONS

Further examples of crustal velocity inversions can be found in the literature [e.g., *Pavlenkova*, 1968; *Lukk et al.*, 1970; *Bloch et al.*, 1969]. Similarly, examples of crustal resistivity inversions have been widely reported [*Keller et al.*, 1966; Kraev et al., 1948, in *Kraev*, 1952; *Petr et al.*, 1964; *van Zijl*, 1969; *Mitchell and Landisman*, 1971a]. In several of these cases, the experimental areas for the electrical and seismic studies were sufficiently contiguous to show that the locations of the low-velocity and low-resistivity zones are at least roughly comparable. The presence of interstitial water in the crust provides a physical mechanism for reductions in both seismic velocity and electrical resistivity. The possible consequences and the origin of a zone of enhanced fluid content are discussed in the following paragraphs. Although a consistent pattern emerges, additional work is needed to study this problem in detail.

Southern Africa is one example of a region in which zones of low velocity and low resistivity may occur at similar depths. A model that includes crustal low-velocity zones is compatible with the multi-mode seismic surface wave dispersion results for southern Africa [*Bloch et al.*, 1969]. The depths interpreted for these zones can be related to those reported for a low-resistivity layer [*van Zijl*, 1969], if appropriate corrections are made for regional uplift in the southerly portions of South Africa [*Matthews*, 1959; *Nicolaysen and Burger*, 1965]. This resistivity experiment is especially important for several reasons. Problems of lateral inhomogeneity were minimized, since a 350-km Schlumberger line was entirely located within a nearly rainless granite-gneiss outcrop area. Both the absence of surficial sediments and the length of the line enhanced the ability of the measurement to penetrate to great depths. Further, the use of an artificial direct current source in the Schlumberger method avoids possible ambiguities related to the distribution of telluric currents from natural sources.

Several independent groups have conducted laboratory research of the effect of interstitial water on the physical properties of rocks under crustal conditions. *Nur and Simmons* [1969, Figure 3] have shown that a granite with a porosity of only 0.7% exhibits a 10% reduction in compressional velocity and a 35% reduction in shear velocity at external pressures of several kilobars, when the fluid pressure in the rock is permitted to approach the external pressure. *Gordon and Davis* [1968] have related low-velocity, attenuating regions in the earth to interstitial fluid at high pore pressures; they consider this process essential to the mechanism responsible for the energy loss and reduction of velocity. It is perhaps significant that *Long and*

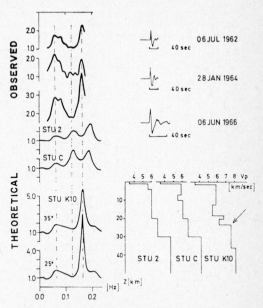

Fig. 8. Observed (upper left) and theoretical (lower left) crustal transfer functions for the region near Stuttgart in West Germany. The model STU K10 was taken from the refraction and reflection interpretation in Figure 7 (reproduced from *Bonjer and Fuchs* [1970], with permission of Schweizerbart'sche Verlag).

Berg [1969] interpreted tape recordings of the Gnome event in terms of a crustal model containing an attenuating zone centered at about 15 km, a depth which is roughly comparable to that postulated by *Mitchell and Landisman* [1971a] for a low-velocity, low-resistivity layer in the same general area. The work of *Brace et al.* [1965] has shown the effectiveness of minor concentrations of water in reducing the electrical resistivity of rocks. Thus, the low velocities and low resistivities found by Mitchell and Landisman and the attenuation measured by *Long and Berg* [1969] may all be caused by higher concentrations of interstitial water of hydration in a thin zone deep within the crust.

The low-velocity zone postulated for moderate crustal depths in eastern New Mexico [*Mitchell and Landisman*, 1971a] is possibly related to a region in Oklahoma that is characterized by enhanced mobility in response to tectonic stress [*Mitchell and Landisman*, 1970]. The enhancement of crustal mobility in a zone that may be the site of minor concentrations of fluid recalls studies such as those by *Hubbert and Rubey* [1959], *Raleigh and Paterson* [1965], and *Heard and Rubey* [1966]. These studies showed that interstitial fluids may reduce the cohesive strength of rocks, since pore pressure may act to reduce the effective pressure, thus lowering the shear stress needed to initiate sliding.

Hyndman and Hyndman [1968], *Berdichevskiy et al.* [1969], and *Mitchell and Landisman* [1971a] have suggested that the presence of electrically conductive layers, sometimes interpreted at moderate depths within the crust, can be explained in terms of certain dehydration reactions resulting in the release of free water into surrounding pore spaces. *Berdichevskiy et al.* [1969] and *Mitchell and Landisman* [1971a] have suggested that the conductive zone and the accompanying dehydrated mineral assemblages lie at moderate depths, above the more basic crustal layers. The temperatures and pressures at mid-crustal depths may well favor dehydration of the more acidic mineral assemblages, whereas higher temperatures and correspondingly greater depths are required for dehydration of more basic rocks. *Berckhemer* [1969] also favors the view that rocks corresponding to mid-crustal depths have been dehydrated, as a result of his geological field studies of the Alpine Ivrea zone, which is postulated to be a section of the crust and uppermost mantle that has been carried to the surface.

Interstitial water at moderate crustal depths can contribute profoundly to the tectonic processes of selective melting, fractionation, and recrystallization, even when present in only small concentrations. *Borchert and Boettcher* [1967] have proposed that an important reaction zone for the production of granitic and flood-basaltic magmas may occur near depths of about 25 km.

Critical-distance refraction amplitudes related to the presence of water of hydration. The first result of the proposed presence of interstitial water in the crust comes from studies of critical-distance refraction amplitudes. The relatively small amplitudes observed for the maximums in the vicinity of the critical points for the Gnome profile, shown in Figure 7, p. 85 in this volume, can be compared to the

remarkably energetic critical arrivals recorded for the profile Delta-West (Figure 6). *Healy and Warren* [1969] presented and discussed the amplitude maximums recorded for a pair of profiles from eastern and western Colorado, in the Great Plains and Rocky Mountain areas, respectively. The apparent velocities for the P_g, P_b (or P^*), and P_n branches were approximately the same for both experiments, but the relative amplitudes near the critical points differed by a large factor. Far stronger signals were observed for the experiment in western Colorado. Similar results are seen when comparing the profile Delta-West with the profiles in the Great Plains.

If the velocity contrasts at the corresponding layer boundaries in the two Colorado profiles are nearly the same, as is reported by *Healy and Warren* [1969], then the wave amplitudes developed near the interference maximums should also be similar. The critical amplitudes observed for these two profiles are quite different, however. This sort of disparity has been described as typical of the results for many, though not all, surveys in the eastern and western United States [*Healy and Warren*, 1969].

A possible explanation may be drawn from theoretical calculations by *Červený* [1966], whose amplitudes for finite-wavelength (6-Hz) signals are reproduced in Figure 9. Amplitudes in the vicinity of the interference maximums are shown as a function of distance, with index of refraction as a parameter. Small indices of refraction (large contrast) are characterized by large, well focused maximums at small distances from the optical (zero wavelength) critical point, denoted by a plus sign to the left of the highest point of each curve. Conversely, refractive indices approaching unity (small contrast) produce weak, poorly focused maximums that migrate farther from the zero-wavelength critical point. There is a striking resemblance of the amplitude-distance curves for the high-contrast case to the illustrative examples taken from the western United States and of the lower contrast case to observations from the eastern United States. If the large-velocity decrease shown in *Nur and Simmons* [1969, Figure 3] is recalled, we suggest that it is not unreasonable to ascribe the amplitude differences between profiles from eastern and western Colorado to poorly developed and well developed velocity inversions above the corresponding refracting horizons. If the suggestions of *Hyndman and Hyndman* [1968], *Berdichevskiy et al.* [1969], and *Mitchell and Landisman* [1971a] are tenable, the inversions may be associated with free water of hydration released by dehydrating reactions at moderate crustal depths. The existence of this water would be favored in regions marked by higher crustal temperatures. It is interesting to consider the possibility that the well focused wave groups near the interference maximums, as sometimes observed in the western United States, might be viewed as an indication of the higher temperatures likely to be found at mid- and lower-crustal depths in this region.

Tectonic implications of zones of free water of hydration. Non-meteoritic water, related to dehydration reactions at depth, might also influence seismic and aseismic deformation at shallower depths as the crust responds to tectonic stresses. There exists an increasing body of evidence implying that water is involved in many stages of crustal deformation, although the process is complex. Field studies of earthquakes in Matsushiro, Japan, have produced evidence that east-west compressive stress 'was released by brittle fracturing where the rock strength was lowered by fluid at the center of the epicentral area; the increasing number of

hypocenters gradually became shallower with the inflation of the surface and extension in a north-south direction of the inflated area. The increase of the fluid pressure culminated in the areal outflow of [about 10^7 m^3 of] the underground water' [*Nakamura*, 1969]. The swelling subsided, and the epicenters moved to another area [*Nakamura*, 1969; *Harada*, 1969]. Similar sequences, though less well documented, have been reported in other regions, such as the Gediz, Turkey, events of March 1970.

In order to discuss the influence of interstitial water of hydration on crustal deformation and shallow earthquakes, the major factors that favor and inhibit these natural phenomena must be considered. Movement often seems to occur along existing faults and their extensions and can take the form of either stable sliding or sudden stick-slip. *Byerlee* [1967] has proposed that the average apical angle of the asperities along the fault can provide a means of estimating which type of motion will prevail. *Byerlee and Brace* [1968, 1969] emphasize the fact that the pressure (i.e., the depth) must reach a minimal value before stick-slip can occur.

Mineralogy is also an important factor; small amounts (as little as 3%) of serpentine [*Brace*, 1969] or altered gabbroic rocks [*Byerlee and Brace*, 1968] are sufficient to terminate the

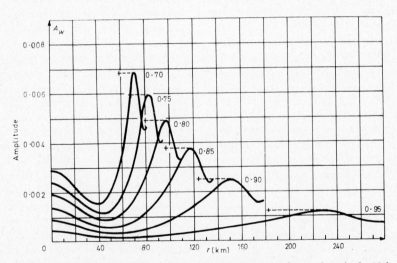

Fig. 9. Amplitude curves for 6-Hz waves reflected from the base of a single 30-km thick layer overlying a half space. The numbers indicate different values for the refractive index at the discontinuity, and the plus signs denote the positions of corresponding critical points from ray theory (reproduced from *Červený* [1966], with permission of the *Geophysical Journal*).

stick-slip process, which then reverts to stable sliding. It is possible that the rather sudden termination of seismicity with depth, as observed in a number of continental earthquake areas other than those associated with deep-plunging Benioff zones [e.g., *Cleary et al.*, 1964; *Stauder and Ryall*, 1967; *Kisslinger*, 1968], may be attributed to more basic rocks which occur in the lower portions of the crust [*Scholz et al.*, 1969]. The depth range for seismicity in the crust is apparently less than that postulated for the span of depths characteristic of the two crustal velocity inversions. It ranges from a few km down to a maximum of the order of twenty-odd kilometers or less, with most of the activity confined to depths as shallow as those proposed for the sialic low-velocity zone.

Laboratory studies indicate that stress corrosion by water is the primary cause of failure in silicates [*Charles*, 1959; *le Roux*, 1965; *Hillig and Charles*, 1965] and that hydration of the silicon-oxygen bond, a thermally accelerated reaction [*Charles*, 1959; *le Roux*, 1965], is responsible for the loss in rock strength. As another result of the proposed hypothesis it is interesting to speculate that water associated with temperature- and pressure-sensitive hydration reactions at moderate depths might slowly migrate upward through zones of weakness in the crust as it moves from regions marked by relatively higher lithostatic and tectonic stresses toward regions of stress release. It might then make its presence known through either seismic or aseismic deformation as the crust adjusts to regional and local forces. Geophysical data related to this speculation are presented in the following section.

Lateral variations of crustal seismicity related to free water of hydration. The seismicity map of North America, which was compiled and discussed by *Woollard* [1969], furnishes important evidence of present tectonic activity in this part of the world. The map, reproduced as Figure 10, contains all epicenters reported in Canada, Mexico, and the United States through 1964. In discussing this map, *Woollard* [1969] notes that the roughly east-west demarcation near the United States-Canada border separates a largely aseismic region to the north from the moderately seismic area to the south, even though there is no appreciable difference in the detectability of earthquakes for the two areas.

Woollard attributes the difference to residual effects of glaciation; the aseismic area generally follows the region that is slowly rebounding after major depression during Pleistocene glaciation, with few exceptions. These include the Pacific margin and the long and narrow seismic region near the Saint Lawrence valley, which has been compared to a rift zone by *Kumarapeli and Saull* [1966]. The sparsely distributed epicenters in northernmost Canada tend to follow the northern edge inferred for the Pleistocene ice sheet. Similarly, northernmost Eurasia does not contribute much seismicity to the maps of *Barazangi and Dorman* [1969]. The Baikal rift zone, running along the Lena River to the Arctic Ocean [*Gutenberg and Richter*, 1954; *Heezen and Ewing*, 1961], may be an area that has similarities to the Saint Lawrence valley, especially as regards its long and narrow pattern of seismicity, which, according to *Petrushevsky* [1969], is a characteristic feature of rift zones. Referring again to Figure 10, the moderately seismic Great Plains of the United States may have a Precambrian basement complex not greatly different from that exposed at the surface of the Canadian Shield, even though the Great Plains are covered by a thin veneer of Paleozoic and younger sediments. The Great Plains area contrasts fairly well with the nearly aseismic borderlands of the Gulf Coast geosyncline, which is noted for its thick blanket of sediments, and the aseismic area of Greenland, which is covered by a continuous ice plateau.

The suggested rationale by which crustal seismicity could be inhibited in these aseismic areas may provide another originally unexpected consequence of the hydration hypothesis. These relatively aseismic regions might be described as areas that either are now, or have recently been, the site of widely distributed crustal loads. The higher pressure associated with crustal loading would affect all relevant mineral reactions in the deeper portions of the crust, inhibit the presence of free water of hydration, and decrease the availability of water for stress-accelerated corrosion; thus it would lead to a lower level of crustal seismicity.

In view of this discussion of the importance of water in crustal tectonics, it is possibly noteworthy that a significant number of hot spring areas are characterized by more than moderate seismicity. Hot springs [*Waring*, 1965] can be

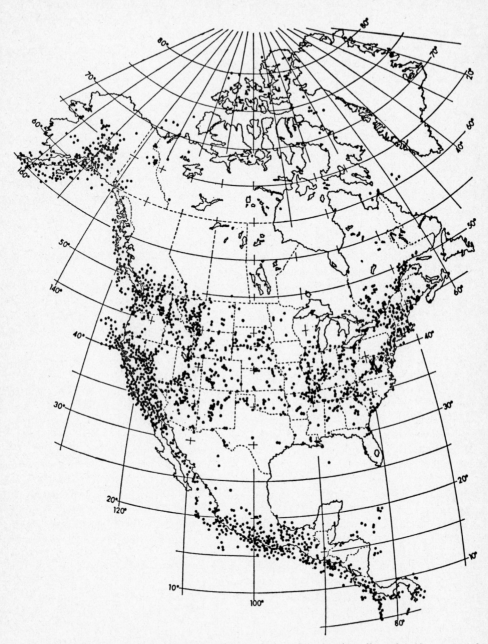

Fig. 10.   Map of earthquake epicenters in North America, showing all earthquakes reported through 1964 by the U.S. Coast and Geodetic Survey, the Dominion Observatory of Canada, and the Instituto de Geofísica of the University of Mexico (reproduced from *Woollard* [1969]).

found along the more seismic rift-like feature which forms an extension of the Saint Lawrence valley that reaches all the way to Arkansas. Hot springs also occur along the parallel seismic trend some 400 km to the southeast, running through Virginia. The greatest concentration of hot springs, though, can be found in the tectonically active western United States, which is also marked by greater crustal seismicity.

SUMMARY

Seismic evidence for inversions of compressional and shear velocity in the continental crust has been re-examined, and results for a few areas have been discussed in some detail. Important evidence for a shallow crustal velocity inversion comes from large secondary P_c refraction signals that arrive about one second after weaker P_g refractions from the crystalline basement at distances of several tens out to at least 100 km from the shot point. The delay of these strong refractions and their occurrence at small distances can be attributed to the presence of the sialic low-velocity zone, which is characterized by a compressional velocity of about 5.5 km/sec, and to an abrupt lower boundary at depths of about 8 to 15 km. The subjacent layers are interpreted to have velocities of approximately 0.2 km/sec or more in excess of those of the overlying basement rocks. Normal-incidence and wide-angle observations of high-frequency intracrustal reflections also provide evidence in support of crustal velocity inversions.

Much of the evidence for the sialic low-velocity zone comes from areas where the basement has been strongly folded or tectonically deformed. Gravity investigations of relatively light, mobile intrusive bodies in tectonic areas give some support to this hypothesis, since the maximum depths inferred for these bodies are concordant with those typical of the sialic low-velocity layer. A band of semi-continuous intrusions would appear to be homogeneous for long-wavelength refraction experiments. Short-wavelength reflection measurements would produce extremely variable results, in accord with numerous observations.

Detailed seismic interpretations are permitted by recent improvements in instrumental technology which generate broad-band recordings on magnetic tape that can be digitized and processed with a digital computer. One of the more useful forms of presentation is a record section consisting of a densely observed series of seismograms that have been digitally processed and plotted. A reduced time scale permits the resolution of small velocity variations.

Seismic results from greater depths within the crust are discussed for areas in the Great Plains and the Basin and Range provinces of North America. Observations of the Gnome event in New Mexico contain reflection and refraction data that are compatible with the interpretation of a deeper crustal velocity inversion at depths of about 20 km [Mitchell and Landisman, 1971a, b]. A corresponding velocity inversion interpreted from seismic data taken in Oklahoma may bear some relation to tectonic uplift in that area [Mitchell and Landisman, 1970]. It is possible that a resistivity inversion inferred from magnetotelluric measurements in northwestern Texas [Mitchell and Landisman, 1971a] and a zone of seismic attenuation derived earlier from the Gnome data [Long and Berg, 1969] are also related to this deeper crustal velocity inversion. Laboratory studies of low-porosity rocks containing interstitial fluid [Gordon and Davis, 1968] and of rocks having saturated pores with pore pressures nearly equal to the external pressure [Nur and Simmons, 1969, Figure 3] indicate that minor concentrations of fluid can significantly reduce the velocity and the amplitude of seismic waves. Minor concentrations of water can also produce greatly reduced values of electrical resistivity [Brace et al., 1965]. It is suggested that the presence of non-meteoritic water governed by the equilibrium boundaries of hydrothermal reactions may be related to the zone of greater electrical conductivity, increased seismic attenuation, decreased seismic velocity, and enhanced tectonic mobility.

The enhancement of the proposed dehydration reaction by increases of temperature suggests a possible result: higher crustal temperatures might produce a well developed velocity inversion sufficient to increase the impedance contrast and produce high-amplitude, critically reflected seismic waves. This process might explain observations of high-amplitude, well focused critical reflections at some locations in the western United States, as compared to small-amplitude reflections recorded east of the Rocky

Mountains, where the inversion is less well developed.

Another possible result of the relation of crustal mobility to the presence of free water of hydration comes from studies of the lateral distribution of crustal seismicity. The retardation of the postulated dehydration reaction by increases of pressure leads to a reduction in the availability of water for corrosive weakening in regions of stress concentration. This restriction in the supply of interstitial water may provide a possible explanation for observed low values of seismicity in formerly glaciated areas covering most of Canada and northern Eurasia and in the geosynclinal borderlands of the Gulf of Mexico.

Acknowledgments. The authors are grateful to many people and organizations for various types of help furnished during the course of this study. Dr. Michael Holdaway, of Southern Methodist University, provided helpful discussions on the petrological aspects of this study. Drs. J. Healy and J. Roller, of the National Center for Earthquake Research, U.S. Geological Survey, permitted the study of analog tape recordings of seismic data collected in North America. Dr. K. Fuchs, of the Geophysical Institute, University of Karlsruhe, Karlsruhe, Germany, furnished his program 'Laufzeiten.' Computer facilities were made available by the Southern Methodist University Computer Laboratory.

Support for this work was provided by the Office of Naval Research under contract N00014-67-0310-0001, by the National Science Foundation under grant GA-25683, and by supplemental funds of the University of Texas at Dallas.

REFERENCES

Ansorge, J., D. Emter, K. Fuchs, J. Lauer, S. Mueller, and E. Peterschmitt, Structure of the crust and upper mantle in the rift system around the Rhinegraben, in *Graben Problems*, edited by H. Illies and S. Mueller, pp. 190–197, Schweizerbart, Stuttgart, Germany, 1970.

Barazangi, M., and J. Dorman, World seismicity maps compiled from ESSA, Coast and Geodetic Survey, epicenter data, 1961–1967, *Bull. Seismol. Soc. Amer., 59,* 369–380, 1969.

Berckhemer, H., Direct evidence for the composition of the lower crust and the Moho, *Tectonophysics, 8,* 97–105, 1969.

Berdichevskiy, M., V. Borisova, V. Bubnov, L. Van' yan, I. Fel'dman, and I. Yakovlev, Anomaly of the electrical conductivity of the Earth's crust in Yakutiya, *Izv., Acad. Sci. USSR, Phys. Solid Earth* (10), 633–637, 1969.

Berg, J., Jr., K. Cook, and W. Dolan, Seismic results of quarry blasts at Lakeside and Promon-tory Point, Utah (abstract), *Geol. Soc. Amer. Bull., 68,* 1858, 1957.

Berg, J., Jr., K. Cook, and H. Narans, Jr., Seismic studies of crustal structure in the eastern Basin and Range province (abstract), *Geol. Soc. Amer. Bull., 70,* 1709, 1959.

Berg, J., Jr., K. Cook, H. Narans, Jr., and W. Dolan, Seismic investigation of crustal structure in the eastern part of the Basin and Range province, *Bull. Seismol. Soc. Amer., 50,* 511–535, 1960.

Bloch, S., A. Hales, and M. Landisman, Velocities in the crust and upper mantle of southern Africa from multimode surface wave dispersion, *Bull. Seismol. Soc. Amer., 59,* 1599–1629, 1969.

Bonjer, K.-P., and K. Fuchs, Crustal structure in southwest Germany from spectral transfer ratios of long-period body waves, in *Graben Problems*, edited by H. Illies and S. Mueller, pp. 198–202, Schweizerbart, Stuttgart, Germany, 1970.

Borchert, H., and W. Boettcher, Zur Petrologie der Lithosphaere in ihrer Beziehung zu geophysikalischen Diskontinuitaeten, auch der Gesamterde, *Gerlands Beitr. Geophys., 76,* 257–277, 1967.

Bott, M., and R. Smith, The estimation of the limiting depth of gravitating bodies, *Geophys. Prospect., 6,* 1–10, 1958.

Bott, M., and S. Smithson, Gravity investigations of subsurface shape and mass distributions of granite batholiths, *Geol. Soc. Amer. Bull., 78,* 859–878, 1967.

Bott, M., A. Day, and D. Masson-Smith, The geological interpretation of gravity and magnetic surveys in Devon and Cornwall, *Phil. Trans. Roy. Soc. London A, 251,* 161–191, 1958.

Brace, W., Laboratory studies pertaining to earthquakes, *Trans. New York Acad. Sci., Ser. 2, 31,* 892–901, 1969.

Brace, W., A. Orange, and T. Madden, The effect of pressure on the electrical resistivity of water-saturated crystalline rocks, *J. Geophys. Res., 70,* 5669–5678, 1965.

Byerlee, J., Theory of friction based on brittle fracture, *J. Appl. Phys., 38,* 2928–2934, 1967.

Byerlee, J., and W. Brace, Stick-slip, stable sliding, and earthquakes, *J. Geophys. Res., 73,* 6031–6037, 1968.

Byerlee, J., and W. Brace, High-pressure mechanical instability in rocks, *Science, 164,* 713–715, 1969.

Červený, V., On dynamic properties of reflected and head waves in the *n*-layered Earth's crust, *Geophys. J. Roy. Astron. Soc., 11,* 139–147, 1966.

Charles, R., The strength of silicate glasses and some crystalline oxides, in *Proc. Int. Conf. Fracture*, pp. 225–250, MIT Press, Cambridge, Mass., 1959.

Cleary, J., H. Doyle, and D. Moye, Seismic activity in the Snowy Mountains region and its relationship to geological structures, *J. Geol. Soc. Aust., 11,* 89–106, 1964.

Diment, W., S. Stewart, and J. Roller, Crustal structure from the Nevada test site to Kingman,

Arizona, from seismic and gravity observations, *J. Geophys. Res., 66,* 201–214, 1961.

Dix, C., Reflection seismic studies, *Geophysics, 30,* 1068–1084, 1965.

Dohr, G., Ein Beitrag der Reflexionsseismik zur Erforschung des tiefern Untergrundes, *Geol. Rundsch., 46,* 17–26, 1957.

Dohr, G., Über die Beobachtungen von Reflexionen aus dem tieferen Untergrunde im Rahmen reflexionsseismischer Messungen, *Z. Geophys., 25,* 280–300, 1959.

Eaton, J., J. Healy, W. Jackson, and L. Pakiser, Upper mantle velocity and crustal structure in the eastern Basin and Range province, determined from SHOAL and chemical explosions near Delta, Utah, *Bull. Seismol. Soc. Amer., 54,* 1567, 1964a.

Eaton, J., J. Healy, W. Jackson, and L. Pakiser, Upper mantle velocity and crustal structure in the eastern Basin and Range province, determined from SHOAL and chemical explosions near Delta, Utah (abstract), *Ann. Meet. Seismol. Soc. Amer. Program,* 30–31, 1964b.

Fuchs, K., The reflection of spherical waves from transition zones with arbitrary depth-dependent elastic moduli and density, *J. Phys. Earth, 16,* 27–41, 1968.

Fuchs, K., On the properties of deep crustal reflectors, *Z. Geophys., 35,* 133–149, 1969.

Fuchs, K., and O. Kappelmeyer, Report on reflection measurements in the Dolomites—September, 1961, *Boll. Geofiz. Teor. Appl., 4,* 133–141, 1962.

Fuchs, K., and M. Landisman, Detailed crustal investigation along a North-South section through the central part of western Germany, in *The Earth beneath the Continents, Geophys. Monogr. Ser.,* vol. 10, edited by J. Steinhart and T. Smith, pp. 433–452, AGU, Washington, D.C., 1966.

Giese, P., Die Geschwindigkeitsverteilung im obersten Bereich des Kristallins, abgeleitet aus Refraktionsbeobachtungen auf dem Profil Böhmischbruck-Eschenlohe, *Z. Geophys., 29,* 197–214, 1963.

Giese, P., The structure of the Earth's crust in central Europe, *10th General Assembly, Europ. Seismol. Comm., Leningrad, 3,* 30 pp., 1968.

Gordon, R., and L. Davis, Velocity and attenuation of seismic waves in imperfectly elastic rock, *J. Geophys. Res., 73,* 3917–3935, 1968.

Gutenberg, B., Structure of the Earth's crust in continents, *Science, 111,* 29–30, 1950.

Gutenberg, B., Crustal layers of continents and oceans, *Geol. Soc. Amer. Bull., 62,* 427–440, 1951.

Gutenberg, B., Effects of low velocity layers, *Geofis. Pura Apl., 28,* 1–10, 1954.

Gutenberg, B., Wave velocities in the Earth's crust, in *Crust of the Earth, Geol. Soc. Amer. Spec. Pap. 62,* edited by A. Poldervaart, pp. 19–34, 1955.

Gutenberg, B., and C. Richter, *Seismicity of the Earth and Associated Phenomena,* 2nd ed., 310 pp., Princeton Univ. Press, Princeton, New Jersey, 1954.

Ham, W., R. Denison, and C. Merritt, Basement rocks and structural evolution of southern Oklahoma, *Okla. Geol. Surv. Bull. 95,* 301 pp., 1964.

Hamilton, W., Mesozoic California and the underflow of Pacific mantle, *Geol. Soc. Amer. Bull., 80,* 2409–2429, 1969.

Harada, Y., Geodetic work in Japan with special reference to the Matsushiro earthquakes (abstract), *Eos, Trans. AGU, 50,* 390, 1969.

Healy, J., and D. Warren, Explosion seismic studies in North America, in *The Earth's Crust and Upper Mantle, Geophys. Monogr. Ser.,* vol. 13, edited by P. Hart, 208–220, AGU, Washington, D.C., 1969.

Heard, H., and W. Rubey, Tectonic implications of gypsum dehydration, *Geol. Soc. Amer. Bull., 77,* 741–760, 1966.

Heezen, B., and M. Ewing, The mid-oceanic ridge and its extension through the Arctic Basin, in *Geology of the Arctic,* pp. 622–642, Univ. Toronto Press, Toronto, Canada, 1961.

Hiller, W., J.-P. Rothé, and G. Schneider, La seismicité du Fosse Rhenan, *Abh. Geol. Landesamt Baden-Wuerttemberg, 6,* 98–100, 1967.

Hillig, W., and R. Charles, Surfaces, stress dependent surface reactions, and strength, pp. 682–705 in *Proc. 2nd Int. Conf. High Strength Materials, Berkeley, California,* edited by V. Zackay, John Wiley, New York, 1965.

Hubbert, M., and W. Rubey, Role of fluid pressure in mechanics of overthrust faulting, *Geol. Soc. Amer. Bull., 70,* 115–205, 1959.

Hyndman, R., and D. Hyndman, Water saturation and high electrical conductivity in the lower continental crust, *Earth Planet. Sci. Lett., 4,* 427–432, 1968.

Illies, J., and S. Mueller (Eds.), *Graben Problems,* 316 pp., Schweizerbart, Stuttgart, Germany, 1970.

Junger, A., Deep reflections in Big Horn County, Montana, *Geophysics, 16,* 499–505, 1951.

Keller, G., L. Anderson, and J. Pritchard, Geological Survey investigations of the electrical properties of the crust and upper mantle, *Geophysics, 31,* 1078–1087, 1966.

Kisslinger, C., Energy density and the development of the source region of the Matsushiro earthquake swarm, *Bull. Earthquake Res. Inst. Japan, 46,* 1207–1223, 1968.

Kraev, A. V., *Osnovi Geoelektriki,* Gostoptekizdat, Moscow, 1952.

Kumarapeli, P., and P. Saull, The St. Lawrence Valley system: A North American equivalent of the East African rift valley system, *Can. J. Earth Sci., 3,* 639–657, 1966.

Landisman, M., and S. Mueller, Seismic studies of the Earth's crust in continents, 2, Analysis of wave propagation in continents and adjacent shelf areas, *Geophys. J. Roy. Astron. Soc., 10,* 539–548, 1966.

Landisman, M., Y. Satô, and T. Usami, Propagation of disturbances in a Gutenberg—Bullen A' spherical Earth model: Travel times and amplitudes of S waves, in The Earth beneath the Continents, Geophys. Monogr. Ser., vol. 10, edited by J. Steinhart and T. Smith, pp. 482–494, AGU, Washington, D.C., 1966.

le Roux, H., The strength of fused quartz in water vapor, Proc. Roy. Soc. London A, 286, 390–401, 1965.

Liebscher, H., Reflexionshorizonte der tieferen Erdkruste im Bayerischen Alpenvorland, abgeleitet aus Ergebnissen der Reflexionsseismik, Z. Geophys., 28, 162–184, 1962.

Liebscher, H., Deutungsversuche für die Struktur der tieferen Erdkruste nach reflexionsseismischen und gravimetrischen Messungen im deutschen Alpenvorland, Z. Geophys., 30, 51–96 and 115–126, 1964.

Long, L., and J. Berg, Transmission and attenuation of the primary seismic wave, 100 to 600 km, Seismol. Soc. Amer. Bull., 59, 131–146, 1969.

Lukk, A., I. Nersesov, and L. Chepkunas, Procedure for identification of the low velocity layer in the Earth's crust and mantle, Izv., Acad. Sci. USSR, Phys. Solid Earth, (2), 77–80, 1970.

Matthews, P., The metamorphism and tectonics of the Pre-Cape formations in the Post-Ntingwe thrust-belt, S.W. Zululand, Natal, Trans. Proc. Geol. Soc. South Africa, 62, 257–322, 1959.

Mead, J., Preliminary results of deep crustal reflection studies (abstract), Eos, Trans. AGU, 31, 324, 1950.

Mitchell, B., and M. Landisman, Interpretation of a crustal section across Oklahoma, Geol. Soc. Amer. Bull., 81, 2647–2656, 1970.

Mitchell, B., and M. Landisman, Electrical and seismic properties of the Earth's crust in the southwestern Great Plains of the U.S.A., Geophysics, 36, 363–381, 1971a.

Mitchell, B., and M. Landisman, Geophysical measurements in the southern Great Plains, in The Structure and Physical Properties of the Earth's Crust, Geophys. Monogr. Ser. vol. 14, edited by J. G. Heacock, AGU, Washington, D. C., this volume, 1971b.

Mueller, S., Low velocity layers within the Earth's crust and mantle, Proc. 10th Assembly Eur. Seismol. Comm. (Leningrad), in press, 1968.

Mueller, S., and K. Fuchs, Investigations on the non-elastic behavior of the upper mantle, Ann. Summary Rep. 1, Contr. AF 61 (052)-861, 52 pp., European Office of Air Force Office of Scientific Research, 1966.

Mueller, S., and M. Landisman, Detailed crustal studies in continental areas, Geophys. Prospect., 13, 498–499, 1965.

Mueller, S., and M. Landisman, Seismic studies of the Earth's crust in continents, 1, Evidence for a low-velocity zone in the upper part of the lithosphere, Geophys. J. Roy. Astron. Soc., 10, 525–538, 1966.

Mueller, S., and M. Landisman, An example of the unified method of interpretation for crustal seismic data, Geophys. J. Roy. Astron. Soc., in press, 1971.

Mueller, S., E. Peterschmitt, K. Fuchs, and J. Ansorge, The rift structure of the crust and upper mantle beneath the Rhine graben, Abh. Geol. Landesamt Baden-Wuerttemberg, 6, 108–113, 1967.

Mueller, S., E. Peterschmitt, K. Fuchs, and J. Ansorge, Crustal structure beneath the Rhine graben from seismic refraction measurements, Tectonophysics, 8, 529–542, 1969.

Nakamura, K., Surface faulting during the Matsushiro earthquakes (abseract), Eos Trans. Amer. Geophys. Union, 50, 389–390, 1969.

Narans, H., Jr., J. Berg, Jr., and K. Cook, Sub-basement seismic reflections in northern Utah, J. Geophys. Res., 66, 599–603, 1961.

Nicolaysen, L., and A. Burger, Note on an extensive zone of 1000 million-year old metamorphic and igneous rocks in southern Africa, Sci. Terre, 10, 497–518, 1965.

Nur, A., and G. Simmons, The effect of saturation on velocity in low porosity rocks, Earth Planet. Sci. Lett., 7, 183–193, 1969.

Pakiser, L., Structure of the crust and upper mantle in the western United States, J. Geophys. Res., 68, 5747–5756, 1963.

Pavlenkova, N., Methods of velocity determination from seismic crustal studies, Inst. Geophys. Acad. Sci. Ukrainian S. S. R., 13 pp., 1968.

Perkins, W., Deep crustal reflections on land and at sea, Ph.D. thesis, 127 pp., Princeton Univ., Princeton, New Jersey, 1970.

Petr, V., J. Pěčová, and O. Praus, A study of the electrical conductivity of the Earth's mantle by magnetotelluric measurement at Srobárová Czechoslovakia), Trav. Inst. Geophys. Acad. Tcheco. Sci., 208, 407–447, 1964.

Petrushevsky, B., Earthquakes and tectonics, in The Earth's Crust and Upper Mantle, Geophys. Monogr. Ser., vol. 13, edited by P. Hart, pp. 279–282, AGU, Washington, D.C., 1969.

Phinney, R., Structure of the Earth's crust from spectral behavior of long-period body waves, J. Geophys. Res., 69, 2997–3017, 1964.

Prodehl, C., Auswertung von Refraktionsbeobachtungen im bayerischen Alpenvorland (Steinbruchsprengungen bei Eschenlohe 1958–1961) im Hinblick auf die Tiefenlage des Grundgebirges, Z. Geophys., 30, 161–184, 1964.

Prodehl, C., Seismic refraction study of crustal structure in the western United States, Geol. Soc. Amer. Bull., 81, 2629–2646, 1970.

Raleigh, C., and M. Paterson, Experimental deformation of serpentinite and its tectonic implications, J. Geophys. Res., 70, 3965–3985, 1965.

Scholz, C., M. Wyss, and S. Smith, Seismic and aseismic slip on the San Andreas fault, J. Geophys. Res., 74, 2049–2069, 1969.

Schulz, G., Reflexionen aus dem kristallinen Untergrund des Pfälzer Berglandes, Zh. Geophys., 23, 225–235, 1957.

Seidl, D., H. Reichenbach, and S. Mueller, Dispersion investigations of Rayleigh waves in the Rhinegraben rift system, in *Graben Problems*, edited by H. Illies and S. Mueller, pp. 203–206, Schweizerbart, Stuttgart, Germany, 1970.

Shor, G., Deep reflections from southern California blasts, *Eos, Trans. AGU, 36,* 133–138, 1955.

Simonen, A., Batholiths and their orogenic setting, in *The Earth's Crust and Upper Mantle, Geophys. Monogr. Ser.,* vol. 13, edited by P. Hart, pp. 483–489, AGU, Washington, D.C., 1969.

Smithson, S., The nature of the 'granitic' layer of the crust in the southern Norwegian Precambrian, *Nor. Geol. Tidsskr., 45,* 113–133, 1965.

Smithson, S., and I. Ramberg, Geophysical profile bearing on the origin of the Jotun Nappe in the Norwegian Caledonides, *Geol. Soc. Amer. Bull., 81,* 1571–1576, 1970.

Stauder, W., and A. Ryall, Spatial distribution and source mechanism of microearthquakes in central Nevada, *Bull. Seismol. Soc. Amer., 57,* 1317–1345, 1967.

Steinhart, J., and R. Meyer, Explosion studies of continental structure, 409 pp., *Carnegie Inst. Wash. Publ. 622,* Washington, D. C., 1961.

Stewart, S., and L. Pakiser, Crustal structure in eastern New Mexico interpreted from the Gnome

explosion, *Bull. Seismol. Soc. Amer., 52,* 1017–1030, 1962.

Talwani, M., G. Sutton, and J. Worzel, A crustal section across the Puerto Rico trench, *J. Geophys. Res., 64,* 1545–1555, 1959.

Thompson, G., and M. Talwani, Crustal structure from Pacific Basin to central Nevada, *J. Geophys. Res., 69,* 4813–4837, 1964.

Tuve, M., and H. Tatel, Seismic sounding, *Carnegie Inst. Yr. Book* (47), 60 pp. 1948.

van Zijl, J., A deep Schlumberger sounding to investigate the electrical structure of the crust and upper mantle in South Africa, *Geophysics, 34,* 450–462, 1969.

Waring, G., Thermal springs of the United States and other countries of the world—A summary, *U.S. Geol. Surv. Prof. Pap. 492,* 383 pp., 1965.

Westhusing, K., Project Gnome volunteer seismological teams, *Tech. Rep. 62-5,* 57 pp., The Geotechnical Corp., Garland, Texas, 1962.

Woollard, G., Tectonic activity in North America as indicated by earthquakes, in *The Earth's Crust and Upper Mantle, Geophys. Monogr. Ser.,* vol. 13, edited by P. Hart, pp. 125–133, AGU, Washington, D.C., 1969.

DISCUSSION

Brace: If temperature is as important as you suggest, why don't you see a large difference in the seismicity of the two extreme regions in the U.S., namely the Basin and Range and the Sierra Nevada, where temperatures differ by a factor of three at any given depth?

Landisman: While the temperature at depth may often be correlated with tectonic activity, I think mineralogy and perhaps water content are at least as important. Clarence Allen has shown the importance of mineralogy along the San Andreas Fault system, and your studies of stick-slip have shown that there is much more sliding and less locking with serpentines than with other materials.

Brace: Is there a velocity inversion in California at the depth at which the earthquakes die out?

Landisman: There has not been enough work done to say for sure. Stewart has published refraction lines from the Gabilan Mountains, in

the Stanford report on the San Andreas Fault Symposium, in which the basement arrival falls off as the sixth power of distance and is gone at 80 km. This may indicate a strong negative velocity gradient at a depth of 10 km or less.

Levin: If the temperature is high enough to permit chemical decomposition would it not also be high enough to permit deformation of the rock minerals and hence preclude the accumulation of stress?

Landisman: We will take the questions in order. First, temperatures of only a few hundred degrees should be sufficient for dehydration reactions to occur. Plastic deformation of crustal silicates would require much higher temperatures. As for the question of the stress in the crust, an earthquake is, in fact, one of the methods for the release of stress, and stress is being released more or less continuously in the active regions.

A Comment on the Evidence for a Worldwide Zone of Low Seismic Velocity at Shallow Depths in the Earth's Crust

J. H. HEALY

National Center for Earthquake Research, U. S. Geological Survey, Menlo Park, California 94025

Abstract. Seismic data recorded on the North American continent reveal that the properties and thicknesses that are derived from these data must be regarded as statistical numbers which represent the structure of the earth's crust with varying degrees of reliability, depending on the validity of the assumptions. For example, seismic waves traveling between distances of 10 to 20 km from the source almost always exhibit velocities of 6 km/sec. However, in detail, the rocks along this profile may have a wide range of seismic velocities, and it is only in an average sense that we can describe the top of the crust as a layer with a velocity of 6km/sec. The data recorded at the Lasa detection network in Montana are used to illustrate the degree of scatter in seismic properties that can be measured in a limited area and to point out the difficulties this statistical scatter presents to methods of interpretation commonly used to define low-velocity layers within the crust.

One objective of the symposium at which this paper was presented was to discuss the evidence for a possible zone of low electrical conductivity at shallow depths within the crust. A layer of low seismic velocity such as is proposed by Landisman et al. in this volume may influence the electrical properties of crustal rocks. The composition and temperature of crustal rocks might be deduced from measurements of seismic velocity. Electrical conductivity can be estimated if the composition, temperature, and fluid content of the rocks are known. If there is a channel of low seismic velocity in the crust, then a similar channel of either high or low conductivity is also possible.

It is our contention that the evidence for a channel of low seismic velocity underlying the continents is not compelling; further evidence is required to resolve this question. The primary difficulty with the evidence of low-velocity layers within the crust lies with the 'statistical' character of the crustal 'layers.' On the average, the earth's crust is layered, and it is useful to describe the average properties of crustal layers. However, in detail, each layer may have a complex structure and contain rocks with a wide range of seismic velocity and density.

A low-velocity zone can be detected under ideal conditions by comparing the travel times of first arrivals with the travel times of reflected phases. First arrivals on a seismic-refraction profile can be used to deduce a velocity-depth structure. This structure is unique if the velocity increases with depth so that some energy from each layer appears as a first arrival at some point along the profile. If there are low-velocity layers, they will be 'masked' and the depths calculated for deeper layers will be too small.

The average velocity and depth to a reflecting horizon can be calculated from the shape of a time-distance plot of a reflected phase. If the depth to the reflecting horizon calculated from the reflection data is greater than the depth to the horizon calculated from the refracted arrivals, a low-velocity zone is required in the model. The velocity and thickness of the low-velocity layer can be estimated by comparing the average velocity computed from the reflected phase with the average velocity computed from the refracted phase.

A flat-layered earth is an implicit assumption in the approach described above. If there is a complex structure along a seismic profile that cannot be corrected to an equivalent flat-layer model, the approach outlined above will not work, because the shape of the reflected travel time graph will not be a sure guide to the average velocity and depth to a reflecting horizon.

There are now sufficient data in the continental United States to define the degree of variation of seismic properties. A good example that illustrates most of the points we wish to make is provided by the recent crustal calibration of the large-aperture seismic array in Montana. This area was the site of some of the first good seismic measurements of the properties of the earth's crust in the United States; they were made by workers from the University of Wisconsin and the Carnegie Institute of Washington. Later, the world's largest seismic array for detecting underground nuclear explosions was installed in this area and, as might have been predicted from the complexity of the seismic structure in early seismic refraction measurements, the array revealed substantial anomalies in travel times of arrivals from distant seismic events.

In 1966 and in 1968, the U.S. Geological Survey conducted extensive seismic surveys in this region in an attempt to define the details of the structure of the earth's crust. A composite travel time curve (Figure 1) consisting of first arrivals from 15 shot points at 50 recording locations (Figure 2) illustrates most of the interesting statistical properties that we believe are important to this discussion.

The surface of the earth is composed of a great variety of rocks with low seismic velocity and complex structure. These rocks are either sedimentary rocks, whose velocities may increase rather regularly with depth, or fractured crystalline rocks, which may have an extremely erratic velocity structure. Although this uppermost layer is not very thick, its low velocity and its extreme complexity produce effects that can be compared to a frost on the surface of an optical system which scatters the wave energy and makes it difficult to focus precisely on the details below.

Beneath this inhomogeneous zone, the rocks of the earth's crust display their most consistent behavior. The velocity of compressional waves is usually between 5.9 and 6.1 km/sec. This is illustrated in the distance range between 10 and 100 km on the composite travel time plot of Figure 1, which in this respect is a typical travel time-distance plot. In some areas, at distances greater than about 100 km, there is evidence in the first-arrival travel time data for layers with higher velocities in the earth's crust. In a few cases, particularly in the eastern part of the United States, velocities of about 6.4 or 6.5 km/sec can be detected. More frequently, we see evidence for seismic velocities of between 6.7 and 7.0 km/sec in the portion of the travel time curve between 100 and 150 km. On many profiles we see no evidence at all in the first-arrival data for velocities in this intermediate range, and the velocity of first arrivals changes from 5.9–6.1 to 7.8–8.2 km/sec as waves refracted in the upper mantle overtake the waves propagating in the upper crust. The composite travel time curve for Lasa shows an increasing scatter in the first-arrival data in the zone between 100 and 200 km. More detailed examination of the data shows evidence for velocities of between 6.4 and 6.6 km/sec.

For the greater distance ranges, if the sources are large enough, we can always record waves with seismic velocity above 7.8 km/sec. Despite a few reports to the contrary, we believe that there are no good data for velocities of upper mantle rocks under continents that are less than about 7.8 km/sec. Upper mantle rocks with low seismic velocities (7.8 km/sec) are found in the western United States, particularly in the Basin and Range province where the heat flow is high. Upper mantle rocks with seismic velocities of 8.0 to 8.2 km/sec are found in the older, more stable portions of the continent in the eastern United States. Referring once more to the composite curve in Figure 1, we can see the source of some confusion in the precise delineation of upper mantle velocities. First ar-

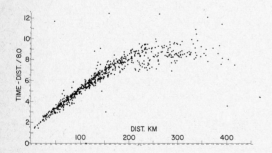

Fig. 1. Composite travel time plot combining all data recorded in the vicinity of the Lasa array. For locations of shot points and recording positions, see Figure 2.

rivals at distances of about 200 km are weak and have passed through major structures at the base of the earth's crust.

We could present many more examples to illustrate statements made above, but it will suffice to say that the situation observed at Lasa is not atypical of the degree of scatter observed in seismic data. From first-arrival data alone it is difficult to determine the detailed structure of the earth's crust in an unambiguous way.

Most seismologists agree that much more information might be obtained by a careful study of the secondary phases. Unfortunately, the secondary phases are also confused and erratic in nature and present difficulties equal to or

greater than those encountered in the interpretation of first-arrival data. In a search of more than 500 seismograms recorded in the Lasa region, it was possible to pick and time about 50 secondary phases that could be used to map the M discontinuity. Because of the very large quantity of data available in the Lasa study, it was possible to map some of the complexities of crustal structure. In other areas, where we have comparatively few data points spaced along a profile, it becomes extremely difficult to make accurate statements about the structure and the velocities of rocks in the lower portion of the earth's crust.

To study the nature of the correlatable secondary arrivals, we designed a computer pro-

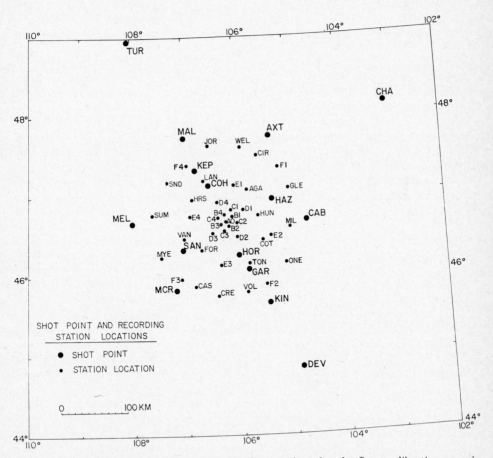

Fig. 2. Locations of shot points and recording stations for the Lasa calibration experiment. Each shot point was also used as a recording location for at least some of the other shots. Recording locations indicated by a single letter and a single number, such as F4, are located at the centers of Lasa sub arrays. Other stations locations were occupied by portable seismic stations.

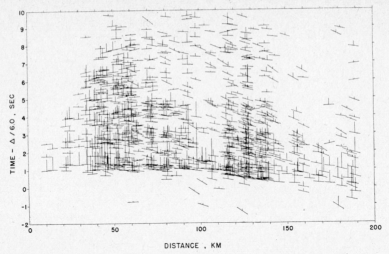

Fig. 3.   Composite travel time plot of correlatable phases picked from the Lasa sub arrays
by an automatic phase picker. The approximately horizontal bars indicate the onset of a
phase and its apparent velocity; the vertical bars indicate the duration of the correlatable
energy in each phase.

gram to function as an automatic phase picker. This picker detects correlatable energy on the seismogram and displays each 'pick' as a line on a record section. A cross line is plotted to indicate the apparent velocity of the phase and this cross-line is followed by a line on the time axis that indicates the duration of the phase. A composite 'phase section' (Figure 3) was made up of all the high-explosive shots recorded on all sub arrays at Lasa (Figure 2).

The details of the picker are complex and will be described in a later paper. For our purpose, it is sufficient to state that the picker does a fair job of identifying and displaying the correlatable seismic phases. There are many interesting features in the phases plotted in Figure 3 that can provide information about the crust. However, we think that the data illustrate our point about the difficulties of using secondary phases.

To summarize then, we can present a velocity-depth profile for the typical continental crust: An inhomogeneous layer of low-velocity rocks at the surface of the earth; a comparatively thick granitic layer with a velocity of about 6 km/sec, increasing slightly with depth; in some areas, the 'granitic' layer is underlain by a layer with velocities of between 6.4 and 6.5 km/sec; in most areas, there is evidence for a layer with velocity of between 6.7 and 7 km/

sec in the lower part of the crust; a possible transition zone above the mantle with seismic velocity of 7.3 (?) km/sec; an upper mantle with seismic velocity of between 7.8 and 8.2 km/sec.

The depths to the various seismic zones or layers can vary considerably, but for purposes of illustration, an average crustal model for Lasa is given in Figure 4.

Seismic refraction profiles in North America are usually recorded along a line 300 to 400 km in length with 2 to 4 shot points spaced at intervals along this line. Seismic waves recorded on these profiles may pass through rocks having a considerable range of physical properties, but usually the structure can be described in terms of layers with an average seismic velocity and thickness.

Attempts to define the crustal structure in greater detail (i.e., the Lasa study) usually reveal greater complexity, and a precise description of the structure always seems to require more data than we have available. This is not surprising, if we think for a moment of the extreme complexity of structure and variation of rock types revealed by surface geology. Perhaps the most surprising aspect of this apparent dilemma is the remarkable consistency of the average properties of crustal rocks, which in detail are so complex.

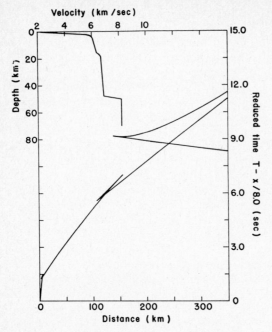

Fig. 4. Average velocity depth structure and travel time plot for the region surrounding the Lasa in Montana.

Workers in the Soviet Union, with support from their mineral industries, have completed thousands of kilometers of extremely detailed crustal surveys, and they too have described problems similar to those we encountered with less detailed surveys. A careful reading of the literature describing these experiments reveals that, even on this fine scale, seismic phases are not continuous over great distances and that there are important lateral variations of velocity in the crust.

The rocks of the earth's crust can be described in an average or statistical sense as being composed of layers. In some ideal situations, we can determine the structure of these layers and relate the structure to tectonic features, but frequently the structure of the crustal layers is too complex to be revealed by available methods.

The implications of this discussion with regard to the possible existence of low-velocity layers within the crust bears primarily on the difficulty of detecting such zones within the confusion presented by the complexities of earth structure. Unlike crustal layers, which can be detected by refracted waves, low-velocity layers depend on the subtleties of the travel time or amplitude data, which can be easily distorted by a complex crustal structure.

I hope that the critical attitudes expressed in this paper will serve to stimulate ideas for the design of experiments to resolve some of the unsolved problems of the structure and composition of the earth's crust. New theoretical techniques coupled with carefully designed experiments offer the possibility of significant advances in this field. In particular, we believe the detailed study of the character of reflected phases and the use of a combination of explosions and earthquake sources offer attractive possibilities for future research.

DISCUSSION

Landisman: I have to agree with much of what you said. The low-velocity layer is a variable feature that you have to look at in tremendous detail. One thing, though, that we do feel is quite important is that you not only have to look at the position of the critical point, but you have to look at the under-critical reflections and you have to look at the normal incidence reflections whenever it is possible to get hold of these. If you neglect these, it is very difficult to come to any positive conclusion.

Healy: I do not know what a reflection is. When you tell me exactly what you do to identify reflection, then we can program it and send you the sections.

Landisman: We have thought along these lines too—trying to find these in a completely unbiased way. I think it is an important subject for future work.

These layers are not continuous, but I do believe the upper low-velocity layer is more often found in regions that are currently active tectonically.

Healy: Some good seismologists have used this qualitative process very successfully and have come up with some very important ideas, and I do not want to run the method down. I think that we seismologists should now describe our process of interpretation in words, and then in mathematical terms, so that if someone disagrees with our interpretations, we can lay out a statistical basis for judgment. I could have come in today with a different presentation and demolished your concept of the low velocity layer—it is easy to do this because it is a matter of data selection. We have to codify what we are doing, so as to come to a systematic proof, or disproof, of these interesting ideas.

Kennedy: What is the sharpness of the Moho and is there a deep low velocity channel, extending down to maybe 100 km, which you would expect as you go from eclogite into the olivine-rich lower layers?

Healy: On so many profiles we get such a major burst of energy coming back at the correct time for a reflection from the Moho that there must be—not a perfectly sharp discontinuity (it does not have that character)—but a relatively abrupt transition to the mantle velocity.

Oliver: If you have to put a low velocity layer somewhere above the core, by far the most evidence puts it 100 or 200 km beneath the top of the mantle. As least in some areas, I find the evidence convincing but it is not sharply defined, and there are many seismologists who would say there is no low velocity layer in that part of the earth.

Brace: If there were a continuous low velocity trough in the crust—say, 5.5 km/sec stretching from Santa Monica to Salt Lake City—would you have seen this, considering the general way you have averaged everything in that part of the world?

Healy: We could easily have missed it. It would not give rise to any prominent arrival unless it had a sharp boundary, but the absence of a sharp boundary does not mean that the trough does not exist.

But we are now getting to the point where we have very accurate locations of earthquakes down to 10 km depth. These are a very powerful source of data which could be used to detect the trough. So far we have not noticed anything in the normal reductions, but there could be a small effect.

R. B. Smith: I know that Prodehl did include a low velocity layer in some of his interpretations of Willdens' profile in southern Utah. What was the evidence for that?

Healy: Prodehl used a method of interpretation described by Giese in German and never adequately in English. Basically, it was a Herglotz-Wiechert approach which is prone to give you a low velocity layer because you work your way downwards and it gives you the highest velocity possible at any particular depth. You are bound to get the velocity too high some place, and you must compensate by putting in a lower velocity at greater depth in order to make the lower arrivals agree. For this reason I am not convinced by his low velocity layer.

(Note: See R. Willden, Seismic refraction measurements of crustal structure between American Falls Reservoir, Idaho, and Flaming Gorge Reservoir, Utah, *U. S. Geological Survey Professional Paper 525-C,* C44-C50, 1965.)

A Reflection Study of the Wind River Uplift, Wyoming

WILLIAM E. PERKINS

Gulf Research and Development Company, Pittsburgh, Pennsylvania 15219

ROBERT A. PHINNEY

Department of Geological and Geophysical Sciences
Princeton University, Princeton, New Jersey 08540

Abstract. Five deep crustal reflection profiles were obtained on a line crossing the Wind River uplift, Wyoming. Geophone and shot arrays were used to suppress surface waves relative to the near-normally incident waves. Reflections were seen at times as great as 14 seconds. A fair correlation of the reflection times (after stripping the sediments) on the two end profiles shows that the crustal structure is nearly identical on either side of the uplift.

It is becoming increasingly apparent that narrow-angle reflections can be recorded from crustal depths as great as 20–40 km. For example, reflections have been routinely reported by workers in Germany [*German Research Group for Explosion Seismology*, 1966]. Experiments specifically designed for this purpose greatly improve the quality of reflection results [*Clowes et al.*, 1968]. Near-vertical reflection profiles, combined with wide-angle and refraction data, make it possible to obtain detailed crustal velocity structure. Furthermore, amplitude studies of reflected and refracted waves have shown that additional inferences can be made regarding the nature of deep interfaces [*Fuchs*, 1969; *Meissner*, 1967; *Helmberger*, 1968]. Narrow-angle reflection data are essential for the removal of ambiguities in refraction data, i.e., for the identification of secondary arrivals and the detection of velocity reversals. Reflection methods are inherently more appropriate for structural studies than refraction methods; the horizontal resolution of detail by reflections is at least ten times better than that resolvable by refraction.

In this paper we report the results of an experiment to test independently the linear shot pattern and geophone array methods described by *Kanasewich and Cumming* [1965]. A reconnaissance profile was shot across the axis of the Wind River Mountains in central Wyoming. Five shot sites were chosen: three in the Wind River basin, one on the anticlinal axis at South pass, and one in the Green River basin (Figure 1). The profile was over 100 km long, with only about 20 km of geophone coverage at the five sites. The degree of correlation between the results from the ends of the profile is sufficiently good to warrant returning to the area to complete the coverage.

EARLIER GEOPHYSICAL WORK AND INTERPRETATIONS

The Wind River Mountains form a long domal uplift (Laramide) striking roughly northwest with an exposed Precambrian core dominated by granites and gneisses. The uplift is asymmetrical, with sediments dipping 10 degrees to the northeast into the Wind River basin. Toward the southwest, sediments of the Green River basin have been steeply folded and overturned by thrusting of the basement. The gently dipping sediments on the Wind River basin side are interrupted only by a long line of thrust folds paralleling the mountains, which have yielded at least eight structures producing oil and gas from the Cretaceous and older sediments. At the northwest end of the range, the Paleozoic and Mesozoic sediments are exposed in the Green River Lakes area, beyond which

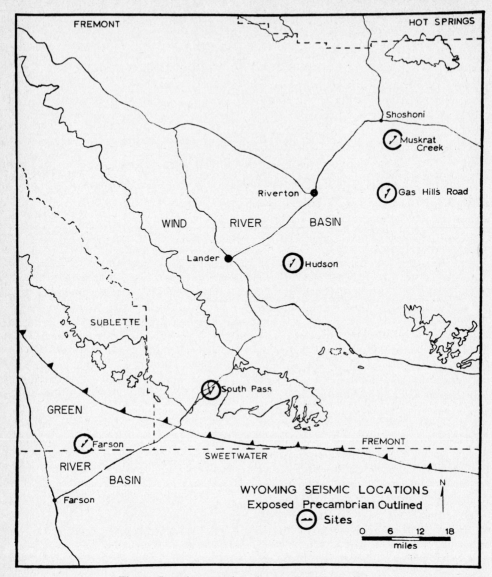

Fig. 1. Location map for seismic reflection profiles.

the range plunges beneath the Tertiary sediments of the Green River basin. To the southeast the range plunges into the region around Bison basin [*Berg,* 1961].

The mode of deformation of the Wind River Range is disputed. The full range of suggestions includes block faulting on high-angle faults [*Chamberlin,* 1945], as well as simple thrusting due to regional compressional forces. Detailed seismic work was done over the thrust fault bordering the mountains on the southwest by

Berg [1961]. He finds that the fault zone itself may be composed of a sheet of Paleozoic rocks that was overturned, thrust, and is now lying parallel to the base of the thrust. He bases his interpretation on the parallel nature of the reflections under the thrust wedge and on the observations of an overturned Paleozoic section under similar thrusts in the Washakie Range and the Owl Creek Mountains to the east. This would imply that there was asymmetrical overturned folding in the upper crust, followed by later thrusting. *Eardley* [1963] has proposed that Laramide uplifts such as this are due to vertical uplift, with secondary thrusting due to lateral deformation caused by flowage and gravity sliding.

The only gravity data over the Wind River Mountains follow a line parallel to our seismic line and very close to it [*Berg and Romberg*, 1966]. Berg and Romberg computed an axial Bouguer anomaly of +100 mgals. They then modeled the southwest flank as a low-angle thrust fault and found, by using $\Delta\rho = 0.3$ gm cm^{-3} for the basin-mountain density contrast, that they could explain the anomaly if the Precambrian thrust wedge has an average thickness of 10,000 feet. This wedge is appreciably thinner than that previously mapped by reflection shooting. A density contrast of 0.3 g cm^{-3} is proper for the surface sedimentary rocks but not for those at depth. Measurements of the sedimentary rocks at depth in the Green River Basin give higher densities, owing to compaction; thus, they indicate that 0.2 g cm^{-3} is a more appropriate average density contrast. If we use this contrast, we find that the central Bouguer anomaly is only 65 mgal (Figure 2). The residual between the observed and computed anomaly is negative on the southwest end of the profile; this fact implies that the model should have more sediments beneath the flank than are indicated by the seismic work. The residual anomaly rises sharply on the southwest end and tapers off to the northeast; it follows closely the trend and shape of the central anomaly. This model implies an upwarped layer or core of extra dense material. This positive excess has been found in other ranges in the Rockies. *Chamberlin* [1935] reports a 'geologic correction' for an assumed mass at depth under the Big Horn Mountains. *Case and Keefer* [1966] show a similar anomaly over the Owl Creek and Granite Mountains.

This excess mass could be due to a discrete intrusive mass at depth or to a local upwelling of the deeper crustal layers. In light of these findings, *Eardley's* [1963] analysis seems germane. He assumed that the domal uplifts with their bordering thrust faults are individual occurrences unrelated to any regional horizontal compression. In these cases the thrust and marginal faults are due to slumping or flow in the basement material. As such, the primary forces have been vertical, with the thrusting occurring only at the margins of those uplifts that expose the Precambrian core. Perhaps his best case for vertical uplift, as contrasted to compressive thrusting, is the Uinta Mountains, a flat-topped anticline as seen in the horizontally bedded Proterozoic sequence on top of the uplift. Granting the primary motion to be vertical, Eardley then assumes that the general broad oval shape of the uplifts implies that there is a large intrusion, a megalaccolith, beneath each feature. To account for the mass excess implied by the gravity, it is assumed to be 'basaltic.'

The gravity profiles in the Wind River basin reflect the configuration of the Precambrian basement and the thickness of sedimentary rocks. The total Wind River anomaly pattern is superimposed on a regional low of -280 mgal. The Wind River Mountains as an individual feature are not isostatically compensated by roots at depth. As a rough measure of isostasy, the free-air anomaly in the flanking basins (-10 mgal) implies that the basins are isostatically compensated, but the Wind River Range ($+170$ mgal) is not. The mountains must be supported by crustal strength or active tectonic forces.

SEISMIC RECORDING SYSTEM AND ANALYSIS

The reflection technique of *Kanasewich and Cumming* [1965] is based on a linear geometry of sources and geophones that discriminates against surface waves in favor of vertically traveling wave energy. This geometry is a simple linear array, whose outputs are phased and summed to be directionally sensitive to waves from a prescribed direction. In this manner the smaller late arrivals, normally masked by surface waves, are detectable. Such a linear array forms an excellent in-line antenna but will accept all off-axis signals without directional discrimination.

For this experiment, two linear arrays were operated, through the collaboration of Princeton and the University of Alberta during the summer of 1969. The total aperture of the two arrays was 10,000 feet. Each array consisted of six subarrays with 400-foot apertures and 1000-foot spacing. Each sub-array consisted of 16 geophones in a tapered pattern, with the outputs summed to provide a single data chan-

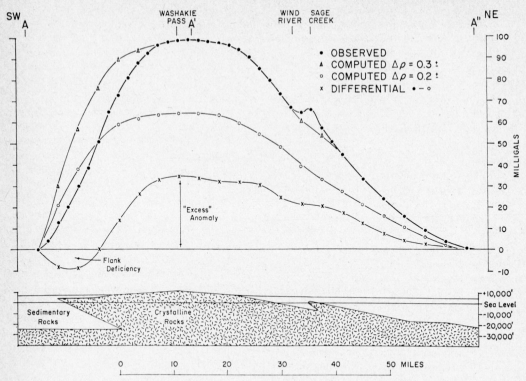

Fig. 2.   Wind River Mountain Bouguer gravity profile, from *Berg* [1966].

nel. At 15 Hz the sub-array sum suppresses ground roll about 18 db with respect to *P* waves incident within 45° of vertical. Data were recorded on both FM magnetic tape and on a galvanometer camera. Amplifiers operated with a passband of from 2 to 23 Hz unless noise conditions made some restriction of the band necessary. After recording, the data were anti-alias filtered, then digitized on the Princeton truck-borne PDP8S computer system. Elementary digital processing of the data was restricted to sign and amplitude normalization, followed by band pass filtering.

The profile: In choosing the sites, care was taken to avoid more complex sedimentary structure on the flanks of the range. The four basin sites were thus on gently dipping sediments, and the site on the axis of the uplift was shot in weathered gneiss. At each site, 14 shot holes were drilled to 100-foot depths. Charge weights ran from 20 to 30 pounds in a tapered pattern (see Figure 3).

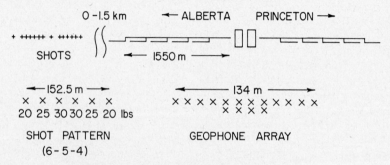

Fig. 3.   Geometry of shot and receiver arrays.

One test shot was fired at each site in order to set the amplifier gains and filters. Since this project was experimental in nature, various combinations of shots were fired at different sites and various filter settings were used. The gains on the linear amplifiers were set as high as possible to bring out the later events and were thus limited by the noise level. This meant that the system was strongly overloaded during the first few seconds of reflection time; loss of the shallow information therefore resulted. This loss is usually avoided by using amplifiers with programmed gain control.

Ideally, all shots would be fired singly, with the synthesis of the shot array being done in later data processing. In practice, we found that the advantage of having good camera records for immediate inspection made it desirable to fire the shots in pairs or groups of six. In general, visibly coherent normal-incidence signals cannot be seen unless at least two holes are simultaneously fired. The simultaneous firing of six holes, which was done at the Hudson and

Farson sites (Figure 1), gives the best camera records and greatest field efficiency but results in lack of flexibility in later data processing.

Any deep reflection must first be considered a candidate for a multiple within the sedimentary section. From *Berg* [1961], we know that the major sedimentary reflectors are the Upper Cretaceous Mesaverde and the Lower Cretaceous Dakota formations. The multiples seen on the records are associated with these two reflection times; the deep events picked are not multiples of these events and are tentatively identified as primary events. The multiples generated in the sedimentary section die out rapidly when we take into account reflection times from 5 to 20 seconds after the shot—times much greater than customary for a reflection profile. The results from South pass, where the shots were detonated directly on the basement surface, have no multiples of the usual sort.

The data are presented in two formats. Portions of the camera records from both the

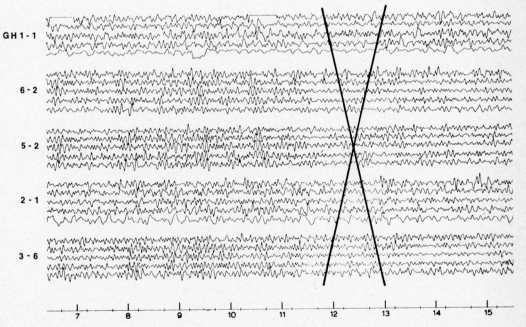

Fig. 4. Composite profile from Gas hills, unfiltered signals. Traces shown are sub-array sums; time is in seconds after the shot. Each group of traces represents a shot pattern consisting of n charges in line at 100 foot intervals; n is, respectively, 1, 2, 2, 1, and 6. The pair of oblique lines dissecting the figure defines the moveout corresponding to an apparent velocity of 8.4 km/sec, which would be obtained from signals arriving at 45° to the vertical in a 6.0-km/sec basement layer.

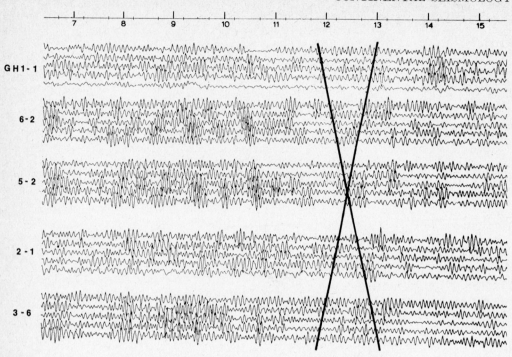

Fig. 5. The same as Figure 4, filtered from 10–20 Hz.

Alberta system and the Princeton system are shown as individual records (Figures 6–8). These are the better examples of deep reflections. In each case there are two sets of six traces at different gains. The low gain traces from the camera records are needed to locate the top of the basement. The middle three or four traces on the records with timing lines are the horizontal geophones (radial and transverse), plus one or two single vertical phones. These channels help verify that simultaneous signals were propagating close to the vertical plane containing the source and receivers. In Figures 4 and 5 are shown processed traces from the Princeton array. Figure 4 shows raw data, while Figure 5 shows the same data digitally band pass filtered from 10 to 20 Hz. For records that were high cut filtered at 15 Hz in the field, these filtered records shows a characteristic lack of bandwith. For most sites, therefore, the filtered records were used to back up the raw data, with the primary picks coming from the raw data and the camera records.

By ascertaining the apparent velocity of an event across the set of sub-array outputs, one can use the full array to determine the direction of incidence of the reflected signals in the line of the array. Of primary interest to us are those signals arriving at the array at angles within 45 degrees of the vertical, within the crust. For a 6.0-km/sec crustal P velocity, the corresponding apparent velocity is 8.4 km/sec, and the time offset across the full array is 0.16 sec. The oblique lines crossing the later portions of Figures 4 and 5 correspond to this case. Nearly all signals that correlate reasonably well across the spread lie within this range. With complete coverage, normal move-out corrections would be applied to the full record, for compositing into a section. We have restricted our consideration to signals arriving within approximately 20 degrees of the vertical.

The Alberta records showed the same major reflection events as did the Princeton records, but the character and wave forms differ significantly. This difference is attributable to differences in filtering and geophone response. We find that better records are obtained when the passband from 15 to 24 Hz is used. This improvement in reflection amplitude with increasing frequency is not characteristic of simple transitional-interface models [*Phinney*, 1970]. It is not clear whether still higher frequencies would be equally suitable for deep sounding purposes. At field sites South Pass and Hudson, high cut filtering at 12 Hz was required, owing to local noise conditions. In future work, it is clear that waiting for reduced noise conditions is more than justified by the improvement in bandwidth and signal strength.

For this experiment, the most realistic way to obtain the seismic parameters of the sedimentary section is from existing stratigraphic and laboratory data, used together with reference to the structure may of the region [*Petroleum Ownership Map Co.*, 1965]. Depth information is derived from the contours on the Lower Cretaceous Dakota formation and from well data below the Dakota formation as given by *Thompson* [1956], *Sharkey* [1956], and *Berg* [1961]. Seismic velocities in the two sedimentary basins are difficult to determine from the literature. *Berg* [1961] gives velocities for the Green River basin that fit well with the reflections we have recorded at Farson. Since the velocities in the basins are a function of lithology and depth, the velocities in the Wind River basin were estimated from Faust's curves [*Clark*, 1966]. In this area, then, the basement arrival time is an estimate, since the system was too overdriven to permit observation of these events. Application of the same procedure to the Farson area led to good agreement between our basement arrival time estimate and that reported by *Berg* [1961].

Interpretation. Records such as those in Figure 5 were used as a basis for the choice of reflection events. The basement arrival times used in reducing data to basement datum are shown below:

Location	Arrival Time, sec
Gas hills	3.2
Hudson	2.2
South Pass	...
Farson	4.4

The basement arrival times were subtracted; the times are shown in Figure 9 as lined up on the present Precambrian surface. For the sites in the sedimentary basins, this process elimi-

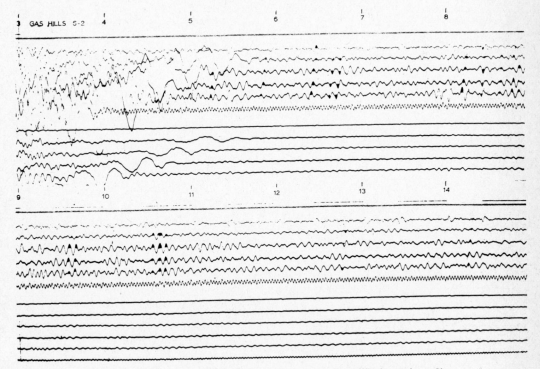

Fig. 6.   Portion of galvanometer camera record, Gas hills shot 6 ($n = 2$).

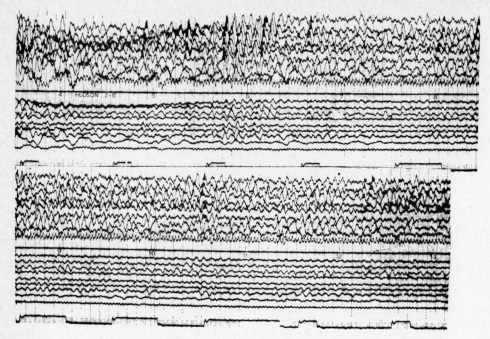

Fig. 7.   Portion of galvanometer camera record, Hudson shot 2 ($n = 6$).

nates the early, overdriven part of the record. At South Pass, the key location in the profile, this reduction runs into difficulty. There, directly over the anticlinal axis of the uplift, the shot holes were drilled through 60 feet of Tertiary gravels into basement rocks; the charges were placed in granite. The basement surface datum should, in this case, refer to the original elevation of the basement after uplift. Erosion of this surface has occurred to some degree. Thus all arrivals plotted in Figure 9 will appear as shallower than were intended.

The first few seconds of subbasement data at South pass are not available for correlation, owing to overdriving of the system, and determination of the correlation with the basin sites must await further field work. The amount of basement section masked by this overdriving is at least 10 km.

Figure 9, however, shows excellent correlation between the Gas hills and Farson sites, 98 km apart. Hudson, with poorer records, shows fewer events, but the significant ones at 3.25 and 3.50 seconds do correlate. We infer that the

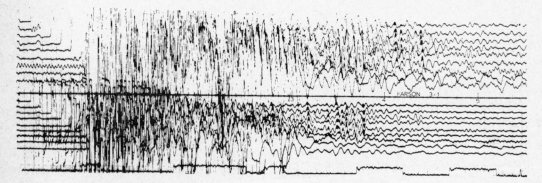

Fig. 8.   Portion of galvanometer camera record, Farson shot 3 ($n = 1$).

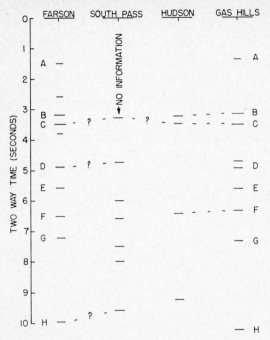

Fig. 9. Compilation of reflection interpretations, with times adjusted to the surface of the Precambrian basement.

crustal section is very similar on both sides of the uplift. It is premature to try to plot a cross section until missing areas are filled in under improved operational practices and velocities are directly measured by wide-angle work. Nonetheless, if the average upper crustal velocity is 6.2 km/sec, event F would correspond to a reflector 20 km below the basement surface. Assuming the correctness of the South Pass correlations marked in Figure 9, we would estimate a loss of 1.2 km of basement by erosion.

The correlation of reflections at Farson and Gas hills is quite good. Events B and C are easily identified and appear to be at the same depth under each basin. For an upper crustal velocity of 6.0 km/sec, the depth to C would be 11 km below basement. The interpretation of these reflectors in terms of geological or other geophysical data is a major objective of this work. The suggestion that the crust contains a sharply defined low velocity zone [*Mueller and Landisman,* 1966] indicates the type of model that must be considered. To make this kind of interpretation possible, wide-angle reflection work is clearly necessary in order to determine velocities and identify interfaces with a negative velocity change [*Mitchell and Landisman,* 1971].

A discussion of the deep structure under the uplift proper would require data from future detailed profiling. The width of the structure at sea level is only about 40 km. This value is closely comparable with normal crustal thickness. It would not be surprising to find that the structure of this block dies out with depth, leaving complete seismic continuity at lower crustal depths. Given the interpretation in Figure 9, with H assigned arbitrarily to Moho, we would need either a shallower Moho or higher velocity material, as suggested by Eardley.

CONCLUSIONS

It is apparent that normal-incidence reflections from deep within the crust can be recorded by using techniques specifically designed for this objective. These techniques are very similar to methods now used in commercial seismic prospecting and differ mainly in the very modest scale of operation to which university research is constrained. In this preliminary profile across the Wind River Mountains, reflections from the crust under the Wind River basin match those from the Green River basin. Results from shooting near the axis of the mountains are inconclusive but warrant more field work. The best reflections come from about 11 km below basement. Their relationship with the proposed crustal low-velocity zone bears inquiry [*Mueller and Landisman,* 1966; *Stewart,* 1968]. Deeper reflections are also recorded, with a possible Moho arrival from approximately 35 km.

More detailed work is needed to fill in the missing information in this profile. Hardware and operational improvements, as mentioned in this paper, are planned, to improve the quality and uniformity of the data and avoid unnecessary loss of information. Wide-angle reflection and refraction data are required, although the large distances (up to 200 km) associated with crustal refraction are not essential. In general, we feel that the finer resolution of reflection methods makes them a desirable adjunct to standard wide-aperture crustal refraction work, which is often fairly ambiguous.

Acknowledgments. We thank R. J. G. Lewis for special assistance in all phases of the project. provided his time and field equipment for the E. R. Kanasewich, of the University of Alberta, duration of these measurements.

This research was supported by NSF grant GA-627. W. E. Perkins was supported as a National Aeronautics and Space Administration trainee.

BIBLIOGRAPHY

Berg, R. R., Laramide tectonics of the Wind River Mountains, in *Wyoming Geological Association Guidebook 16*, 70–80, 1961.

Berg, R. R., and F. E. Romberg, Gravity profile across the Wind River Mountains, Wyoming, *Geol. Soc. Amer. Bull.*, 77, 647–656, 1966.

Case, J. E., and W. R. Keefer, Regional gravity survey, Wind River basin, Wyoming, *US Geol. Surv. Prof. Pap. 550-C*, C120–C128, 1966.

Chamberlin, R. T., Geological analysis of gravity anomalies of the Bighorn-Beartooth region, *Geol. Soc. Amer. Bull.*, 46, 393–408, 1935.

Chamberlin, R. T., Basement control in Rocky Mountain deformation, *Amer. J. Sci.*, 243, 98–116, 1945.

Clark, S. P., Handbook of Physical Constants, *Geol. Soc. Amer. Mem. 97*, 203, 1966.

Clowes, R., E. R. Kanasewich, and G. L. Cumming, Deep crustal seismic reflections at near-vertical incidence, *Geophysics*, 33, 441–451, 1968.

Eardley, A. J., Relations of uplifts to thrusts in the Rocky Mountains, in *Backbone of Americas, Amer. Ass. Petrol. Geol. Mem. 2*, edited by O. E. Childs and B. W. Beebe, 209–219, 1963.

Fuchs, K., On the properties of deep crustal reflectors, *Zh. Geophys.*, 35, 133–149, 1969.

German Research Group for Explosion Seismology, Seismic wide angle measurements in the Bavarian molasse basin, *Geophys. Prospect.*, 14, 1–6, 1966.

Helmberger, D. V., The crust-mantle transition in the Bering Sea, *Bull. Seismol. Soc. Amer.*, 58, 179–214, 1968.

Kanasewich, E. R., and G. L. Cumming, Near-vertical incidence seismic reflections from the 'Conrad' discontinuity, *J. Geophys. Res.*, 70, 3441–3446, 1965.

Meissner, R., Exploring deep interfaces by seismic wide angle measurements, *Geophys. Prospect.*, 15, 598–617, 1967.

Mitchell, B. J., and M. Landisman, Geophysical measurements in the Southern Great Plains, in *The Structure and Physical Properties of the Earth's Crust, Geophys. Monogr. Ser.*, vol. 14, edited by J. Heacock, AGU, Washington, D.C., this volume, 1971.

Mueller, S., and M. Landisman, Seismic studies of the Earth's crust in continents, *Geophys. J.*, 10, 525–538, 1966.

Petroleum Ownership Map Co., Structural map of Wyoming, Petroleum Ownership Map Co., Cooper, Wyoming, 1965.

Phinney, R. A., Reflection of acoustic waves from a continuously varying interfacial region, *Rev. Geophys. Space Phys.*, 8, 517–532, 1970.

Sharkey, H.H.R., Structural control of oil fields in the Wind River basin, Wyoming, *Geol. Rec., AAPG Rocky Mt. Sec.*, 159–170, 1956.

Stewart, S. W., Preliminary comparison of seismic travel times and inferred crustal structure adjacent to the San Andreas fault in the Diablo and Gabilan Ranges of central California, *Stanford Univ. Publ. Geol. Sci.*, 11, 218–229, 1968.

Thompson, R. M., Tectonics of Central Wyoming, *Geol. Rec., AAPG Rocky Mt. Sec.*, 145–152, 1956.

DISCUSSION

Ward: Do seismologists use the concept of discrete scatterers in filtering their data, as we are beginning to develop in electromagnetic theory, or do you just use layered models?

Phinney: Typically, no. It is clear, however, from the amount of apparently incoherent signal on the sub-array sums that three-dimensional scattering is quantitatively important. Also, larger discrete scatterers give rise to the characteristic hyperbolic echoes, which might be computed by Kirchoff integration. A complete theory for the three-dimensional crust in terms of scattering distributions is needed.

Meyer: Is all the noise on your records generated by the signal?

Phinney: Yes, there is no ground noise at all.

R. B. Smith: Have you considered a land air gun as an economical repetitive source?

Phinney: Yes, that would be a very attractive thing.

Crustal and Mantle Inhomogeneities as Defined by Attenuation of Short-Period P Waves

JOSEPH W. BERG, JR.[1], L. TIMOTHY LONG[2], SURYYA K. SARMAH[3], AND LYNN D. TREMBLY[4]

Abstract. Attenuation was calculated by using the slope of the logarithm of the amplitude spectrum of first-arriving P waves at each epicentral distance. This was done for arrivals to distances of 90° from three nuclear explosions. Seismic arrivals from two profiles recorded in eastern New Mexico and Nevada by the U. S. Geological Survey were used to determine attenuation to depths immediately below the crust-mantle boundary. For regional and teleseismic distances, data from permanent stations recording the first arrivals on Benioff (short period) instruments were used. A ray-tracing program was employed to compute attenuation structure at shallow depths (60–70 km). The shallow attenuation structure is more complex than the velocity structure given for the eastern New Mexico and Nevada profiles. It is suggested that this type of information could possibly augment conventional interpretation of seismic refraction arrivals. An analysis of wave type is important to such an interpretation. Along refraction profiles, the average Q for the eastern New Mexico area was computed to be 169 ± 42 at 5 cps (frequency of peak amplitude) and that for the Nevada area was calculated to be 116 ± 38 at 4 cps. Ten models were fitted to the attenuation data for the mantle. The interpreted best fitting model yields: $Q = 200$ for depths to 200 km; $Q = 400$ for depths between 200 and 600 km; and $Q = 2000$ for depths greater than 600 km. The upper-mantle data apply to the western United States. There is indication that horizontal as well as vertical variations may occur in the attenuation structure of the mantle.

Attenuation of body waves from seismic sources is complex, and it is extremely difficult to separate the effects of geometrical spreading, scattering, and absorptive attenuation in observed seismic signals. Additional complications may be encountered when different sources and instruments are used. In some cases, wave-amplitude attenuation is considered to be a 'lumped' quantity and is described as an exponential function of distance [*Romney*, 1959; *Romney et al.*, 1962], and in other cases, specific modes of propagation are assumed in order to fix the geometrical spreading and thus to describe absorptive losses. Even so, scattering effects are usually present [*Willis and DeNoyer*,

1966; *Howell*, 1966]. Some difficulties can be overcome by observing multiply reflected waves arriving at the same observing station, as *Kovach and Anderson* [1964] did for S waves reflected from the core-mantle boundary.

One method of describing the absorptive attenuation of seismic waves is to utilize the dimensionless quantity Q (or the specific dissipation function $1/Q$). Several excellent review papers have been written about the significance of Q measurements and their implications, and the reader is referred to them for a more complete general discussion [*Knopoff*, 1964; *Jackson and Anderson*, 1970]. For the purposes of this work, Q is assumed to be independent of frequency, as is usually observed for the frequencies considered and is defined in terms of a spatial wave attenuation factor for a wave function in the form exp $(-\alpha r)$, where $\alpha = (\pi ft)/(Qr)$ and t and r are time and radial distance from the source, respectively. The amplitude spectrum of a wave at a fixed point in space, r_1, can be expressed by the following

[1] Division of Earth Sciences, National Academy of Sciences, Washington, D.C. 20036.

[2] Mineral Engineering Branch, Georgia Institute of Technology, Atlanta, Georgia.

[3] Department of Physics, Gauhati University, Gauhati, Assam, India.

[4] Marathon Oil Company, Denver Research Center, Physics and Mathematics Department, Littleton, Colorado.

equation:

$$A(f, r_1) = A_0(f)I(f)F(r_1) \exp\left(-\frac{\pi f t}{Q}\right) \quad (1)$$

where: $A(f, r_1)$ is the amplitude spectrum of the recorded arrival; $A_0(f)$ is the source spectrum; $I(f)$ is the instrument response; $F(r_1)$ is the geometrical spreading factor (assumed to be frequency-independent for the narrow bands used in this work); and

$$-\frac{\pi t}{Q} = \frac{\partial}{\partial f} \log_e [A(f, r_1)] - [\text{source factor}] -$$

$$[\text{instrument factor}] \quad (2)$$

which allows the value of Q to be determined from the slope of the logarithm of the amplitude spectrum when the data are corrected for source and instrument factors.

Equations 1 and 2 were used in the analyses of attenuation of P waves generated by nuclear explosions in New Mexico and Nevada, and the results are presented here. The next section of this paper is taken mainly from *Long and Berg* [1969] and deals with the crust and the region immediately beneath it.

ATTENUATION AT SHALLOW DEPTH

Long and Berg [1969] determined apparent Q's of 169 ± 42 (5 cps) and 116 ± 38 (4 cps) by using primary seismic waves recorded along profiles in New Mexico (Gnome) and Nevada-Utah (Shoal). The analysis assumed that the Q and the geometrical spreading factor were both independent of frequency. These values resulted from a least-squares fit of straight lines to a plot of the slopes of the logarithms of the amplitude spectra versus travel times. When the geometrical spreading term was approximated by r^{-n} and values of Q were calculated for different values of n using a modification of (1), a value of $n = 0.5$ yielded Q's most closely corresponding to the above values. In these cases, the first arrival appears to be propagated as a direct wave (cylindrically spreading wave affected by an average positive velocity gradient) rather than as a head wave [*Berg and Long*, 1966]. In addition, the arrivals at different epicentral distances travel different paths, which are dependent on the depth-versus velocity function, and if horizontal layers of constant Q are assumed, the average Q observed at the surface,

$\langle Q \rangle$, can be directly related to the attenuation structure at depth.

Following the above authors, Q is written as an integral along the entire propagation path s for a pulse traveling with velocity c:

$$\frac{\omega}{2} \int \frac{ds}{cQ} = \frac{\omega t}{2\langle Q \rangle} \quad (3)$$

where t is the propagation time, ω is the angular frequency, $\langle Q \rangle$ is the average for the entire path, and

$$\frac{t_i}{\langle Q \rangle_i} = \sum_{j=1}^{N} t_{ij} \frac{1}{Q_j} \quad (4)$$

where (t_{ij}) is the matrix of travel times through the jth layer to the ith station, and

$$\frac{1}{Q_i} = (t_{ij})^{-1}\left(\frac{t_i}{\langle Q \rangle_i}\right) \quad (5)$$

The thicknesses and depths for the layers of constant Q were determined from the maximum depth of penetration of the ray to each station. Thus, the total number of possible layers is equal to the total number of stations along the profile. The values of $\langle Q \rangle_i$ were obtained by assuming a value for one station and then computing the others by using the slopes of spectral ratios for pulses recorded at the different ranges. The attenuation structures resulting from this analysis are shown in Figure 1. The accuracy is directly related to the reliability of the measured values and, as a result, only the trends of structure shown in Figure 1 should be considered reliable.

For comparison, crustal-velocity profiles are shown in Figure 2 for the two seismic refraction lines that were used to compute the attenuation profiles. The continuous velocity-depth profiles given by *Long* [1968] were used for ray-tracing purposes for the attenuation models. The curves showing constant-velocity layers at various depths were taken from *Stewart and Pakiser* [1962] and *Eaton* [1963]. In general, the attenuation profiles are similar to the constant-velocity-layer interpretations, but in both examples, greater detail is derived by using attenuation profiles. Attenuation below the crust-mantle boundary (see Figure 2 at 55 to 60 km, for Gnome, and at 25 to 30 km, for Shoal) appears to increase in both. The Gnome data show this increase for only one station, but the Shoal data show a more complex attenuation structure. Above the crust-mantle boundary

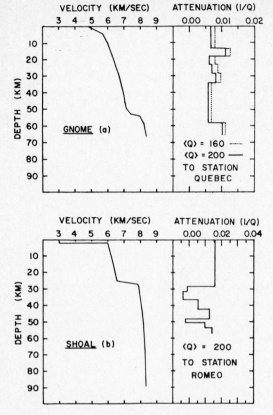

Fig. 1. Velocity versus depth and attenuation versus depth profiles: (a) is Gnome, and (b) is Shoal [*Long and Berg*, 1969].

(Shoal), the seismic arrival at one station shows high attenuation, whereas the attenuation profile for Gnome data is complex, with high attenuation shown at depths of 15 and 30 km. These attenuation trends appear to be real, but other factors that affect the trends, such as interference of arrivals or subtle changes in the velocity structure, may be present.

It would appear that interpretation of data from seismic refraction lines in terms of attenuation structure offers the possibility of obtaining information to augment conventional interpretations. In appropriate cases, it may reveal layers within the earth that would normally be excluded from interpretations using constant-velocity layers. A knowledge of the type of wave (head versus direct wave) being propagated as the initial arrival is very important in determining the supplemental information to be gained from this type of analysis.

ATTENUATION IN THE MANTLE

The basic data used in this section are taken from *Sarmah* [1966]. First arrivals (1½ cycles) of seismic waves from the Shoal and Bilby nuclear explosions recorded between 11° and 90° epicentral distance were used to evaluate attenuation. After spectral analysis was performed, the slopes of straight lines drawn between the amplitudes computed for 0.8 and 1.0 cps were calculated for all pulses. These values are given in Table 1 together with travel times and azimuths to the stations.

All data in Table 1 were obtained from seismographs taken by Benioff instruments. Other data were available, but to keep the instrument factor in (2) as nearly constant as possible and for other reasons, such as noise, only the above data were used in this study.

As an example, Figure 3 shows attenuation ($\pi t/Q$) versus travel time for first arrivals from the Shoal nuclear explosion as recorded at three Canadian stations with Willmore instruments, as well as values at other stations recorded with Benioff instruments. There is a distinct difference in the data. All amplitude spectra used in this research were corrected for instrument response, but such anomalous sets of data are suspicious. The data for Figure 3 were taken from *Sarmah* [1966] and were derived from spectral slopes between 0.7 and 1.0 cps using the method of least squares. Variation in those data was, on the average, ±30%. Analysis

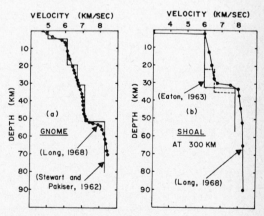

Fig. 2. Crustal structures: (a) shows Gnome velocity versus depth profiles, and (b) shows Shoal velocity versus depth profiles [*Long and Berg*, 1969].

TABLE 1. Data Used to Evaluate Attenuation

Station	Spectral Slope*	Travel Time, sec	Azimuth†
Bilby nuclear explosion			
Longmire	−1.99	156.7	O
Hungry Horse	−1.04	166.6	A
Shamrock	−0.45	187.8	B
Atlanta	−1.20	336.0	B
Berlin	−0.55	363.0	B
State College	−0.97	368.3	B
Caracas	−0.65	547.7	B
Kevo	+0.34	672.5	A
Sodankyla	+0.35	685.1	A
Konsberg	+0.23	695.4	A
Kajaani	+0.27	703.6	A
Nurmijarvi	−0.12	717.3	A
Toledo	−1.85	738.3	A
Honiara	−1.36	785.8	O
Shoal nuclear explosion			
Hungry Horse	+1.36	143.0	A
Manhattan	+0.29	239.1	B
Dallas	+1.19	260.1	B
Rolla	+0.79	283.8	B
Florissant	+1.37	293.9	B
Atlanta	+0.64	352.4	B
Ogdensberg	+0.80	398.6	B
Arequipa	+1.23	678.9	O

* 0.8 to 1.0 cps.
† A is 350° to 50°; B is 51° to 110°; O is other.

showed that the difference in the frequency intervals used in that work and those used in the research presented in this paper is, on the average, ±0.1 units of attenuation. Arrivals at the Canadian stations from other sources (Gnome and Haymaker nuclear explosions) similarly showed large attenuation. Anomalous attenuation with azimuth or with depth cannot be ruled out as an explanation, but, because of the difference in instruments and because of other amplitude data, the data from these Canadian stations have been excluded from this research.

Figure 4 shows slopes of the logarithms of the amplitudes of spectra of distant arrivals versus travel time (see Table 1). There is a distinct difference between the data pertaining to the two sources. Equation 2 shows that the source factor must be considered in the analysis of attenuation. Although estimates of source factor can be obtained from recordings near the source [*Trembly and Berg*, 1966; *Sarmah*, 1966] these data are usually not of sufficient quality to be used in conjunction with the above equation.

A measure of the difference in source factor was obtained by comparing the data for several stations at approximately 400 sec in travel time from the source. The average of the two

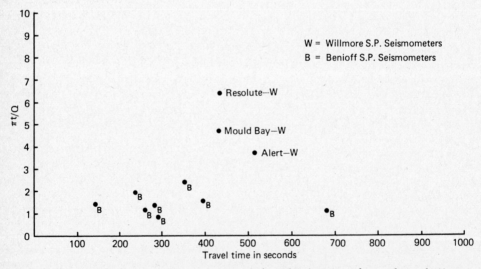

Fig. 3. Sample plot of attenuation versus travel time showing anomalous values of attenuation. Source is Shoal nuclear explosion.

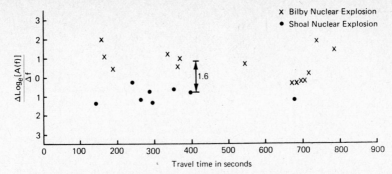

Fig. 4. Spectral slopes (0.8–1.0 cps) versus travel time for first arrivals (1½ cycles) from the Bilby and Shoal nuclear explosions showing source factor difference between the two sets of data.

data points for Pennsylvania compared to that for New Jersey is 1.6 units of attenuation. Since instrument factors for the two sets of data should be represented by the same constant, this estimated difference should be attributable to the source. The scatter of data shown in Figure 4 is considerably reduced by adjusting the data derived from the Bilby source to those derived from the Shoal source by the 1.6 units. The data in the resulting plot are directly re-

lated to attenuation values $\pi t/Q$ by using constant source and instrument factors that do not affect trends in the slopes of the logarithms of the amplitude spectra.

Shown in Figure 5 are ten attenuation models that were computed using a ray-tracing program for comparison with the observed data. Among other quantities, this program computed: attenuation $(\pi t/Q)$; amplitude affected by geometrical spreading; amplitude af-

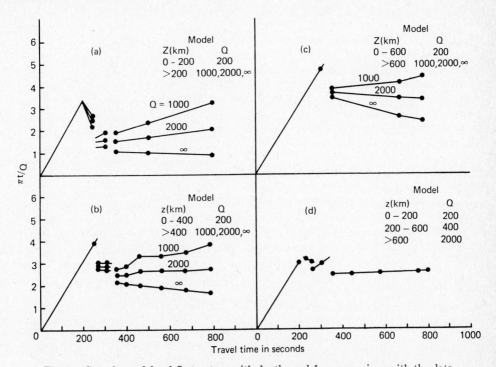

Fig. 5. Sample models of Q structure with depth used for comparison with the data.

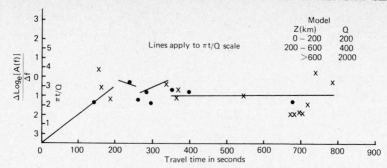

Fig. 6. Data, corrected for source factor difference, plotted with interpreted best fitting model.

fected by geometrical spreading and $\pi t/Q$ losses; maximum depth of penetration; and time of propagation through specified depth intervals. For this study, the values for the depth versus velocity function were taken from three sources, as follows: 0 to 60 km are from *Long* [1968]; 60 to 800 km are from *Niazi and Anderson* [1965]; below 800 km are from *Bullen* [1963]. The exactness of the function is not critical, since minor variations will not strongly affect the values of $\pi t/Q$ and the scatter in the observed values is large.

Figure 6 shows our interpreted best model fitted to the combined data adjusted for difference in source factor. It is important to note that major trends of the data have been fitted by this model and that further refinement does not seem warranted. The other models were rejected because the peaks are too high (Figure 5a), because the peaks are too far displaced in time (Figure 5c), and because there is a lack of correspondence with the data between 200 and 300 sec (Figure 5b).

Amplitude studies support the character of the chosen attenuation model between 30° and 90° epicentral distance. *Cleary* [1967] showed that the observed amplitudes from a set of earthquake data were in substantial agreement between 39° and 90° epicentral distance with an amplitude-distance curve derived from the Cleary-Hales travel time curve.

This was found to be true for amplitudes affected by geometrical spreading only, derived for distances between 30° and 90° from the velocity-depth function used in this work. This implies that the attenuation is nearly constant in this interval, which is the case shown in

Figure 6 (i.e., the attenuation curve is flattened because the wave spends progressively more time below 600 km in the mantle as the epicentral distance increases). In addition, *Hales et al.* [1968] found no evidence for major discontinuities between 24° and 96° in their travel time curve. Lateral inhomogeneities may exist, as was mentioned earlier for the data from the Canadian stations, and it is also a possibility that such inhomogeneities may be indicated by data from the Scandanavian stations grouped in Figure 6 at a travel time of approximately 700 seconds. Further, the data presented here were obtained from explosive sources located in the western part of the United States. As a result, data for the upper mantle from sources in other regions can be expected to differ significantly. For example, P waves generated by explosions in Lake Superior have been observed to have very low attenuation when propagating in the upper mantle beneath the Central lowlands and the Great Plains [*Roller and Jackson*, 1966]. In addition, *Oliver and Isacks* [1967] presented evidence for an anomalous portion of low attenuation material (high Q) protruding into a highly attenuating portion of the upper mantle beneath the Tonga-Kermadec arch. In general, it would appear that the upper mantle is quite heterogeneous in geologic structure, and, as a result, in the attenuation, or Q structure.

Acknowledgments. After September 1966, the data analysis was performed at the Seismic Data Laboratory, in Alexandria, Virginia. This was made possible through the courtesy of Mr. William Best.

This research was supported by the Air Force

Office of Scientific Research under contract AF-AFOSR-49(638)-1403 as part of the Vela Uniform Program directed by the Advanced Research Projects Agency of the Department of Defense.

REFERENCES

Berg, J. W., and L. T. Long, Characteristics of refracted arrivals of seismic waves, *J. Geophys. Res., 71* (10), 2583–2589, 1966.

Bullen, K. E., *An Introduction to the Theory of Seismology*, 3rd ed., Cambridge University Press, London, 1963.

Cleary, J., Analysis of the amplitudes of short-period *P* waves recorded by long range seismic measurements stations in the distance range 30° to 102°, *J. Geophys. Res., 72* (18), 4705–4712, 1967.

Eaton, J. P., Crustal structure from San Francisco, California, to Eureka, Nevada, from seismic refraction measurements, *J. Geophys. Res., 68* (20), 5789–5806, 1963.

Hales, A. L., J. R. Cleary, and J. L. Roberts, The velocity distribution in the lower mantle (abstract), *Eos Trans. AGU, 49* (1), 284–285, 1968.

Howell, B. F., Jr., Lake Superior seismic experiment: Frequency spectra and absorption, in *The Earth Beneath the Continents, Geophys. Monogr. Ser.,* vol. 10, edited by J. S. Steinhart and T. J. Smith, 227–233, AGU, Washington, D. C., 1966.

Jackson, D. D., and D. L. Anderson, Physical mechanisms of seismic-wave attenuation, *Rev. Geophys. Space Phys., 8* (1), 1–63, 1970.

Knopoff, L., *Q, Rev. Geophys. Space Phys., 2* (4), 625–660, 1964.

Kovach, R. L., and D. L. Anderson, Attenuation of shear waves in the upper and lower mantle, *Bull. Seismol. Soc. Amer., 54* (6), 1855–1864, 1964.

Long, L. T., Transmission and attenuation of the primary seismic wave, $\Delta = 100$ to 600 km, Ph.D. thesis, Department of Oceanography, Oregon State University, Corvallis, 1–110, 1968.

Long, L. T., and J. W. Berg, Jr., Transmission and attenuation of the primary seismic wave, 100 to 600 km, *Bull. Seismol. Soc. Amer. 59* (1), 131–146, 1969.

Niazi, M., and D. L. Anderson, Upper mantle structure of western North America from apparent velocities of *P* waves, *J. Geophys. Res., 70* (18), 4633–4640, 1965.

Oliver, J., and B. Isacks, Deep earthquake zones, anomalous structures in the upper mantle, and the lithosphere, *J. Geophys. Res., 72* (16), 4259–4275, 1967.

Roller, J. C., and W. H. Jackson, Seismic-wave propagation in the upper mantle: Lake Superior, Wisconsin to Denver, Colorado, in *The Earth Beneath the Continents, Geophys. Monogr. Ser.,* vol. 10, edited by J. S. Steinhart and T. J. Smith, 270–275, AGU, Washington, D. C. 1966.

Romney, C., Amplitudes of seismic body waves from underground nuclear explosions, *J. Geophys. Res., 64* (10), 1489–1498, 1959.

Romney, C., B. G. Brooks, R. H. Mansfield, D. S. Carder, J. N. Jordan, and D. W. Gordon, Travel times and amplitudes of principal body phases recorded from Gnome, *Bull. Seismol. Soc. Amer., 52* (5), 1057–1074, 1962.

Sarmah, S. K., Attenuation of compressional waves in the earth's mantle, Ph.D. thesis, Department of Oceanography, Oregon State University, Corvallis, 1–80, 1966.

Stewart, S. W., and L. C. Pakiser, Crustal structure in New Mexico interpreted from the Gnome explosion, *Bull. Seismol. Soc. Amer., 52* (5), 1017–1030, 1962.

Trembly, L. D., and J. W. Berg, Jr., Amplitudes and energies of primary seismic waves near the Hardhat, Haymaker, and Shoal nuclear explosions, *Bull. Seismol. Soc. Amer., 56* (3), 643–653, 1966.

Willis, D. E., and J. M. DeNoyer, Seismic attenuation and spectral measurements from the Lake Superior experiment, in *The Earth Beneath the Continents, Geophys. Monogr. Ser.,* vol. 10, edited by J. S. Steinhart and T. J. Smith, 218–226, AGU, Washington, D. C., 1966.

DISCUSSION

Mitchell: What effect would it have on your attenuation model if you used a velocity model with abrupt changes with depth?

Berg: It certainly would have an effect and might result in the computation blowing up when you invert the travel-time matrix. However, attenuation computed for a given station would be the same, and an attenuation model could be calculated.

Madden: Why is it that you can account for the constant T/Q in your data with a constant Q in the bottom layer of your model?

Berg: That is just the way the numbers come out. Rays spend progressively more time in the deeper mantle with increasing epicentral distance.

Oliver: Are all your crusts and upper mantle attenuation data from the Basin and Range province?

Berg: Yes, everything really hangs on the Shoal and Bilby nuclear shots.

Crustal Structure of the Eastern Basin and Range Province and the Northern Colorado Plateau from Phase Velocities of Rayleigh Waves

ROBERT L. BUCHER AND ROBERT B. SMITH

Department of Geological and Geophysical Sciences, University of Utah
Sale Lake City, Utah 84112

Abstract. Phase-velocity dispersion data from Rayleigh waves were used to study crustal and upper mantle structure beneath the eastern part of the Basin and Range and the northern part of the Colorado Plateau of the western United States. Seismograms of 29 earthquakes were analyzed in this study, most with epicenters in the southwest Pacific Ocean area. These were recorded by 14 long-period vertical seismographs. Geophysical investigations provided initial control for the theoretical models. Two models from more than 100 were selected as the most probable for the eastern Basin and Range and the northern Colorado Plateau. The two most significant features determined from this study were confirmation of crustal thicknesses obtained from seismic refraction profiles and the determination of low upper-mantle shear velocities. Under the northern Colorado Plateau the crust was modeled to be 40 km thick; under the eastern Basin and Range it was modeled to be 32 km thick. Both models show a compressional velocity of 7.8 km/sec in the uppermost mantle, which is notably low. For the two models, respective upper mantle shear velocities of 4.25 and 4.17 km/sec are also low. In each case, these velocities corroborate the results of other geophysical studies in this region which suggest that the upper mantle may include a partial melt zone.

During recent years, important geophysical studies of the crust and upper mantle have been made in the western United States. These include seismic, gravity, magnetic, electrical conductivity, and heat flow investigations, which when interpreted together give reasonably complete earth models. The most revealing of these studies has been extensive seismic refraction profiling [*Berg et al.*, 1960; *Eaton*, 1963; *Pakiser and Hill*, 1963; *Roller*, 1965; *Ryall and Stuart*, 1963]. Further information about crustal and upper mantle structure in the western United States has been obtained from dispersive surface-wave data [*Ewing and Press*, 1959; *Alexander*, 1963; *Wickens and Pec*, 1968]. These studies have partially corroborated results of the seismic refraction work, but more importantly, they have proved a useful method for determining shear velocities and hence Poisson's ratio.

In the early 1960's, mobile seismograph stations operated extensively throughout the western United States as part of the Long Range Seismic Measurement (LRSM) program of the Vela Uniform Project. Twelve of these portable stations, together with the Worldwide Standard Seismic (WWSS) station at Dugway, Utah, and the Uinta Basin Seismological Observatory near Vernal, Utah, provided a network of long-period seismographs in the eastern part of the Basin and Range and northern part of the Colorado Plateau (Figure 1). It is the purpose of this investigation to use Rayleigh-wave data from this network to study crustal and upper mantle structure in the eastern Basin and Range and the northern Colorado Plateau. Information on the seismograph stations used in this study is given in Table 1.

Earthquake data analysis. Long-period records were searched for suitable Rayleigh waves from distant earthquakes. These events occurred during 1962 and 1963, when most of the LRSM stations in the Basin and Range and the Colorado Plateau were operating. Only the fundamental modes were picked, since higher Rayleigh modes and Love waves were not suitable for the analysis used in this study. From more than 100 earthquakes, 29 were selected because

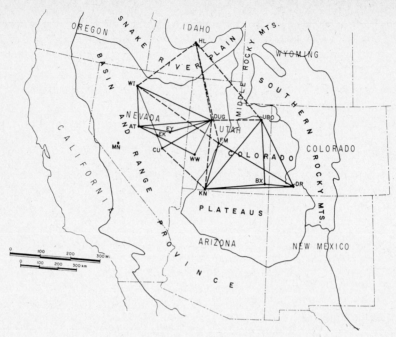

Fig. 1. Index map of seismograph stations with primary and secondary arrays indicated by continuous and broken lines, respectively.

they could be correlated well at three or more stations in the network. Most of these events originated in the southwest Pacific at an average distance of 7000 km.

Phase velocity computation. For an array of three stations, phase velocities can be computed by the tripartite method [*Evernden,* 1953; *Press,* 1956]. However, for an array of more than three stations (a multipartite array) phase velocities and directions of approach are better

TABLE 1. Seismograph Station Information

Station	Symbol	Type
Austin, Nev.	AT-NV	LRSM
Blanding, Ut.	BX-UT	LRSM
Currant, Nev.	CU-NV	LRSM
Dugway, Ut.	DUG	WWSS
Durango, Co.	DR-CO	LRSM
Ely, Nev.	EY-NV	LRSM
Eureka, Nev.	EK-NV	LRSM
Fillmore, Ut.	FM-UT	LRSM
Hailey, Id.	HL-ID	LRSM
Kanab, Ut.	KN-UT	LRSM
Mina, Nev.	MN-NV	LRSM
Uinta Basin, Ut.	UBO	Geneva
Wah Wah Mts., Ut.	WW-UT	LRSM
Winnemucca, Nev.	WI-NV	LRSM

determined by a least-squares technique [*Aki,* 1961]. To compute phase velocities for both tripartite and multipartite arrays, a method following Aki's was programmed for use on a digital computer.

The program read in successive arrival times of correlative peaks and troughs for each station in an array; these times were then corrected for instrumental phase shift according to the formula of *Hagiwara* [1958]. Although this correction is small or negligible when nearly identical instruments are used, such as those of the LRSM stations, it can be as much as 10 sec when the long-period records from the WWSS or Vela Observatory stations are combined with those from LRSM stations.

Corrected arrival times, distances, and azimuths between stations were then used to compute the phase velocities. The period associated with each phase velocity was the computed average of the periods observed at each station in the array. After phase velocities were computed, weighted averaging of these data reduced the scatter and reduced the large number of plotted points for the shorter-period dispersion data.

Error analysis. Significant errors in computing phase velocities for an array may result from inaccuracies in picked arrival times, in phase shift corrections, and in distances and azimuths between stations. The combined uncertainty of these experimental errors for most phase velocities computed for our tripartite arrays was usually less than 1.5 per cent, or 0.06 km/sec; for multipartite arrays it was usually less than 1 per cent, or 0.04 km/sec. Part of the scatter in the dispersion data may have been the result of these errors, but the remainder was probably due to lateral inhomogeneities in the earth itself.

Modeling parameters. After the observed dispersion data were plotted, theoretical dispersion curves were computed by using *Dorman's* [1962] PV-7 program for flat-layered earth models. Shear velocities, layer thicknesses, and compressional velocities were altered until an acceptable fit between the observed and the theoretical dispersion curves was obtained. To relate the compressional and shear velocities, we used Poisson's ratio, hereafter designated by σ, where

$$\sigma = \frac{1 - \frac{1}{2}(\alpha^2/\beta^2)}{1 - (\alpha^2/\beta^2)}$$

and α is the compressional velocity and β is the shear velocity.

Density distributions given by *Jackson and Pakiser* [1965] for refraction models in the western United States were used for the theoretical crustal models; it is recognized that these densities may differ by 0.1 g/cc from those given by *Birch* [1961]. These differences in density do not produce significant changes in the phase velocities. For example, increasing the density by 0.1 g/cc decreased the phase velocity in general by only about 0.01 km/sec, or 0.3 per cent.

To estimate the depths at which differences in the model parameters change phase velocities that are less than a few hundredths km/sec over the observed period ranges, a simple rule is useful. The depth at which the wave amplitude reduces to a small fraction of its near-surface value is about one-half its wavelength. Below this depth, the effect of the material on the particular wave is small. Hence, a wavelength with a period of 45 sec effectively samples the material from the surface to a depth of about 85 km. For the period range 12 to 45 sec, which is used principally in this study, altering theoretical model parameters below a depth of 100 km produces little effect. *Der et al.* [1970]

calculated errors of this nature and have also suggested that upper mantle velocities may be inaccurate when determined from Rayleigh-wave data in the period range 15 to 80 sec.

Northern Colorado Plateau interpretations. Rayleigh-wave dispersion data were recorded in the northern Colorado Plateau for three arrays: KN-FM-DR, KN-DR-UBO, and KN-BX-DR-UBO (Figure 1). Because differences in average phase velocities for the three arrays were no greater than 0.05 km/sec, theoretical curves for varied crustal models were visually fitted to all dispersion data for the three northern Colorado Plateau arrays (Figure 2).

Roller's [1965] reversed refraction profile from Hanksville, Utah, to Chinle, Arizona, is near the center of the above arrays (Figure 1). By using his velocity and thickness determinations at Hanksville, theoretical dispersion curves for two- and three-layered crustal models were computed on the assumption that Poisson's ratio is 0.25 (models C1 and C2, Table 2). The theoretical curve for the three-layered crustal model C1 lies within the scatter of the observed phase velocities for all periods (Figure 2), but it is somewhat higher than the average observed phase velocities for the longer periods. The theoretical dispersion curve for the two-layered model, C2, is below the average phase velocity observed for periods less than 27 sec and above the average for periods greater than 27 sec (Figure 2). Both models suggest that the upper mantle shear velocity must be lower than 4.5 km/sec, which is the value computed for a compressional velocity of 7.8 km/sec.

Other theoretical dispersion curves were computed for different crustal models (Table 2) in an attempt to match the observed dispersion data for the northern Colorado Plateau. Assuming $\sigma = 0.25$, a three-layered crustal model (model C3, Figure 2) must have a lower layer 28 km thick. This gives a total crustal thickness of 55 km.

Model C4 has a three-layered crust with $\sigma = 0.25$ and an upper mantle layer with $\sigma = 0.28$. Model C5 has a Poisson's ratio of 0.27 for both the lower crustal and upper mantle layers. Both models have crustal thicknesses of 40 km and low shear velocities in the upper mantle of 4.31 and 4.37 km/sec, respectively. These produce a reasonable fit between the theoretical and the observed dispersion data (Figure 2).

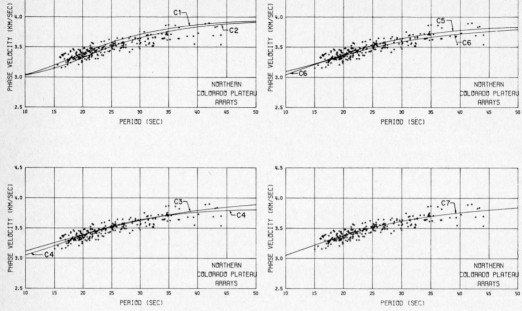

Fig. 2. Theoretical Rayleigh-wave dispersion curves for northern Colorado Plateau models: C1 and C2 (top left); C3 and C4 (bottom left); C5 and C6 (top right); and C7 (bottom right).

Model C6 includes an 8.2-km/sec layer at a depth of 85 km, which was suggested by *Jackson and Pakiser* [1965]. The crust is modeled at 40 km thick and the upper mantle has a finite 7.8-km/sec layer 45 km thick overlying a semi-infinite 8.2-km/sec layer. To fit the observed dispersion data, upper mantle shear velocities of 4.17 and 4.38 km/sec were used (Figure 2). These give a high Poisson's ratio of 0.30, whereas in the crust it remains 0.25.

Model C7 is similar to model C6 with two exceptions: the lower crustal shear velocity is 3.87 instead of 3.93 km/sec and the uppermost mantle shear velocity is 4.25 instead of 4.17 km/sec. The theoretical curve for this model matches observed phase velocities for the intermediate periods better than the curve for model C6 (Figure 2).

All of the above three-layered crustal models are reasonable for the northern Colorado Plateau, although models C3, C4, C5, and C7 have theoretical curves which match the observed data more closely over the period range 15 to 45 sec. Model C7, which includes a layer at a depth of 85 km, is preferred to model C5 because it is more consistent with the observed dispersion

data in the intermediate period range. It is preferred over model C3, which has a crustal thickness of 55 km. This is at least 10 km more than the 40- to 43-km thickness determined by *Roller* [1965, p. 107], *Ryall and Stuart* [1963, p. 5821], and *Jackson and Pakiser* [1965, p. 89].

The Dugway-Tucson shear-velocity model of *Wickens and Pec* [1968, p. 1827], which was determined from Love-wave phase-velocity dispersion, includes parts of both the eastern Basin and Range and western Colorado Plateau and is similar to model C5. This model has a 35-km crust with an upper mantle shear velocity of 3.20 km/sec which compares with the 40-km crust and 3.56-km/sec shear velocity of model C7. The upper mantle shear velocity in the Dugway-Tucson model increases from 4.40 to 4.50 km/sec at a depth of 80 km and corresponds closest to model C7, which increases from 4.25 to 4.48 km/sec at a depth of 85 km.

If the crustal shear velocity is actually higher, as suggested by *Wickens and Pec* [1968], then upper mantle shear velocities for the Dugway-Tucson model must be lowered to fit the data. These would be in better agreement with the upper mantle velocities of model C7. In any event, the most significant aspect of both models is evidence for low velocities at the top of the mantle.

These low shear velocities determined for the upper mantle of the northern Colorado Plateau can also be deduced from low P_n velocities, low average crustal and upper mantle densities, and the absence of long-wavelength magnetic anomalies. Roller's [1965, p. 107] low P_n velocity of 7.80 km/sec along the top of the upper mantle in southeastern Utah and northeastern Arizona can be associated with a low upper-mantle density of 3.3 g/cc by using the velocity-density relationships given by Nafe and Drake [Talwani et al., 1959].

Cook et al. [1966a] determined density models from a Bouguer gravity profile extending from the Nevada test site to the Kansas-Colorado border. Across the Colorado Plateau of southern Utah they determined crustal thicknesses of 42 km for the western half and 38 km for the eastern half. These thicknesses do not include approximately 2 km of 2.67 g/cc material above sea level. All crustal thicknesses for the Rayleigh-wave models are measured downward from the average elevation of the region. They used densities of 2.80 g/cc for an upper crustal layer, 3.00 g/cc for a lower crustal layer, and 3.40 g/cc for the upper mantle. These correspond well to respective densities of 2.83, 2.99, and 3.30 g/cc for the northern Colorado Plateau models used in this study.

Pakiser and Zietz [1965] have interpreted a transcontinental aeromagnetic profile which crossed southern Nevada and Utah. For the portion west of the Colorado Plateau, the profile shows no long-wavelength magnetic anomalies. On this basis, they suggest that the Curie temperature is reached at or above the upper mantle boundary for this part of the profile. Moderate heat flow values between 1.0 and 1.5 μcal/cm^2 sec [Blackwell, 1969, p. 1000] have been measured in the northern Colorado Plateau. These values do not indicate an extremely high upper mantle temperature below the thick 40 km crust determined for the northern Colorado Plateau. However, a moderately high upper mantle temperature can explain the lower than normal mantle velocities and densities as well as the absence of long-wavelength magnetic anomalies.

Eastern Basin and Range interpretations.

Theoretical dispersion curves were computed for three eastern Basin and Range models which are based on the refraction results of Eaton [1963], Hill and Pakiser [1967], and Berg et al. [1960]. These are models B1, B2, and B3 (Table 2) with shear velocities determined for a Poisson's ratio of 0.25. A fourth theoretical curve, B4, was computed for the Basin and Range surface-wave model of Alexander [1963].

The most surprising result from these theoretical curves (Figure 3) is that the phase velocities of the observed longer-period data are significantly higher than those determined for the four models above. This difference suggests that the theoretical models for these arrays should be assigned higher shear velocities for the upper mantle or shallower crustal depths.

Theoretical dispersion curves were computed for numerous crustal models which included high upper-mantle shear velocities. Models B5, B6, B7, and B8 have theoretical dispersion curves which fit the observed phase velocities (Figure 3). These models use crustal and upper mantle compressional velocities determined by Eaton [1963] and Hill and Pakiser [1967] for the northern Nevada area.

Model B5 is a three-layered crustal model with a thickness of 30 km. The uppermost mantle layer with a compressional velocity of 7.9 km/sec is 20 km thick. It overlies a semi-infinite 8.3-km/sec layer with a high shear velocity of 5.0 km/sec. Poisson's ratio is 0.22 for this semi-infinite layer and 0.25 for the others. The theoretical curve for this model fits the average long-period phase velocities but is below it for periods of less than 26 sec (Figure 3).

Model B6 is similar to B5 except that it has a two-layered crust with no near-surface layer and an 8.3-km/sec layer beginning 10 km lower at a depth of 60 km. The theoretical curve for this model is below the average observed phase velocities for all periods (Figure 3). These models indicate the crust is less than 30 km thick with upper mantle shear velocities between 4.6 and 5.0 km/sec.

B7 and B8 are three-layered crustal models that have thicknesses of 25 km. The significant features of these models are upper mantle shear velocities of 4.65 and 4.73 km/sec for model B7 and 4.60 and 4.95 km/sec for model B8. The theoretical curve for model B7 matches the average observed short-period velocities and

TABLE 2.   Crustal and Upper Mantle Models

Model	Layer Thickness, h, km	Compressional Velocity, α, km/sec	Shear Velocity, β, km/sec	Density, ρ, g/cc	Poisson's ratio, σ
northern Colorado Plateau					
C1	2.5	3.00	1.73	2.40	.25
	24.5	6.20	3.58	2.83	.25
	13.0	6.80	3.93	2.99	.25
		7.80	4.50	3.30	.25
C2	2.5	3.00	1.73	2.40	.25
	35.0	6.20	3.58	2.83	.25
		7.80	4.50	3.30	.25
C3	2.5	3.00	1.73	2.40	.25
	24.5	6.20	3.58	2.83	.25
	28.0	6.80	3.93	2.99	.25
		7.80	4.50	3.30	.25
C4	2.5	3.00	1.73	2.40	.25
	24.5	6.20	3.58	2.83	.25
	13.0	6.80	3.93	2.99	.25
		7.80	4.31	3.30	.28
C5	2.0	3.00	1.73	2.40	.25
	25.0	6.20	3.58	2.83	.25
	13.0	6.80	3.81	2.99	.27
		7.80	4.37	3.30	.27
C6	2.5	3.00	1.73	2.40	.25
	24.5	6.20	3.58	2.83	.25
	13.0	6.80	3.93	2.99	.25
	45.0	7.80	4.17	3.30	.30
		8.20	4.38	3.43	.30
C7	2.5	3.00	1.73	2.40	.25
	24.5	6.20	3.58	2.83	.25
	13.0	6.80	3.87	2.99	.26
	45.0	7.80	4.25	3.30	.29
		8.20	4.38	3.43	.30
eastern Basin and Range					
B1	2.0	4.50	2.60	2.50	.25
	15.0	6.00	3.46	2.78	.25
	7.0	6.60	3.81	2.93	.25
	70.0	7.80	4.50	3.30	.25
		8.20	4.73	3.43	.25
B2	3.0	4.50	2.60	2.50	.25
	20.0	6.00	3.46	2.78	.25
	10.0	6.70	3.87	2.96	.25
	60.0	7.90	4.56	3.30	.25
		8.20	4.73	3.43	.25
B3	9.0	5.73	3.31	2.67	.25
	16.0	6.33	3.66	2.84	.25
	47.0	7.59	4.38	3.18	.25
		7.97	4.60	3.27	.25
B4	2.0	3.80	1.75	2.20	.36
	13.0	6.10	3.60	2.83	.23
	10.0	6.20	3.65	2.83	.23
	24.0	7.70	4.10	3.30	.30
	15.0	8.15	4.60	3.40	.27
	20.0	8.00	4.50	3.45	.27
	20.0	7.90	4.45	3.45	.27
		7.80	4.40	3.45	.27
B5	2.0	4.50	2.60	2.50	.25
	18.0	6.00	3.46	2.78	.25
	10.0	6.70	3.87	2.96	.25

TABLE 2.   (*Continued*)

Model	Layer Thickness, h, km	Compressional Velocity, α, km/sec	Shear Velocity, β, km/sec	Density, ρ, g/cc	Poisson's ratio, σ
	20.0	7.90	4.56	3.30	.25
		8.30	5.00	3.43	.22
B6	20.0	6.00	3.46	2.78	.25
	10.0	6.70	3.87	2.96	.25
	30.0	7.90	4.56	3.30	.25
		8.30	5.00	3.43	.22
B7	3.0	4.50	2.60	2.50	.25
	15.0	6.00	3.46	2.78	.25
	7.0	6.70	3.90	2.96	.24
	25.0	7.90	4.65	3.30	.24
		8.20	4.73	3.43	.25
B8	1.5	4.50	2.60	2.50	.25
	15.0	6.00	3.46	2.78	.25
	8.5	6.60	3.81	2.93	.25
	35.0	7.90	4.60	3.30	.24
		8.30	4.95	3.43	.23
BC2*	48.0	6.10	3.51	2.80	.25
		8.10	4.68	3.50	.25
BC3*	2.5	4.50	2.60	2.50	.25
	20.0	6.00	3.41	2.78	.26
	10.0	6.70	3.71	2.96	.28
	45.0	7.80	4.17	3.30	.30
		8.20	4.38	3.43	.30
eastern Basin and Range and northern Colorado Plateau					
BC1	2.0	3.50	2.03	2.45	.25
	20.0	6.10	3.52	2.80	.25
	13.0	6.70	3.87	2.96	.25
		7.80	4.25	3.30	.29

* Data from different Basin and Range-Snake River Plain arrays were used (see text).

lies along the lower limit of phase-velocity scatter for the longer periods (Figure 3).

The theoretical curve for model B8 best fits the observed dispersion data (Figure 3), although an upper mantle shear velocity of 4.95 km/sec is extremely high. This best fitting model would suggest that the average crust in the eastern part of the Basin and Range province is about 25 km thick.

Poisson's ratio of 0.25 for the crust is compatible with the observed dispersion data. An uppermost mantle layer 35 km thick, which has a shear velocity of 4.60 km/sec and Poisson's ratio of 0.25, is also compatible with the observed data. However, the semi-infinite 8.3 km/sec layer has a high shear velocity of about 4.95 km/sec but is beyond the depth of good resolution for the range of periods used.

The average crustal thickness of 25 km for the northern Basin and Range is reasonable when compared with the 24- to 32-km thickness determined for refraction models and is the same as the thickness determined by *Alexander* [1963] from group-velocity data over a larger area of the Basin and Range. In addition, compressional velocity and density distributions are not extremely different from those determined for the refraction models.

Normal shear velocities may reach approximately 4.80 km/sec before they begin to decrease with depth. This decrease is true of the Basin and Range model of *Alexander* [1963], which has a maximum shear velocity of 4.60 km/sec and decreases below a depth of 64 km. Although model B7 has a maximum shear velocity of 4.73 km/sec, which is more logical than 4.95 km/sec, it does not match the observed dispersion data. In this case, waves of 40-sec

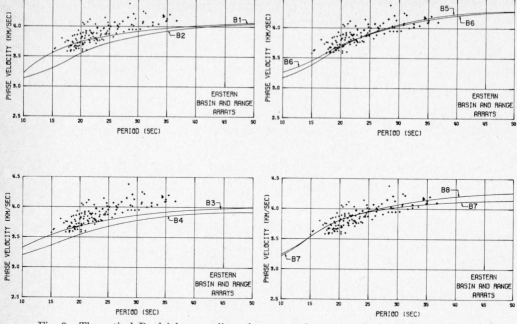

Fig. 3.   Theoretical Rayleigh-wave dispersion curves for eastern Basin and Range models:
B1 and B2 (top left); B3 and B4 (bottom left); B5 and B6 (top right); and B7 and B8
(bottom right).

period or greater are required to confirm these abnormal shear velocities but were not available in our data.

Observed dispersion data for smaller arrays which cross physiographic province boundaries were computed as well. The arrays WI-HL-DUG and WI-HL-FM are representative of several considered. These arrays included the central part of the Snake River Plain with the eastern part of the Basin and Range province. These phase-velocity dispersion data show results that are similar to those of the eastern Basin and Range province given by *Ewing and Press* [1959] for their Eureka-Butte-Salt Lake City and Eureka-Salt Lake City-Boulder City arrays.

Ewing and Press [1959, p. 243] found an average crustal thickness of 48 km for their Basin and Range arrays (model BC2, Table 2). *Smith's* [1962, p. 1031] later reinterpretation of their data indicated a 30-km crust. If *Jackson and Pakiser's* [1965, p. 91] approximate layer thicknesses, compressional velocities, and densities are used for a theoretical Basin and Range model, the resultant crustal thickness is 32 km. The corresponding shear velocities are 2.6, 3.41, 3.71,

and 4.17 km/sec in the upper crustal, lower crustal, and upper mantle layers, respectively (model BC3). Theoretical dispersion curves for these models in Figure 4 indicate that BC3 is the better model. Model BC3 has a compressional velocity of 6.0 km/sec in the upper crustal layer and a crustal thickness of 32.5 km. Compressional velocities in this area have been determined by *Eaton* [1963] and *Hill and Pakiser* [1967]. *Prodehl's* [1970] reinterpretation of the United States Geological Survey seismic refraction data indicated a 30- to 34-km crustal thickness in the eastern part of the Basin and Range province.

Cook et al. [1966a, p. 193] determined a crustal layer 26.5 to 27.5 km thick with a density of 2.80 g/cc from the Nevada test site eastward into the Colorado Plateau. *Cook et al.* [1966b, p. 192] determined a 31.5-km crust with a density of 2.80 g/cc in the eastern Basin and Range of western Utah. Both models require an upper mantle density of 3.30 g/cc and a density of 3.40 g/cc below 60 to 70 km. Eastern Basin and Range model B8, determined here, has a slightly thinner crust of 25 km, with crustal densities of 2.78 and 2.93 g/cc. Model BC3 has a thicker crust of 32.5 km, which includes an upper crustal layer of 2.78 g/cc and a lower crustal layer with a density of 2.96 g/cc.

Kovach and Robinson [1969] determined an upper mantle shear-velocity distribution for the Basin and Range province of Arizona which is different from both. Their model shows a 4.50-km/sec lid zone 9 km thick at the top of the mantle, which begins at a depth of 33 km. Below 42 km they show a shear-velocity decrease from 4.50 to 4.40 km/sec at 100 km. *Archambeau et al.* [1969] have interpreted a similar low-velocity zone for compressional waves in the upper mantle of the eastern Basin and Range. Their model has a velocity of 7.72 km/sec in the uppermost mantle which extends from a depth of 28 to 45 km. Below this possible lid zone, the compressional velocity decreases very slightly to 7.71 before gradually increasing to 7.90 km/sec at 146 km. At this depth it increases to 8.33 km/sec. Because model BC3 has similar low upper mantle shear and compressional velocities, as determined from other surveys, it

probably is more reasonable than model B8 on the basis of these findings.

High average heat flow of 2.0 μcal/cm^2 sec [*Roy et al.*, 1968; *Blackwell*, 1969, p. 1003] for the eastern Basin and Range province suggests abnormally high temperatures in the upper mantle which may possibly extend into the lower crust. As a result, *Roy et al.* [1970, p. 31] have proposed a partial melt zone from approximately 60 to 170 km below the Basin and Range province. The *P*-wave delay studies made by *Hales and Herrin* [1970, p. 24] in the Basin and Range lend further support to this hypothesis. Low shear velocities in the lower crust and upper mantle of the eastern Basin and Range would be expected for higher than normal temperatures in these regions. A sharp increase in mantle conductivity [*Reitzel et al.*, 1970, p. 233] under the Wasatch front of the eastern Basin and Range may also imply lower than normal shear velocities at these depths. The low compressional velocities in the upper mantle of the Basin and Range have been interpreted by

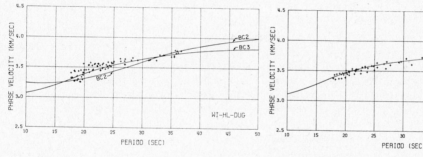

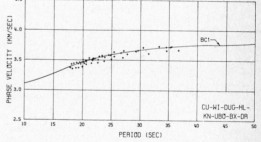

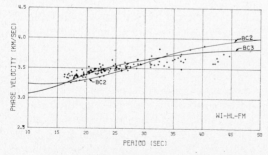

Fig. 4. Rayleigh-wave dispersion data for the array WI-HL-DUG with theoretical dispersion curves for models BC2 and BC3 are shown at top left; Rayleigh-wave dispersion data for the array WI-HL-FM with theoretical dispersion curves for models BC2 and BC3 are shown at bottom left; and Rayleigh-wave dispersion data for the eight station array CU-WI-DUG-HL-KN-UBO-BX-DR with theoretical dispersion curve for model BC1 are shown at top right.

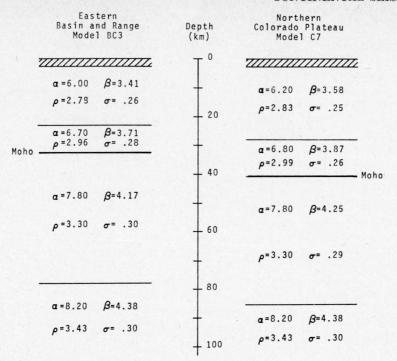

Fig. 5. Crustal and upper-mantle models for the eastern part of the Basin and Range province and for the northern part of the Colorado Plateau. Cross-hatched area indicates near-surface layers.

Cook [1962] as an anomalous layer designated as the mantle-crust mix. In conclusion, a low-velocity high-temperature relation indicated by most geophysical data best supports the low shear-velocity model BC3.

Composite interpretation of Basin and Range and Colorado Plateau. An eight-station array, CU-WI-DUG-HL-KN-UBO-BX-DR, included stations in three physiographic areas: the central Snake River Plain, the eastern Basin and Range province, and the northern Colorado Plateau (Figure 4). Phase velocities for this large array were computed from least-square fitted arrival times at the eight stations. As shown, the phase-velocity scatter is 0.1 km/sec or less over the period range of 17 to 37 sec (Figure 4).

Average values of the phase velocities are substantially lower than those determined for the eastern Basin and Range but just slightly higher than those for the northern Colorado Plateau. This implies that the average structure is similar to that of the Colorado Plateau. A theoretical dispersion curve for a three-layered crustal model (BC1, Table 2) was fitted to the observed dispersion data from the three-province array. However, the great lateral changes over such a large region give it little real significance for local crustal investigations.

CONCLUSIONS

Phase velocity dispersion of Rayleigh waves has been used to study the crust and upper mantle structure beneath the eastern Basin and Range and the northern Colorado Plateau. From several plausible models, two were selected as the most probable (Figure 5).

Model C7 for the northern Colorado Plateau confirms a crustal thickness of 40 km and exhibits a relatively low upper mantle shear velocity of 4.25 km/sec. The eastern Basin and Range model BC3 has a 32-km crust with a shear velocity of 4.17 km/sec in the upper mantle. These low shear velocities support the hypothesis of partial melting in the upper mantle.

The results of this investigation generally confirm the findings and implications of other geophysical studies of this region. A thin crust, high heat flow, low upper-mantle seismic velocities, high attenuation of seismic waves, high electrical conductivities, and lack of long-wavelength magnetic anomalies characterize the eastern part of the Basin and Range province. Thicker crust, moderate heat flow, and low upper-mantle seismic velocities characterize the northern Colorado Plateau.

Severe lateral refraction of Rayleigh waves across the Basin and Range-Colorado Plateau boundary strongly suggests a sharp transition zone in the upper mantle. This boundary transition is at least 60 km deep and correlates with the location of the southern end of the intermountain seismic belt [*Smith and Sbar*, 1970]. The continuity and structure of crustal and upper mantle layers in this abrupt boundary zone are poorly determined from surface wave dispersion studies. Further experiments to describe the crustal structure in the zone should include high-resolution refraction profiling incorporated with normal incidence reflection studies.

A major objective of the ONR-CIRES Symposium on physical properties of the crust, reported in this volume, was to determine the extent of suggested 'sialic' low velocity layers across different geologic regimens. The phase-velocity dispersion technique provides a method whereby thin low-velocity layers can be detected if sufficiently short periods can be recorded. However, our results do not indicate the presence of major low velocity layers in the crust of the eastern Basin and Range or the northern Colorado Plateau. Thin low-velocity layers suggested by *Landisman et al.* [1971] for the Basin and Range could not be confirmed by our study because of the insensitivity of the phase velocity method that results when periods greater than 15 sec are used.

Acknowledgments. Data for this study were obtained through the cooperation of the U. S. Air Force, Vela Seismological Center, Washington, D. C. Dugway seismograms were available at the Department of Geological and Geophysical Science, University of Utah. We are indebted to Professor E. S. Robinson (formerly at the Department of Geophysics, University of Utah, now at Virginia Polytechnic Institute), who suggested this study. Professors R. T. Shuey, K. L. Cook, H. D. Goode, and F. W. Christiansen, at the Department of Geological and Geophysical Sciences, University of Utah, also made suggestions.

The Research Corporation, Burlingame, California, granted funds for the research.

REFERENCES

Aki, K., Crustal structure of Japan from the phase velocity of Rayleigh waves, *Bull. Earth Res. Tokyo Univ.*, *39*, 255–283, 1961.

Alexander, S. S., Crustal structure in the Western United States from multi-mode surface wave dispersion, Ph.D. thesis, California Institute of Technology, Pasadena, 229 pp., 1963.

Archambeau, C. B., E. A. Flinn, and D. G. Lambert, Fine structure of the upper mantle, *J. Geophys. Res.*, *74*, 5825–5865, 1969.

Berg, J. W., Jr., K. L. Cook, H. D. Narans, Jr., and W. M. Dolan, Seismic investigation of crustal structure in the eastern part of the Basin and Range province, *Bull. Seismol. Soc. Amer.*, *50*, 511–535, 1960.

Birch, F., The velocity of compressional waves in rocks to 10 kilobars, 2, *J. Geophys. Res.*, *66*, 2199–2224, 1961.

Blackwell, D. D., Heat-flow determinations in the northwestern United States, *J. Geophys. Res.*, *74*, 992–1007, 1969.

Cook, K. L., The problem of the mantle-crust mix: Lateral inhomogeneity in the uppermost part of the earth's mantle, *Advan. Geophys.*, *9*, 295–360, 1962.

Cook, K. L., R. C. Gray, and J. K. Costain, Crustal and upper mantle structure from the Nevada Test Site area to the Colorado-Kansas border as determined by a gravity profile (abstract), *Eos, Trans. AGU*, *47*, 192–193, 1966a.

Cook, K. L., D. A. Zimbeck, and J. K. Costain, Crustal and upper mantle structure as determined by northward-trending gravity profiles across the boundary between the Basin and Range province and the southern and southwestern parts of the Colorado Plateau (abstract), *Eos, Trans. AGU*, *47*, 192, 1966b.

Der, Z., R. Masse, and M. Landisman, Effects of observational errors on the resolution of surface waves at intermediate distances, *J. Geophys. Res.*, *75*, 3399–3409, 1970.

Dorman, J., Period equation for waves of Rayleigh type on a layered, liquid-solid half space, *Bull. Seismol. Soc. Amer.*, *52*, 389–398, 1962.

Eaton, J. P., Crustal structure from San Francisco, California, to Eureka, Nevada, from seismic-refraction measurements, *J. Geophys. Res.*, *68*, 5789–5806, 1963.

Evernden, J. F., Direction of approach of Rayleigh waves and related problems, 1, *Bull. Seismol. Soc. Amer.*, *43*, 335–374, 1953.

Ewing, M., and F. Press, Determination of crustal structure from phase velocity of Rayleigh waves, 3, The United States, *Bull. Geol. Soc. Amer.*, *70*, 229–244, 1959.

Hagiwara, T., A note on the theory of the electromagnetic seismograph, *Bull. Earthquake Res. Inst., Univ. Tokyo*, *36*, 139–164, 1958.

Hales, A. L., and E. T. Herrin, Discussions of the travel times of P and S waves (abstract), pp. 22–24, in *The nature of the solid earth—A symposium in honor of Professor Francis Birch*, Harvard Univ., Cambridge, Mass., 22–24, 1970.

Hill, D. P., and L. C. Pakiser, Seismic-refraction study of crustal structure between the Nevada Test Site and Boise, Idaho, *Bull. Geol. Soc. Amer., 78*, 685–704, 1967.

Jackson, W. H., and L. C. Pakiser, Seismic study of crustal structure in the Southern Rocky Mountains, *U.S. Geol. Surv. Prof. Pap. 525-D*, D85–D92, 1965.

Kovach, R. L., and R. Robinson. Upper mantle structure in the Basin and Range province, North America, from the apparent velocities of S waves, *Bull. Seismol. Soc. Amer., 59*, 1653–1665, 1969.

Landisman, M., S. Mueller, and B. J. Mitchell, Review of evidence for velocity inversions in the continental crust, in *The Structure and Physical Properties of the Earth's Crust, Geophys. Monogr. Ser.*, vol. 14, edited by J. G. Heacock, AGU, Washington, D. C., this volume, 1971.

Pakiser, L. C., and D. P. Hill, Crustal structure in Nevada and Southern Idaho from nuclear explosions, *J. Geophys. Res., 68*, 5757–5766, 1963.

Pakiser, L. C., and I. Zietz, Transcontinental upper-mantle structure, *Rev. Geophys. Space Phys., 3*, 505–520, 1965.

Press, F., Determination of crustal structure from phase velocity of Rayleigh waves, 1, Southern California, *Bull Geol. Soc. Amer., 67*, 1647–1658, 1956.

Prodehl, C., Seismic refraction study of crustal

structure in the western United States, *Bull. Geol. Soc. Amer., 81*, 2629–2646, 1970.

Reitzel, J. S., D. I. Gough, H. Porath, and C. W. Anderson, III, Geomagnetic deep sounding and upper mantle structure in the western United States. *Geophys. J. Roy. Astron. Soc., 19*, 213–235, 1970.

Roller, J. C., Crustal structure in the eastern Colorado Plateaus province from seismic-refraction measurements, *Bull. Seismol. Soc. Amer., 55*, 107–119, 1965.

Roy, R. F., E. R. Decker, D. D. Blackwell, and F. Birch, Heat flow in the United States, *J. Geophys. Res., 73*, 5207–5221, 1968.

Roy, R. F., D. D. Blackwell, and E. R. Decker, Continental heat flow (abstract), pp. 31–32, in *The nature of the solid earth—A symposium in honor of Professor Francis Birch*, Harvard Univ., Cambridge, Mass., 1970.

Ryall, A., and D. J. Stuart, Travel times and amplitudes from nuclear explosions, Nevada test site to Ordway, Colorado, *J. Geophys. Res., 68*, 5821–5835, 1963.

Smith, R. B., and M. Sbar, Tectonics of the inter-mountain seismic belt, Western United States, 2, Focal mechanisms of major earthquakes (abstract), *1970 Ann. Meet. Geol. Soc. Amer., Circular 2*, 687–688, 1970.

Smith, S. W., A reinterpretation of phase velocity data based on the Gnome travel time curves, *Bull. Seismol. Soc. Amer., 52*, 1031–1035, 1962.

Talwani, M., G. H. Sutton, and J. L. Worzel, A crustal section across the Puerto Rico trench, *J. Geophys. Res., 64*, 1545–1555, 1959.

Wickens, A. J., and K. Pec, A crust-mantle profile from Mould Bay, Canada, to Tucson, Arizona, *Bull. Seismol. Soc. Amer., 58*, 1821–1831, 1968.

DISCUSSION

Kennedy: How much variation can you allow in the sub-Moho velocity of 7.8 km/sec?

Healy: We define the Moho here as that layer immediately beyond this most prominent break in the travel-time curve. On that basis, the NTS shots give velocities of 7.8 to 7.9 km/sec in the Basin and Range province. Recent work in the San Francisco Bay area gives a velocity below 8 km/sec for the NTS shots after correcting for the depth of the Moho, and this suggests a velocity of about 7.8 km/sec down to some considerable depth—around 80 km, I guess. On another profile in Utah we are still fitting the data with a velocity of 7.8 km/sec somewhere near Salt Lake City. However, we do not have data exactly on the path Smith has been discussing.

Landisman: The sensitivity to P-wave ve-

locity or to density for surface wave measurements is less than the sensitivity to shear wave velocity by a factor of 5 or 10. Really, using surface waves you are only investigating the distribution of shear velocity. More details are given in the paper 'Effects of observational errors on the resolution of surface waves at intermediate distances' by Der, Massé, and Landisman in the June 10, 1970, issue of the *Journal of Geophysical Research*, which defines the limitation on the accuracy achievable from surface wave inversion studies. Of course, refraction lines give you a better determination of the P-wave velocity.

Sutton: Is there any correlation of the scattering in your surface-wave data with azimuth?

R. B. Smith: No. All the events come within about a 30-degree range in azimuth.

Velocity Gradients in the Continental Crust from Head-Wave Amplitudes[1]

DAVID P. HILL

Seismological Laboratory, California Institute of Technology, and U. S. Geological Survey, Pasadena 91109

Abstract. Small velocity gradients in a refracting horizon have a pronounced effect on the spectral amplitudes of head waves. Negative velocity gradients and anelasticity (Q^{-1}) result in a similar amplitude decay with distance for narrow-bandwidth data. Positive velocity gradients result in a net amplitude gain with distance compared with the head wave from a homogeneous, perfectly elastic refractor. Wave-theoretical expressions for these effects applied to published amplitude data for the major crustal refraction branches, P_g and P^*, suggest that the 'granitic' crust in the Basin and Range province has either negative velocity gradients of the order of 10^{-2} km/sec/km or an anelastic Q of the order of 400, whereas the 'granitic' crust in the eastern United States and on the California coast has slightly positive velocity gradients. Similarly, the 'basaltic' intermediate layer appears to have a negative gradient of the order of 10^{-2} sec^{-1} under the Snake River plain and null or slightly positive gradients under Lake Superior and Mississippi. Velocity gradients inferred from laboratory measurements on granite and basic igneous rocks, together with published geothermal gradients, are generally consistent with the gradients inferred from amplitude data.

VELOCITY GRADIENTS AND AMPLITUDES

Recent wave-theoretical studies show that the spectral amplitudes of critically refracted waves (head waves) are quite sensitive to small velocity gradients in the refracting horizon [*Červený and Jansky*, 1967; *Hill*, 1971]. Negative gradients and anelasticity (Q^{-1}) result in a similar amplitude decay with distance for narrow bandwidth data; positive gradients result in a net amplitude gain with distance with respect to theoretical amplitudes for head waves refracted from a homogeneous medium with infinite Q. These theoretical effects of velocity gradients on head-wave amplitudes are illustrated in Figure 1. The negative gradient case is based on asymptotic solutions obtained by *Hill* [1971]; the positive gradient case is adapted from *Červený and Jansky* [1967].

In a preliminary attempt to determine the existence and distribution of velocity gradients in well-established crustal horizons, these theo-retical results have been applied to published amplitude data for the major crustal travel time branches P_g and P^*. In this study, only first-arrival data for these branches are included.

RELATION BETWEEN Q' AND VELOCITY GRADIENTS

Effective Q values (hereafter referred to as Q') were computed for each amplitude data set by determining in a least squares sense the net gain ($-Q'$) or decay ($+Q'$) of the observed data with respect to theoretical head-wave amplitudes from homogeneous, infinite Q refractors. The results are interpreted in terms of negative velocity gradients for Q' values smaller than 10^3, null gradients for Q' values greater than 10^3, and positive gradients for negative Q' values as summarized in Table 1. Four examples of the amplitude data showing the least squares fit for Q' are given in Figure 2.

Q' VERSUS REGIONAL HEAT FLOW

The distribution of Q' values for the data considered is shown in Figure 3, together with regional heat flow contours presented by *Archambeau et al.* [1968]. From the figure, we see that

[1] Division of Geological and Planetary Sciences, California Institute of Technology, Pasadena, California, Contribution 1914. Publication authorized by the Director, U. S. Geological Survey.

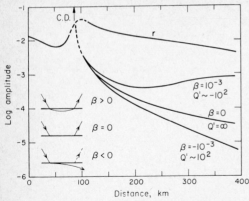

Fig. 1. Amplitude curves illustrating effect of positive ($\beta = 10^{-3}$ sec^{-1}), null ($\beta = 0$), and negative ($\beta = -10^{-3}$ sec^{-1}) velocity gradients on head-wave amplitudes, together with effective Q values (Q') with respect to the $\beta = 0$, $Q = \infty$ case. The reflected wave amplitude r, which is insensitive to small gradients in the refracting medium, is shown for reference. The arrow at C.D. indicates the ray-theoretical critical point. These curves were generated for 6-Hz waves in a 6.4-km/sec layer 30 km thick over an 8.0-km/sec half-space. The positive gradient case is adapted from Červený and Jansky [1967].

there is a general inverse correlation between Q' values and heat flow. Q' values in the 'normal' heat flow provinces of the eastern United States and west coast are greater than 10^3 or are negative; these values imply null or small positive velocity gradients for both the upper portions of the crystalline crust ('granitic layer') and the intermediate layer in these regions. Q' values in the Basin and Range high heat-flow province, which probably includes the Snake River plain, are generally smaller than 10^3. These low Q' values common to the Basin and Range province may be due to: 1, scattering of the critically refracted waves by relief on the refractor associated with Basin and Range faulting; 2, a temperature-dependent anelastic Q; or 3, negative velocity gradients in the upper portions of the crystalline crust associated with high geothermal gradients.

Available information is insufficient to quantitatively assess effects of the first two factors, and they remain as possible contributing factors. The third factor can be assessed quantitatively and is found to be consistent with the correlation between Q' and heat flow mentioned above. Heat flow studies suggest that the crustal geo-

thermal gradients in the Basin and Range heat flow province is 30°/km or possibly somewhat higher [Roy et al., 1968; Lachenbruch, 1970; Minster and Archambeau, 1970]. This geothermal gradient, together with partial derivative data for P-wave velocities in granites with respect to pressure, $(\partial V_\alpha / \partial P)_T$, and temperature, $(\partial V_\alpha / \partial T)_P$, measured in the laboratory [Hughes and Maurette, 1956], suggests negative velocity gradients of

$$\beta \sim -0.8 \times 10^{-2} \text{ km/sec/km}$$

in the 6.0-km/sec 'granitic' horizon of the Basin and Range province. The lower geothermal gradients (about 15°/km) that are associated with the 'normal' heat flow province, when combined with the same partial derivative data, suggest positive velocity gradients of

$$\beta \sim 0.5 \times 10^{-2} \text{ km/sec/km}$$

in the upper crystalline crust. Thus, the differences in velocity gradients in the upper part of the crystalline crust (P_g refractor) between the Basin and Range and eastern United States-west coast heat flow provinces inferred separately from P_g amplitudes and geothermal gradients are reasonably consistent.

Similar calculations using the partial derivative data for P waves in basic igneous rocks reported by Hughes and Maurette [1957] indicate that the 'basaltic' 6.7-km/sec intermediate layer has negative velocity gradients of

$$\beta \sim -1.6 \times 10^{-3} \text{ km/sec/km}$$

beneath the Snake River plain and

$$\beta \sim -0.3 \times 10^{-2} \text{ km/sec/km}$$

beneath Lake Superior and Mississippi. As before, a geothermal gradient of 30°/km is assumed for the high heat flow region (the Snake River plain) and 15°/km is assumed for the 'normal' heat flow regions (Lake Superior and Mississippi). These velocity gradients inferred from geothermal data are more negative than those implied by the Q' values obtained from $P*$ amplitude data in these regions (Table 1). If we accept this result at face value, then a compositional gradient (increasingly mafic with depth) is required in the upper parts of these intermediate layers to bring the velocity gradient estimated from the geothermal gradient

TABLE 1. Summary of Q' and Velocity Gradients from Crustal Amplitudes Data

Standard error refers to Q'^{-1} fit to amplitude data, β is velocity gradient, and 'null' and '$+$' β indicate zero and positive gradients, respectively.

Region and Profile	Phase	Q'	Q'^{-1a}	Std. Error in Q'^{-1a}	$\beta(\sec^{-1})^a$	Source
Basin and Range						
Fallon-Eureka	P_g	471	2.10	0.91	-15	*Eaton* [1963]
Fallon-S.F.	P_g	760	1.22	0.36	-9.7	*Eaton* [1963]
Fallon-Owens V.	P_g	-446	-2.2	1.16	$+$	*Eaton* [1963]
Eureka-Fallon	P_g	972	1.03	0.47	-7.5	*Eaton* [1963]
Eureka-North	P_g	-1290	-0.77	0.88	$+$	*Hill & Pakiser* [1966]
Mt. City-South	P_g	403	2.48	3.71	-18	*Hill & Pakiser* [1966]
NTS-East	P_g	117	8.50	0.87	-26	*Ryall & Stuart* [1963]
California						
S.F.-Fallon	P_g	3810	0.262	0.076	null	*Eaton* [1963]
S.F.-S. Monica	P_g	-1580	-0.63	1.13	$+$	*Healy* [1963]
Camp Roberts	P_g	2860	0.349	0.87	null	*Healy* [1963]
S. Monica-L. Mead	P_g	230	4.34	2.17	-15	*Roller & Healy* [1963]
San Juan (6.06)	P_g	-47	$-21.$	4.9	$+$	*Stewart* [1968b]
San Juan (6.35)	P_g?	54	18.	3.7	-20	*Stewart* [1968b]
Colorado Plateau						
Hanksville	P_g	1260	0.795	1.02	null	*Roller* [1965]
Chinle	P_g	221	4.53	1.33	-16	*Roller* [1965]
Missouri						
Hannibal	P_g	-613	-1.63	0.86	$+$	*Stewart* [1968a]
Swan L.-Hannibal	P_g	-549	-1.82	0.54	$+$	*Stewart* [1968a]
Swan L.-St. Joseph	P_g	-1960	-0.511	0.633	$+$	*Stewart* [1968a]
St. Joseph	P_g	-475	-2.10	0.63	$+$	*Stewart* [1968a]
Mississippi	$P*$	-382	-2.62	0.92	$+$	*Warren et al.* [1966]
Lake Superior	$P*$	31,700	0.0315	0.604	null	*O'Brien* [1968]
Snake River plain						
Boise-South	$P*$	337	2.97	0.84	-4.4	*Hill & Pakiser* [1966]

^{a} 1×10^{-3}.

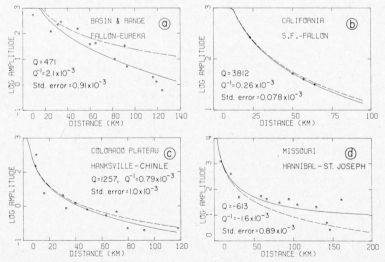

Fig. 2. Examples of P_g amplitude data in four regions: a, Fallon toward Eureka in the Basin and Range [*Eaton*, 1963]; b, San Francisco east in California [*Eaton*, 1963]; c, Hanksville north in the Colorado Plateau [*Roller*, 1965]; and d, Hannibal west in Missouri [*Stewart*, 1968a]. Continuous line is least squares fit for Q' through the amplitude data. Broken line shows amplitude decay with $Q' = \infty$ for comparison.

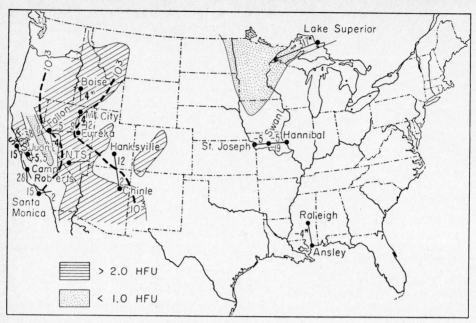

Fig. 3.   Location of seismic profiles and the relation of Q' to heat flow. Number beside each profile is $Q' \times 10^{-2}$. The dark broken contour line separates the region in which Q' is less than 10^3 from that in which it is either greater than 10^3 or negative. Regional heat flow patterns are adapted from *Archambeau et al.* [1968]. Numbers associated with heat flow contours have units $\mu cal/cm^2/sec$.

(e.g., -16×10^{-3} sec^{-1} in the Snake River plain) to the level of the velocity gradient estimated from Q' (4×10^{-3} sec^{-1} in the Snake River plain). However, neither the amplitude nor the thermal data are of sufficient accuracy to provide much confidence for this suggested compositional gradient in the intermediate layers.

CONCLUSIONS

Thus, on the basis of P_g amplitudes and thermal data, we conclude that a slight P-wave low velocity zone may exist in the upper crystalline crust in the Basin and Range high heat flow province and that such zones are unlikely in the eastern United States and west coast normal heat flow provinces. In the Basin and Range province, the crustal low velocity zone would have the form of a gradual decrease in velocity from the top of the crystalline crust downward (at a maximum rate of about 1×10^{-2} km/sec/km) and would terminate rather abruptly at the top of the intermediate layer. The abrupt, pronounced crustal low velocity zone at depths of about 10 km of the type proposed by *Mueller*

and Landisman [1966] could be present in either the Basin and Range province or the eastern United States, but such a zone would be difficult to detect using the data and methods described in this paper.

Acknowledgment. I would like to express my appreciation to C. B. Archambeau and D. L. Anderson for many helpful discussions.

This research was supported in part by the Advanced Research Projects Agency and was monitored by the Air Force Office of Scientific Research under contract (F44620-69-C-0067).

REFERENCES

Archambeau, C. B., R. F. Roy, D. D. Blackwell, D. L. Anderson, L. Johnson, and B. Julian, A generalized study of continental structure (abstract), *Eos Trans. AGU, 49,* 328, 1968.

Červený, V., and J. Jansky, On some dynamic properties of the diving wave, *Proc. 7th Assembly European Seismol. Comm., Copenhagen,* 397–402, 1967.

Eaton, J. P., Crustal structure from San Francisco, California, to Eureka, Nevada, from seismic-refraction measurements, *J. Geophys. Res., 68,* 5789–5806, 1963.

Healy, J. H., Crustal structure along the coast of California from seismic refraction measurements, *J. Geophys. Res., 68,* 5777–5787, 1963.

Hill, D. P., Velocity gradients and anelasticity from crustal body wave amplitudes, *J. Geophys. Res., 76*(14), 3309–3325, 1971.

Hill, D. P., and L. C. Pakiser, Crustal structure between the Nevada Test Site and Boise, Idaho, from seismic-refraction measurements, in *The Earth Beneath the Continents, Geophys. Monogr. Ser.,* vol. 10, edited by J. S. Steinhart and T. J. Smith, pp. 391–419, AGU, Washington, D.C., 1966.

Hughes, D. S., and G. Maurette, Variation of elastic wave velocities in granites with pressure and temperature, *Geophysics, 21,* 277–284, 1956.

Hughes, D. S., and G. Maurette, Variation of elastic wave velocities in basic igneous rocks with pressure and temperature, *Geophysics, 22,* 23–31, 1957.

Lachenbruch, A. H., Crustal temperature and heat production: Implications of the linear heat-flow relation, *J. Geophys. Res., 75,* 3291–3300, 1970.

Minster, J. B., and C. B. Archambeau, Systematic inversion of continental heat flow and temperature data, (abstract), *Eos Trans. AGU, 51,* 824, 1970.

Mueller, S., and M. Landisman, Seismic studies of the earth's crust in continents, 1, Evidence for a low-velocity zone in the upper part of the lithosphere, *Geophys. J. Roy. Astron. Soc., 10,* 525–538, 1966.

O'Brien, P. N. S., Lake Superior crustal structure—A reinterpretation of the 1963 seismic experiment, *J. Geophys. Res., 73,* 2669–2689, 1968.

Roller, J. C., Crustal structure in the eastern Colorado plateaus province from seismic refraction measurements, *Bull. Seismol. Soc. Amer., 55,* 107–110, 1965.

Roller, J. C., and J. Healy, Seismic refraction measurements of crustal structure between Santa Monica Bay and Lake Mead, *J. Geophys. Res., 68,* 5837, 1963.

Roy, F. R., D. D. Blackwell, and F. Birch, Heat generation of plutonic rocks and continental heat flow provinces, *Earth Planet. Sci. Lett., 5,* 1–12, 1968.

Ryall, A., and D. J. Stuart, Travel times and amplitudes from nuclear explosions, Nevada Test Site to Ordway, Colorado, *J. Geophys. Res., 68,* 5821–5835, 1963.

Stewart, S. W., Crustal structure in Missouri by seismic-refraction methods, *Bull. Seismol. Soc. Amer., 58,* 291–323, 1968*a*.

Stewart, S. W., Preliminary comparison of seismic travel times and inferred crustal structure adjacent to the San Andreas fault in the Diabb and Gabilan ranges of central California, in *Proceedings of Conference on Geologic Problems of San Andreas Fault System, Stanford Univ. Pub., Geol. Sci., 11,* edited by W. R. Dickinson and A. Grantz, pp. 218–230, Stanford Univ., Stanford, California, 1968*b*.

Warren, D. H., J. H. Healy, and W. H. Jackson, Crustal seismic measurements in southern Mississippi, *J. Geophys. Res., 71,* 3437–3458, 1966.

DISCUSSION

Porath: You have a uniform 6.4-km/sec velocity for the Columbia plateau. Might you not expect a lower velocity at depth because the plateau basalts probably cover an ancient granitic crust?

Hill: The data only give the time it takes a *P* wave to travel from the Moho to the surface; there is no way of dividing the crust up into layers, so this is just an average velocity.

Meyer: Did you consider using a lower *Q* and positive gradient in the Lake Superior region?

Hill: All I did was to fit the homogeneous head-wave potential, derived from the amplitude data published by O'Brien, and make a least square solution for *Q*. This gives an estimate of the lower limit for the gradient and an upper limit for *Q*.

Higgins: What is the physical significance of a negative *Q*?

Hill: That is an artifact; in reality there must be a positive *Q* and a positive velocity gradient to give the amplitude increase.

Geophysical Measurements in the Southern Great Plains[1]

B. J. MITCHELL AND M. LANDISMAN

Geosciences Division, University of Texas at Dallas, Dallas, Texas 75230

Abstract. Detailed seismic velocity models for Oklahoma and eastern New Mexico are compared in order to study lateral variations of the crust within a single geologic province. The lower three crustal layers in each model are similar in thickness and velocity and appear to be continuous across a major fault zone. Detailed study of the reversed refraction profile in Oklahoma favors an interpretation in which the shallowest continuous seismic interface is encountered at a depth of about 18 km, just beneath a crustal low-velocity layer; this feature apparently extends to eastern New Mexico, where recordings of the Gnome explosion can be interpreted in terms of a crustal velocity reversal with a lower boundary approximately 23 km beneath the surface. The total crustal thickness increases from 46 to 52 km between Oklahoma and eastern New Mexico; the increase is confined largely to the upper portions of the crust. Observations of shear wave travel times, crustal transfer functions determined from teleseismic body waves, and observed gravity differences support the compressional velocity models derived earlier. The low-velocity layer in eastern New Mexico may correlate with a zone of reduced electrical resistivity in northwestern Texas.

Several recent analyses of seismic refraction and reflection observations in North America [*Mitchell and Landisman,* 1970, 1971; *Mueller and Landisman,* 1971] and Europe [*Mueller et al.,* 1969; *Ansorge et al.,* 1970] have yielded multi-layered models characterized by one or more velocity reversals. Detailed interpretations like these require the consideration of secondary refracted arrivals and wide-angle reflections, as well as the more commonly used first arrivals. Both 'detailed' and 'first arrival' analyses indicate substantial lateral variations of velocity from one geologic province to another. Lateral differences of velocity between dissimilar geologic provinces and within individual provinces must be carefully investigated in order to improve current understanding of the behavior and properties of the continental crust. In this paper, the results from two detailed seismic studies in the southern Great Plains [*Mitchell and Landisman,* 1970, 1971] are compared in an attempt to delineate intra-province lateral variations in that region.

Shear wave record sections were constructed for both profiles, and crustal transfer functions for teleseismic body waves were determined from seismograms recorded at nearby seismological observatories. These new data, as well as observed gravity differences, support the compressional velocity distributions interpreted earlier.

Further insight into the nature of the crust can be gained from a comparison of these velocity models with the distribution of other geophysical properties in the region. *Mitchell and Landisman* [1971] have derived an electrical resistivity model for the crust in northwestern Texas and have compared it with their seismic interpretation for eastern New Mexico. An interesting correlation occurred between a zone of reduced seismic velocity and one of reduced electrical resistivity. These anomalous crustal zones may possibly be correlated with a zone of high seismic attenuation centered at a depth of approximately 15 km, which is a feature of the *Long and Berg* [1969] interpretation of the Gnome data for eastern New Mexico. *Mitchell* [1970] and *Mitchell and Landisman* [1971] have suggested the possibility that interstitial water of hydration might be a common causative mechanism for all these anomalous crustal features in the southern Great Plains.

[1] Contribution 182, Geosciences Division, University of Texas at Dallas.

SEISMIC RESULTS

Refraction and reflection data.   Mitchell and Landisman [1970, 1971] made detailed analyses of previously interpreted refraction profiles made across Oklahoma [*Tryggvason and Qualls*, 1967] and eastern New Mexico [*Stewart and Pakiser*, 1962]. Seismograms from the original eastern New Mexico study were supplemented with recordings of the Gnome volunteer teams [*Westhusing*, 1962]. The locations of both profiles are indicated in Figure 1. The two shot points of the reversed Oklahoma profile were designated Chelsea and Manitou [*Qualls*, 1965]. The Gnome nuclear event provided the source for the eastern New Mexico profile.

The reversed profile in Oklahoma traversed a deep basin and a major fault system. Figures 2 and 3 are record sections that include digitally processed and plotted versions of seismograms recorded for each refraction line. Several of the traces were either bandpass filtered or deconvolved in order to improve the character of the arrivals. The expanded time scale in Figures 2, 3, and 7 was obtained by reducing the travel times

with a velocity of 6.0 km/sec. This reduction provides greater resolution for those refraction branches that have apparent velocities near 6.0 km/sec. The theoretical arrival times for the model in Figure 6 are plotted as continuous lines that intersect the seismic traces on the record sections. Critical points for the refraction branches are indicated by a small *X*. The model is not only concordant with the seismic refraction data but, as indicated in Figure 6, the theoretical gravity anomaly values derived from a two-dimensional computation satisfy the regional gravity measurements reported by *Lyons* [1964]. Field geological observations [*Ham et al.*, 1964; *Harlton*, 1963; *Jordan*, 1964] and well-log data [*Douze*, 1964] serve as constraints for the upper portion of the model.

Figures 4 and 5 include those intervals of the Oklahoma seismograms that correspond to theoretical arrival times for the shear waves. A reduction velocity of 3.5 km/sec was used in these figures. The theoretical shear wave travel times, shown on the record sections by continuous lines, are those that would be expected for the model

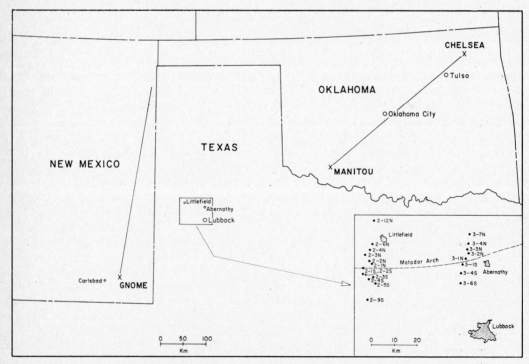

Fig. 1.   Map of Oklahoma and portions of Texas and New Mexico, including the locations of seismic refraction profiles, seismological observatories, and the array of magnetotelluric stations discussed in the present study.

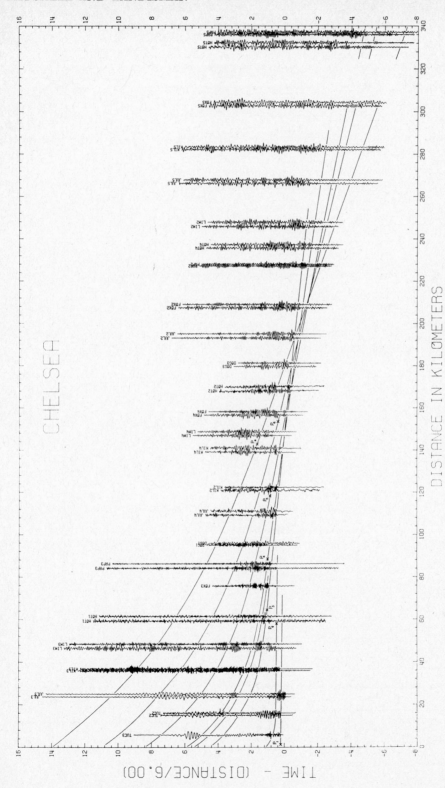

Fig. 2. Record section of the portions of seismograms that include the compressional phases received from the Chelsea shot point. Theoretical refraction branches and under-critical reflection curves (continuous lines) computed from the Chelsea velocity model in Figure 6 are superposed on the record section. The symbol X indicates critical points predicted by ray theory. A gap appears over the distance interval occupied by the Wichita fault system (see Figure 6), and an offset occurs at a distance of about 325 km because of a correction for near-surface structural differences (reproduced from *Mitchell and Landisman* [1970], with the permission of *Geol. Soc. Amer. Bull.*).

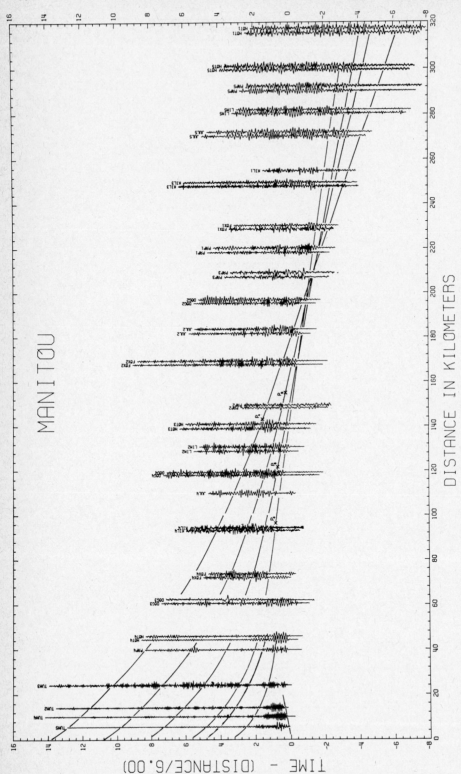

Fig. 3. Record section of the portions of seismograms that include the compressional phases received from the Manitou shot point. Theoretical refraction branches and under-critical reflection curves for the Manitou velocity model in Figure 6 are shown for distances less than 43 km from the shot point. The positions of the lines at greater distances have been corrected for differences between the Chelsea and Manitou models in Figure 6. A small additional correction has been made for the increased sedimentary thickness between 63 and 100 km from the source. A gap appears at the position of the Wichita fault system. X's denote critical points, and continuous lines indicate arrival times predicted by ray theory (reproduced from *Mitchell and Landisman* [1970], with the permission of *Geol. Soc. Amer. Bull.*).

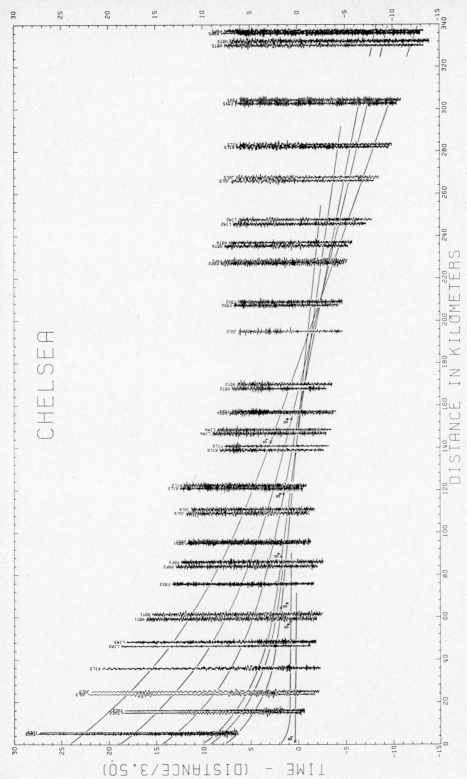

▶ Fig. 4. Record section of the portions of seismograms that include the shear phases received from the Chelsea shot point. The theoretical times were determined from the compressional arrival times in Figure 2, with the assumption that Poisson's ratio has a value of 0.25. X's denote critical points, and continuous lines indicate arrival times predicted by ray theory.

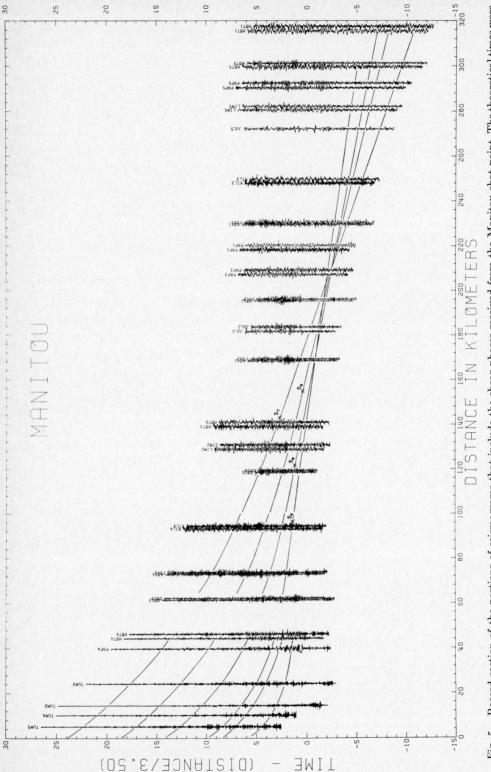

Fig. 5. Record section of the portions of seismograms that include the shear phases received from the Manitou shot point. The theoretical times were determined from the compressional arrival times in Figure 3, with the assumption that Poisson's ratio has a value of 0.25. X's denote critical points, and continuous lines indicate arrival times predicted by ray theory.

in Figure 6, if a value of 0.25 is assumed for Poisson's ratio. The arrival times on the P and S record sections follow similar patterns, since the reduction velocities (6.0 km/sec for the P and 3.5 km/sec for the S record sections) correspond to a Poisson's ratio of 0.242, which is very close to the assumed value of 0.25. This similarity has been emphasized on the record sections by using identical distance scales and a shear wave time scale shortened by the factor $\sqrt{3}$. ($Vp/Vs = \sqrt{3}$ when Poisson's ratio is 0.25.) Weak but continuous phases occur at the theoretically predicted arrival times for each shear wave branch. The largest amplitudes usually occur near the critical points for both

shear and compressional wave refractions. These maximums appear at distances somewhat greater than the zero-wavelength critical distance in most cases, as predicted by the theory of Červený [1966]. Reflections are usually weak or undiscernable at near-vertical incidence, but they increase in amplitude at wider angles approaching the critical angle. Transition zones consisting of gradients or thin laminations have been suggested as causes for this pattern of arrivals [*Fuchs and Kappelmeyer*, 1962; *Meissner*, 1967; *Fuchs*, 1969].

The shear wave observations generally support the earlier model derived from the analysis of the compressional wave arrivals. The

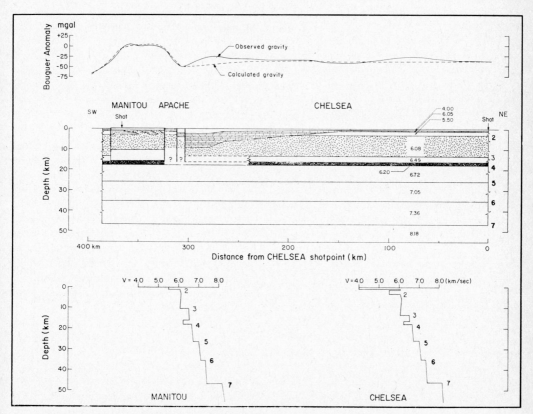

Fig. 6. The top diagram shows theoretical and observed Bouguer gravity anomaly values along the Oklahoma profile. The middle diagram shows a crustal model with average compressional velocity values indicated for each layer. Dotted lines indicate uncertain positions of boundaries. The limestone symbol only implies sediments with a velocity greater than 6.0 km/sec and does not necessarily indicate limestone or horizontal layering. The numbers to the right of the model refer to the refraction branches in Figures 2, 3, 4, and 5 (vertical exaggeration is ×2). The bottom diagram shows velocity-depth curves near the Chelsea and Manitou shot points. The numbers to the right of each discontinuity refer to the refraction branches in Figures 2, 3, 4, and 5 (reproduced from *Mitchell and Landisman* [1970], with the permission of *Geol. Soc. Amer. Bull.*).

continuity of compressional and shear wave refractions from the four deepest refractors in Figures 2, 3, 4, and 5 suggests that those horizons are continuous or nearly so over the length of the profile and are uninterrupted by the fault zone. The presence of refracted energy from shallower horizons on the Chelsea line (Figures 2 and 4) and its absence on the Manitou record section (Figures 3 and 5) have been taken as evidence for the disruption of the shallower horizons by the normal fault system associated with the uplifted region. Finally, a crustal low-velocity zone just above the shallowest continuous interface shown in the cross section (Figure 6) is suggested by reflection arrivals slightly earlier than those corresponding to the P_4 (or S_4) undercritical reflection branches in Figures 2, 3, 4, and 5.

A shallower 'sialic' low-velocity zone has been observed in several continental regions [*Mueller and Landisman*, 1966; *Landisman and Mueller*, 1966; *Ansorge et al.*, 1970]. However, the data of the present study permit no conclusive statements regarding the existence of this feature in the southern Great Plains.

The Gnome record section, presented by *Mitchell and Landisman* [1971], and the model derived for eastern New Mexico are shown in Figures 7 and 11, respectively. Figure 8 includes the corresponding shear wave record section, as well as the theoretically predicted refraction and under-critical reflection times; a Poisson's ratio of 0.25 was also assumed in this case. The most prominent shear arrivals are those that just penetrate the upper mantle (S_n). They correspond to the predicted arrival times of S_7 on the record section. The arrivals from the other horizons, although continuous, are weaker. This pattern is to be expected, since shear conversion near the source is more efficient at steeper angles. The times of these arrivals accord well with the theoretically predicted values for the model in Figure 11, except for the S_2 (or S_g) branch. This branch, which appears only on some of the Gnome volunteer team records, has the predicted apparent velocity, but its intercept time is earlier than that of the theoretical S_2 branch. One possible, though fairly unlikely, explanation is that the shear velocity in the shallowest unconsolidated sediments is higher than the value used in the calculation, based on a Poisson's ratio of 0.25. It is considered more

likely, however, that conversion from compressional to shear waves occurred at some appropriate distance from the source rather than in the immediate vicinity of the shot, as was assumed in the computation. The early intercept time observed for S_g would then indicate that the seismic energy propagated some distance as a compressional wave before conversion.

The velocity model for eastern New Mexico (Figure 11) is strikingly similar to the Oklahoma model (Figure 6), except for the uppermost layers. The velocities for each crustal layer from the basement refractions (P_2 and S_2 in Figures 2, 3, 4, 5, 7, and 8) down to those for the upper mantle (P_7 and S_7) are nearly the same for both areas; they differ by only a few hundredths of a km/sec. The thicknesses of the three deepest crustal layers are also quite similar. A deep low-velocity zone occurs in the same relative position for both models, but it is 2 to 3 km thicker beneath eastern New Mexico. The other shallower crustal layers and the sediments are also thicker beneath eastern New Mexico; this pattern leads to a total crustal thickness of about 52 km for this region as compared to about 46 km beneath Oklahoma. The average velocities and thicknesses for these models are given in Table 1.

Long and Berg [1969] derived a Q distribution for the crust in eastern New Mexico by considering the amplitudes of the Gnome recordings. The model of Long and Berg included a zone of high attenuation at a depth similar to that inferred for the low-velocity zone postulated by *Mitchell and Landisman* [1971]. The significance of this correlation will be discussed near the end of this paper.

Crustal transfer functions. Long-period recordings from seismological observatories at Leonard (near Tulsa), Oklahoma (TUL), and Lubbock, Texas (LUB), which are situated near the refraction profiles (see Figure 1), have been used to obtain the crustal transfer functions for teleseismic P waves [*Phinney*, 1964]. These results provide a further check on the derived models. Deep earthquakes (with focal depths greater than 200 km and epicentral distances less than 70°) are preferable for such studies, because reflected phases such as pP and PcP do not interfere with the crustal reverberations [*Bonjer and Fuchs*, 1970]. A search of seismograms from January 1963 through December 1968 yielded

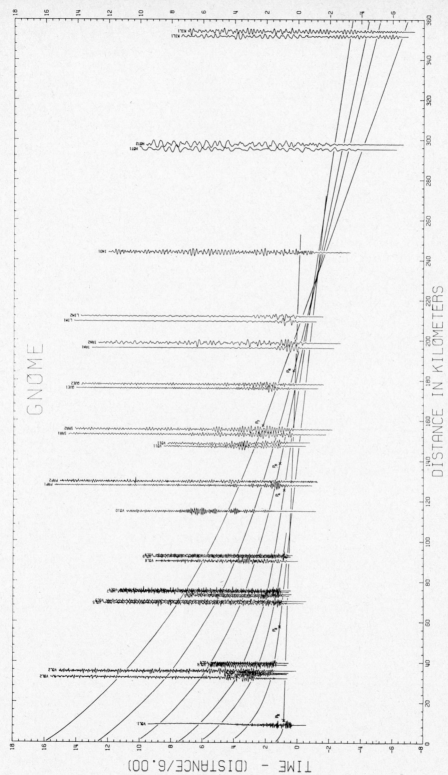

Fig. 7. Record section of the portions of seismograms that include the compressional phases received from the Gnome event. Theoretical refraction branches and under-critical reflection curves computed for the seismic model in Figure 11 are superposed on the record section. X's denote critical points and continuous lines indicate arrival times predicted by ray theory (reproduced from *Mitchell and Landisman* [1971], with the permission of *Geophysics*).

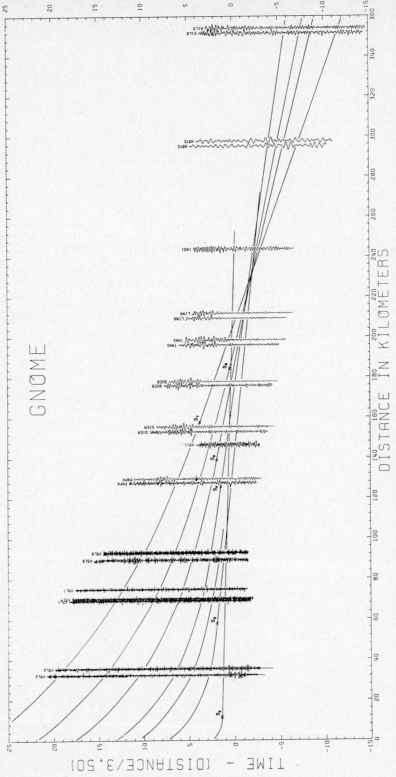

Fig. 8. Record section for the portions of seismograms that include the shear phases received from the Gnome event. The theoretical times were determined from the compressional wave arrival times in Figure 7, with the assumption that Poisson's ratio has a value of 0.25. X's denote critical points and continuous lines indicate arrival times predicted by ray theory.

only one outstandingly well recorded event that satisfied the above criteria at both stations. That earthquake occurred in the Peru-Brazil border region on November 3, 1965. The long-period seismograms from both stations were digitized over a time interval of 35 seconds, and the ratios between the Fourier transformed vertical and horizontal components of ground motion were plotted as functions of frequency (Figure 9). The theoretical ratios in the same figure were computed by using the formalism of *Haskell* [1962]. The theoretical ratios were normalized to unity at zero frequency and the observed ratios were scaled to make the maximum value equal to 5.0. The only significant unexplained observed peak occurs near 0.18 cycle/sec at both stations. This frequency corresponds to that of the microseismic noise observed on the seismograms. The positions of the other observed peaks at TUL and LUB correspond to the positions of the theoretical peaks for the

Chelsea and eastern New Mexico models, respectively, for the frequency range from 0.03 to 0.2 cycle/sec.

Although there is fair agreement between the observed and theoretical transfer functions, if the observed peak near 0.18 cycle/sec is ascribed to microseismic noise, this agreement only implies that the gross properties of the two theoretical models are correct. Other features, such as the relatively thin low-velocity zones that are included in the models, have little observable effect on the theoretical ratios. The ratios at the two stations differ mainly in that the frequencies of the minimum near 0.1 cycle/sec and the two adjacent maximums are slightly lower at LUB than at TUL. This shift toward lower frequencies is predicted by the theoretical models and supports the inference made earlier that the significant crustal differences between Oklahoma and eastern New Mexico occur in the upper few kilometers of the crust.

TABLE 1. Average Crustal Velocities and Thicknesses*

Eastern New Mexico				Manitou (SW Oklahoma)				Chelsea (NE Oklahoma)			
Thick-ness km	Depth km	Compres-sional Velocity km/sec	Shear Velocity km/sec	Thick-ness km	Depth km	Compres-sional Velocity km/sec	Shear Velocity km/sec	Thick-ness km	Depth km	Compres-sional Velocity km/sec	Shear Velocity km/sec
								0.6	0.0	4.00	2.31
								0.4	0.6	6.05	3.49
3.5	0.0	4.93	2.85	1.0	0.0	5.50	3.18	2.1	1.0	5.50	3.18
8.8	3.5	6.14	3.55	9.5	1.0	6.08	3.51	10.3	3.1	6.08	3.51
6.3	12.3	6.52	3.76	5.1	10.5	6.49	3.75	3.0	13.4	6.49	3.75
4.4	18.6	6.20	3.58	2.5	15.6	6.20	3.58	1.5	16.4	6.20	3.58
8.8	23.0	6.72	3.88	8.2	18.1	6.72	3.88	8.2	17.9	6.72	3.88
8.9	31.8	7.10	4.10	9.1	26.3	7.05	4.07	9.1	26.1	7.05	4.07
11.9	40.7	7.35	4.24	11.1	35.4	7.36	4.25	11.1	35.2	7.36	4.25
	52.6	8.23	4.75		46.5	8.18	4.72		46.3	8.18	4.72

*The compressional velocities were derived from the record sections in Figures 2, 3, and 7. Shear velocities predicted for a Poisson's ratio value of 0.25 agree well with observed arrivals on the record sections (Figures 4, 5, and 8) in most cases. The solid horizontal lines imply that the interfaces may be continuous. Depths were measured from the surface for each model. The surface elevations average about 0.8 km for the eastern New Mexico model, 0.6 km for the Manitou model, and 0.4 km for the Chelsea model.

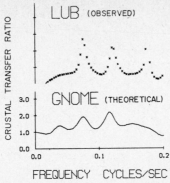
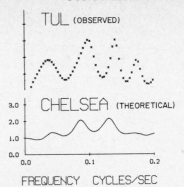

Fig. 9. Crustal transfer ratios from long-period teleseismic P waves recorded by the seismological observatories at Lubbock, Texas (LUB), and Tulsa, Oklahoma (TUL), compared to theoretical ratios expected for the eastern New Mexico (Gnome) and northeastern Oklahoma (Chelsea) models, respectively.

GRAVITY

The upper portion of Figure 6 presents a comparison of theoretical and observed gravity curves for the Oklahoma profile. The density values for the model were derived from well-log data and seismic refraction measurements, converted as prescribed by the Nafe-Drake empirical velocity-density relationship [*Talwani et al.*, 1959]. The observed gravity values were taken from the Bouguer gravity map compiled by *Lyons* [1964]. Significant discrepancies between the observed and theoretical curves occur only where lateral disturbances on the Bouguer gravity map seem to indicate that off-profile density variations have disturbed the observations along the profile.

A rough investigation of the gross differences in the density distributions between Oklahoma and eastern New Mexico was made by assuming one horizontally layered model for each region. The two flat-layered density models were juxtaposed, and the gravity values at great distances from the contact were computed. The difference between these computed values was compared with the observed gravity difference between northeastern Oklahoma and eastern New Mexico [*Woollard and Joesting*, 1964]. Both differences were found to be about 100 mgals; this value indicates that the lateral variation in density predicted by the two theoretical models satisfies the regional gravity data.

ELECTRICAL RESISTIVITY

Mitchell and Landisman [1971] interpreted a resistivity model for a region in northwestern

Texas, using data from 20 magnetotelluric stations shown in Figure 1. The data furnished for that study had been collected and reduced by personnel of Mobil Research and Development Corporation's Field Research Laboratory in Dallas, Texas. The techniques used for the reduction of these data are similar to those described by *Sims and Bostick* [1969].

Mitchell and Landisman [1971] find that the magnetotelluric data for all of the 20 sites can be adequately explained by slight variations of a single one-dimensional model. They attribute differences in the apparent resistivity data from station to station, especially those in the short to intermediate period range, to minor changes in the uppermost 0.5 km. When corrections for these changes are applied, the observations at periods greater than 3 seconds exhibit great uniformity for both components of apparent resistivity, as indicated in Figure 10a. In addition, they ascribe uniform differences in the two apparent resistivity curves, at moderate to long periods for each station, to the influence of a 2-km thick anisotropic low-resistivity layer at a depth slightly greater than 20 km. These differences are exhibited in Figure 10b, in which the ratios of the two corrected apparent resistivity components are presented as functions of period. Laboratory measurements of electrical resistivity [*Hill*, 1969] indicate that resistivity anisotropy is a characteristic feature of metamorphic rocks.

Mitchell and Landisman [1971] discuss the non-uniqueness inherent in their magnetotelluric interpretation and particularly emphasize departures from one-dimensionality as an alternate explanation for the differences in the two apparent resistivity curves at each station. Recent computations for a graben model [*Wright*, 1970] and a basin model [*Porath*, 1971] suggest that a lateral change in resistivity can cause fluctuations in the apparent resistivity curves at great distances from an anomalous structure and can produce fairly uniform apparent resistivity curves over an array of stations relatively far from the anomaly. The graben model of *Wright* [1970] exhibits the greater lateral resistivity contrast of the two. Wright's graben model produces a ratio of nearly 1.5 between the apparent resistivities determined from E perpendicular (E vector perpendicular to the strike of the structure) and those determined from E parallel, when the determination was made 30 km from the edge of the structure. The data from northwestern Texas produce ratios of 2 and larger

over an array of stations separated from one another by as much as 40 km. If differences between the two apparent-resistivity curves are caused by departures from one-dimensionality in northwestern Texas, then these departures must result from a more complicated resistivity distribution than that of a distant lateral change in a single direction from the array of magnetotelluric stations.

NATURE OF THE CRUST IN THE SOUTHERN GREAT PLAINS

Perhaps the most interesting result of the present study is the similarity of the lower crust in both Oklahoma and eastern New Mexico. The thicknesses and seismic velocities of the lower three crustal layers appear to be surprisingly uniform and seem to be continuous across a major fault zone. The major lateral inhomogeneities occur above the shallowest continuous

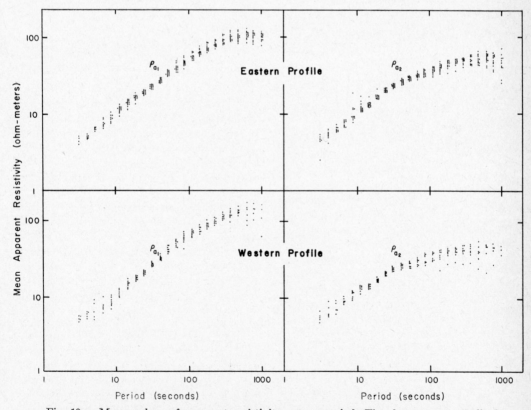

Fig. 10*a*. Mean values of apparent resistivity versus period. The data were normalized such that a least-squares line through magnetotelluric observations in the period range from 10 to 32 sec always assumed a value of 12 ohm-m at a period of 24 sec (reproduced from *Mitchell and Landisman* [1971], with the permission of *Geophysics*).

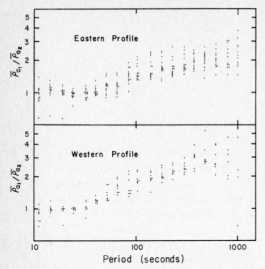

Fig. 10*b*. Ratios of mean apparent resistivity versus period. The values were obtained by fitting a least-squares line through the data for the period range 18 to 32 sec and normalizing the least-squares value of the ratio to unity at a period of 24 sec (reproduced from *Mitchell and Landisman* [1971], with the permission of *Geophysics*).

interface which also forms the base of a low-velocity layer in both Oklahoma and eastern New Mexico.

The depth of the low-velocity zone in eastern New Mexico correlates well with that of a zone of low resistivity interpreted from a one-dimensional analysis of magnetotelluric data from northwestern Texas. If this correlation is valid, it may indicate that a common causal mechanism has produced both features. The seismic velocity just below this zone is about 6.7 km/sec, which is a value commonly taken as indicative of a preponderantly basic mineral assemblage in the rock layer. The base of the low-velocity, low-resistivity zone may correspond, then, to the transition from intermediate to basic rocks in the crust. The apparent continuity of the crust below this transition in Oklahoma seems to indicate that tectonic movements in this region have been restricted to the upper crust. It is interesting to speculate that these movements might be related to the presence of the low-velocity, low-resistivity zone, which might well be characterized by a higher degree of mobility than the surrounding rock.

Hyndman and Hyndman [1968] suggested that water of hydration causes the reduced re-

sistivities sometimes interpreted for the lower crust. *Berdichevskiy et al.* [1969], *Mitchell* [1970], and *Mitchell and Landisman* [1971] have proposed that water of hydration produces zones having reduced values of both electrical resistivity and seismic velocity at moderate depths in the crust. Velocity reductions caused by interstitial fluids which increase the pore pressure in crystalline rocks have been observed in the laboratory by *Gordon and Davis* [1968] and by *Nur and Simmons* [1969, Figure 3; note the lines labeled $P_{pore} = P_{ext}$].

Mitchell [1970] and *Mitchell and Landisman* [1971] have suggested that zones of low resistivity and low velocity are produced by the intersection of the crustal temperature-depth curve with the equilibrium boundary of a hydrothermal reaction. If this is true, then the greater depth to the low-velocity zone in eastern New Mexico as compared to that in Oklahoma may indicate that the geothermal gradient in eastern New Mexico is lower than that in Oklahoma.

CONCLUSIONS

The major results of the present study are the following:

1. The lower layers of the crust exhibit only minor lateral velocity variations and appear to be continuous over a wide region in the southern Great Plains.

2. Major lateral inhomogeneities and faults are restricted to the upper 15 or 20 km of the crust in this area.

3. The shallowest continuous interface forms the base of a low-velocity zone in the crust, and the velocity below this interface corresponds to that which is commonly used to characterize basic rock materials.

4. The low-velocity zone may correlate in depth with a zone of reduced electrical resistivity.

These results suggest that tectonic movements in the interiors of continental plates may be related to a zone having reduced values of seismic velocity and electrical resistivity at moderate depths. Water of hydration is suggested as a possible causal mechanism for the reduced values observed for both velocity and resistivity and for local enhancement of seismic attenuation and increased mobility in response to tectonic

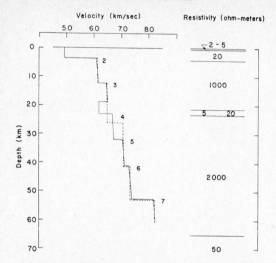

Fig. 11. Comparison of seismic velocity model for eastern New Mexico and electrical resistivity model for northwestern Texas. The broken line represents a seismic model derived from a simple analysis of apparent velocities and intercept times. The continuous line is the velocity distribution obtained from combined observations of refracted and reflected arrivals. The numbers to the right of each discontinuity refer to the refraction branches in Figures 7 and 8 (reproduced from *Mitchell and Landisman* [1971], with the permission of *Geophysics*).

forces. The fluid content of the crust may be governed by the intersection of the crustal temperature-depth curve and the equilibrium boundary of a hydrothermal reaction. If this is so, then the positions of the upper surfaces of low-resistivity low-velocity zones are specified by the pressure and temperature at the point of intersection. The bases of such zones are produced by mineralogical changes to more basic material, in which hydrothermal reactions occur at higher temperatures.

We favor the seismic and resistivity interpretations presented in Figures 6 and 11, for reasons discussed in the preceding text. However, non-uniqueness is inherent in both the seismic and electrical methods, especially for laterally varying structures. For this reason, it is important that intensive multi-disciplinary studies be made in a few specially selected regions, with a view toward increasing our detailed knowledge of the earth's crust.

Acknowledgments. The authors gratefully acknowledge appreciation to Drs. J. Healy and J. Roller of the National Center for Earthquake Research, U.S. Geological Survey, Dr. E. Tryggvason of the University of Tulsa, and Mr. R. Reakes of Geotech, a Teledyne Company, for providing seismic data from the Oklahoma and Gnome experiments, and Dr. Gustave Hoehn of Mobil Research and Development Corporation's Field Research Laboratory for providing the magnetotelluric data used in this study. Dr. R. DuBois of the University of Oklahoma generously furnished long-period seismograms recorded by the observatory in Leonard, Oklahoma. Prof. E. Herrin of Southern Methodist University kindly arranged for computing facilities at The Dallas Geophysical Observatory.

The work was supported by the Office of Naval Research under contract N00014-67-0310-0001, by the National Science Foundation under grant GA-25683, and by supplemental funds of the University of Texas at Dallas.

REFERENCES

Ansorge, J., D. Emter, K. Fuchs, J. Lauer, S. Mueller, and E. Peterschmitt, Structure of the crust and upper mantle in the rift system around the Rhinegraben, in *Graben Problems,* edited by H. Illies and S. Mueller, pp. 190–197, Schweizerbart, Stuttgart, Germany, 1970.

Berdichevskiy, M., V. Borisova, V. Bubnov, L. Van'yan, I. Fel'dman, and I. Yakovlev, Anomaly of the electrical conductivity of the Earth's crust in Yakutiya, *Izv. Akad. Nauk SSSR, Phys. Solid Earth,* (10), 633–637, 1969.

Bonjer, C., and K. Fuchs, Crustal structure in southwest Germany from spectral transfer ratios of long-period body waves, in *Graben Problems,* edited by H. Illies and S. Mueller, pp. 198–202, Schweizerbart, Stuttgart, Germany, 1970.

Červený, V., On dynamic properties of reflected and head waves in the *n*-layered Earth's crust, *Geophys. J. Roy. Astron. Soc., 11,* 139–147, 1966.

Douze, E., Rayleigh waves in short-period seismic noise, *Bull. Seismol. Soc. Amer., 54,* 1197–1212, 1964.

Fuchs, K., and O. Kappelmeyer, Report on reflection measurements in the Dolomites—September, 1961, *Boll. Geofiz. Teor. Appl., 4,* 133–141, 1962.

Fuchs, K., On the properties of deep crustal reflectors, *Z. Geophys., 35,* 133–149, 1969.

Gordon, R., and L. Davis, Velocity and attenuation of seismic waves in imperfectly elastic rock, *J. Geophys. Res., 73,* 3917–3935, 1968.

Ham, W., R. Denison, and C. Merritt, Basement rocks and structural evolution of southern Oklahoma, *Okla. Geol. Surv. Bull., 95,* 301 pp., 1964.

Harlton, B., Frontal Wichita fault System of southwestern Oklahoma, *Bull. Amer. Assoc. Petrol. Geol., 47,* 1552–1580, 1963.

Haskell, N., Crustal reflection of plane *P* and *SV* waves, *J. Geophys. Res., 67,* 4751–4767, 1962.

Hill, D., A laboratory investigation of conductivity and dielectric constant tensors of rocks,

Ph.D. thesis, 228 pp., Michigan State Univ., East Lansing, 1969.

Hyndman, R., and D. Hyndman, Water saturation and high electrical conductivity in the lower continental crust, *Earth Planet. Sci. Lett., 4,* 427–432, 1968.

Jordan, L., Geology of Oklahoma—A summary, *Proc. Geophys. Soc. Tulsa, 8,* 27–41, 1964.

Landisman, M., and S. Mueller, Seismic studies of the earth's crust in continents, 2, Analysis of wave propagation in continents and adjacent shelf areas, *Geophys. J. Roy. Astron. Soc., 10,* 539–548, 1966.

Long, L., and J. Berg, Jr., Transmission and attenuation of the primary seismic wave, 100 to 600 km, *Bull. Seismol. Soc. Amer., 59,* 131–146, 1969.

Lyons, P., Gravity map of Oklahoma, *Proc. Geophys. Soc. Tulsa, 8,* 53–63, 1964.

Meissner, R., Exploring deep interfaces by seismic wide angle measurements, *Geophys. Prospect., 15,* 598–617, 1967.

Mitchell, B., *Electrical and Seismic Properties of the Earth's Crust in the Southwestern Great Plains,* Ph.D. thesis, 53 pp., Southern Methodist Univ., Dallas, Texas, 1970.

Mitchell, B., and M. Landisman, An interpretation of a crustal section across Oklahoma, *Bull. Geol. Soc. Amer., 81,* 2647–2656, 1970.

Mitchell, B., and M. Landisman, Electrical and seismic properties of the Earth's crust in the southwestern Great Plains of the U.S.A., *Geophysics, 36,* 363–381, 1971.

Mueller, S., and M. Landisman, Seismic studies of the Earth's crust in continents, 1, Evidence for a low-velocity zone in the upper part of the lithosphere, *Geophys. J. Roy. Astron. Soc., 10,* 525–538, 1966.

Mueller, S., and M. Landisman, An example of the unified method of interpretation for crustal seismic data, *Geophys. J. Roy. Astron. Soc.,* in press, 1971.

Mueller, S., E. Peterschmitt, K. Fuchs, and J. Ansorge, Crustal structure beneath the Rhinegraben from seismic refraction and reflection measurements, *Tectonophysics, 8,* 529–542, 1969.

Nur, A., and G. Simmons, The effect of saturation on velocity in low porosity rocks, *Earth Planet. Sci. Lett., 7,* 183–193, 1969.

Phinney, R., Structure of the earth's crust from spectral behavior of long-period body waves, *J. Geophys. Res., 69,* 2997–3017, 1964.

Porath, H., A review of the evidence on low-resistivity layers in the Earth's crust, in *The Structure and Physical Properties of the Earth's Crust, Geophys. Monogr. Ser.,* vol. 14, edited by J. G. Heacock, this volume, AGU, Washington, D.C., 1971.

Qualls, B., *Crustal Study of Oklahoma,* M.S. thesis, 82 pp., University of Tulsa, Oklahoma, 1965.

Sims, W., and F. Bostick, Methods of magnetotelluric analysis, *Tech. Rep. 58,* 86 pp., Electronics Res. Ctr., Univ. Texas, Austin, 1969.

Stewart, S., and L. Pakiser, Crustal structure in eastern New Mexico interpreted from the Gnome explosion, *Bull. Seismol. Soc. Amer., 52,* 1017–1030, 1962.

Talwani, M., G. Sutton, and J. Worzel, A crustal section across the Puerto Rico trench, *J. Geophys. Res., 64,* 1545–1555, 1959.

Tryggvason, E., and B. Qualls, Seismic refraction measurements of crustal structure in Oklahoma, *J. Geophys. Res., 72,* 3738–3740, 1967.

Westhusing, K., Project Gnome volunteer seismological teams, *Tech. Rep. 62–5,* 57 pp., The Geotechnical Corp., Garland, Texas, 1962.

Woollard, G., and H. Joesting, Bouguer gravity anomaly map of the United States, AGU and U.S. Geol. Surv., Washington, D.C., 1964.

Wright, J., Anisotropic apparent resistivities arising from non-homogeneous, two-dimensional structures, *Can. J. Earth Sci., 7,* 527–531, 1970.

DISCUSSION

Ward: Could a two-dimensional computation have eliminated the deeper highly conductive layer in Figure 11?

Mitchell: Both apparent resistivity components assume uniform values at periods of about 30 seconds and greater for the entire array of stations, after a correction has been made for differences in the conductivity of the surficial sediments. It will be difficult to explain both this consistency and the magnitude of the difference between the two components at each station with any simple type of two-dimensional model. The possibility that a complex two- or three-dimensional configuration could produce the same results cannot be ruled out, however.

Ward: Would you comment on the order of fit of the curves.

Mitchell: We tried to fit the data at each of the 20 measurement sites with a model which included the fewest possible changes from site to site. We know that lateral changes occur in the sediments, and we tried to match the apparent resistivity curves by varying the resistivity in only the uppermost half kilometer in our models. Uniform differences between the two components for periods greater than 30 seconds at every site required that the fourth layer be anisotropic and have low values of resistivity.

Kelly: You had gravity anomalies indicating structure off to the side of the profiles. Could this structure have given rise to multi-path problems in the seismic data?

Mitchell: I do not think it would have a noticeable effect.

Oliver: The section in Figure 6 seems incredible to me. How can you have flat-lying layers in view of all the structure?

Mitchell: The upper part is taken from direct observations in a well. I am not saying that the layers are perfectly flat—only that the plane-layer approximation adequately explains the data. We have attempted to determine the depth at which major crustal movements in this region originated. The best explanation of the entire body of data indicates that these movements originated in the deepest granitic rocks at depths less than 20 kilometers.

Oliver: It does seem to fit.

Smith: Is your model sensitive to the conducting layer?

Mitchell: Yes, it is sensitive to that layer, since it is a conducting layer sandwiched between highly resistive layers.

The Resolving Power of Geoelectric Measurements for Delineating Resistive Zones within the Crust

THEODORE R. MADDEN

Department of Earth and Planetary Sciences
Massachusetts Institute of Technology, Cambridge, Massachusetts   02139

Abstract. Theoretical solutions for simple geometries and for layered mediums are examined to investigate the resolving power of surface electrical measurements for measuring crustal resistivities. It is shown that galvanic-resistivity measurements can determine only the resistivity-thickness product of the resistive zone and that magnetotelluric measurements can determine only the thickness. Neither method is practical for studying the resistive crust in areas where appreciable conductivity exists near the surface. Near-surface lateral variations are shown to give a large scatter to the measurements, but in areas of high surface resistivities one can still recognize the existence of extreme crustal resistivities, provided the resistive zone has a large enough lateral extent.

It is widely recognized that the electrical resistivity of the upper crust is controlled by the free-water content of the rocks. The electrical properties of water in the pressure and temperature range of the crust are also known. The mechanical and thermodynamic conditions that exist at depths of 10 to 30 km are enough in doubt, however, that estimates of probable electrical-resistivity values at these depths have varied from 10^4 to 10^{11} ohm-m. Surface electrical and electromagnetic measurements are therefore important factors in any attempt to establish reasonable estimates of the actual electrical properties existing at these depths. There are very strong limitations on the interpretation of such surface measurements, however, especially when the region of interest is a poor conductor relative to the surrounding areas. In this study we wish to review certain classic results concerning simple geometries and then (by using numerical techniques) to look at model results for more realistic geometries in order to establish the possibility of detecting and evaluating high-resistivity zones in the crust by means of surface measurements.

Two different types of measurements are usually used to study crustal electrical properties. One type of measurement, called simply resistivity measurements, consists of applying current into the earth with a long horizontal grounded dipole and measuring the electric field strength at varying distances from the current source. The other measurement technique is called the magnetotelluric method, and it consists of measuring the ratio of the horizontal electric and magnetic fields at the earth's surface at varying frequencies. The exciting fields are the natural fluctuations of the earth's magnetic field, and they are assumed to be homogeneous over distances that are large compared to the depth of interest. A third measurement technique has also been used that is a cross between magnetotelluric and resistivity measurements. It consists of measuring the electric and magnetic field ratios for the field set up by a dipole source. The magnetic-variation method, which uses the vertical and horizontal magnetic fluctuations over an area to determine the internal induced electromagnetic fields, has often been used to study deeper electrical properties, but it is not well suited for studying resistive crustal zones. The resistivity method appears to be the most applicable method. We shall concern ourselves mostly with this method, but we shall look at the magnetotelluric method for comparison.

SIMPLE GEOMETRIES

Some of the difficulties of detecting resistive zones are demonstrated by very simple geometries. If we have only a single plane boundary between mediums of different resistivities, the secondary fields set up by the mediums in response to dc current sources can be described as the fields of image sources. The strength of these image sources depends on a contrast factor K

$$K = (\rho_2 - \rho_1)/(\rho_2 + \rho_1) \qquad (1)$$

The contrast factor is limited to values between ± 1 and exhibits a saturation behavior for large contrasts. Thus it might appear that one could never determine accurately the resistivity of the second medium from measurements in the first medium when their resistivities have a large contrast. For the typical measurement geometries, however, this limitation applies only if the second medium is more resistive. This comes about because K is negative when the second medium is conductive, and if one is close enough to the boundary, the image source can largely cancel out the direct source term. In this case, the small deviation of K from -1 is directly measurable. If the normal derivative of the electric potential very close to the boundary could be measured, one could determine small deviations of K from $+1$, but this measurement is denied one when the target zone is buried and flat lying.

The same saturation effects are obtained from target zones of finite size. As an example we can consider a sphere of radius R_o and resistivity ρ_2 immersed in an otherwise uniform medium of resistivity ρ_1. If we apply a uniform incident electric field of magnitude E_0, the resultant potential in the outside medium [Stratton, 1941, p. 205] can be written as

$$\Phi_1 = -E_0 r \cos \theta$$

$$\cdot \{1 + [(\rho_2 - \rho_1)/(2\rho_2 + \rho_1)]R_0^3/r^3)\} \qquad (2)$$

where θ is the angle between the r position vector and E_0. The secondary field tends to

$$\Phi_{\text{secondary}} = -E_0 r \cos \theta R_0^3/2r^3$$

$$\text{for} \quad \rho_2 \gg \rho_1 \qquad (3)$$

or

$$\Phi_{\text{secondary}} = +E_0 r \cos \theta R_0^3/r^3$$

$$\text{for} \quad \rho_2 \ll \rho_1 \qquad (4)$$

Again, in a fashion analogous to the plane boundary case, E field measurements close to the sphere are sensitive to conductive spheres if the measurement is parallel to the boundary, and they are sensitive to resistive spheres if the measurement is perpendicular to the boundary. For a buried sphere, one could only approach the former measurement, however.

LAYERED-MEDIUM GEOMETRY

These results pertaining to the influence of a resistive target are modified in the actual earth situation by the presence of the air-earth interface and by the fact that the air is a much better insulator than any naturally occurring rock formations. Theoretical solutions now must involve layered mediums. Standard curves for resistivity measurements over 2- and 3-layered models are available [Mooney and Wetzel, 1957; Compagnie Generale de Geophysique, 1955]. In these geometries, the properties of the bottom layer can always be determined even when that layer is resistive, provided the measurements can be made across sufficiently large spreads. Unfortunately for the crustal-resistivity problem, the resistive layers cannot be considered as infinitely thick, since the temperature increase with depth produces an appreciable conductivity increase in the lower crust and upper mantle. In fact, the zone in the crust likely to be most resistive is small in thickness compared to the distances needed to make the resistivity measurements. Hence it can be treated as a thin layer. This situation allows us to simplify the analysis, and we can clearly define the resolution problem of determining the influence of a resistive zone on the surface-resistivity measurements.

To arrive at the thin-layer approximation, consider the potential about a point source in a layered medium. The potential can be expressed as a Hankel transform

$$\Phi_1 = (I/2\pi) \int_0^\infty \lambda A(\lambda) J_0(\lambda r) \, d\lambda \qquad (5)$$

The kernel function $A(\lambda)$ is the surface impedance of a one-dimensional transmission system that satisfies [Sunde, 1949] the following equation

$$dV/dz = -\rho I \qquad (6a)$$

$$dI/dz = -(\lambda^2/\rho) V \qquad (6b)$$

where ρ is the medium resistivity and z is the vertical dimension. Within a homogeneous layer such a transmission system is characterized by a propagation constant k and a characteristic impedance K.

$$k = i\lambda \qquad (7a)$$

$$K = \rho/\lambda \qquad (7b)$$

The impedance at the top of a layer is given in terms of the impedance at the bottom of the layer, Z_T, and the layer thickness, Δz.

$$Z = K[Z_T - iK \tan (k \Delta z)]$$
$$\div [K - iZ_T \tan (k \Delta z)] \qquad (8)$$

The terminal impedance is given as

$$Z_T = \rho_T/\lambda \qquad (9)$$

If the terminating space is homogeneous, ρ_T is the resistivity of that half space; otherwise it is an apparent resistivity and will depend on λ. A 'thin' layer is one for which $k\Delta z \ll 1$, and since $k = i\lambda$ and λ is the horizontal wave number in the Hankel transform representation, the measuring distance r should be large compared to the layer thickness. In this approximation, $\tan (k\Delta z) \cong k\Delta z$ and

$$Z \cong (\rho/\lambda)(\rho_T + \rho\lambda \Delta z)/(\rho + \rho_T\lambda \Delta z) \qquad (10a)$$

If the layer is resistive, $\rho \gg \rho_T$

$$Z \cong (\rho_T + \lambda\rho \Delta z)/\lambda \qquad (10b)$$

If the layer is conductive, $\rho \ll \rho_T$

$$Z \cong (\rho_T/\lambda)/(1 + \rho_T\lambda\sigma \Delta z) \quad \sigma = 1/\rho \qquad (10c)$$

Thus a thin layer is characterized by a resistivity-thickness product if it is resistive or by a conductivity-thickness product if it is conductive in comparison with the zone below. The actual value of the resistivity or of the thickness is not resolved, however.

There is a further difficulty involved when the resistive layer is beneath a more conductive surface layer. Let us say that

ρ_1 is surface layer resistivity.
Δz_1 is surface-layer thickness.
ρ_2 is resistive-layer resistivity.
Δz_2 is resistive-layer thickness.
ρ_3 is bottom half-space resistivity.

Set

$$\rho_3 \cong 0 \qquad (11)$$

Using the thin-layer approximations, the impedance at the top of the resistive layer is $\rho_2 \Delta z_2$ from (10b), and thus the impedance at the surface is

$$Z_{surface} \cong \rho_1\rho_2 \Delta z_2/(\rho_1 + \lambda^2 \Delta z_1 \Delta z_2 \rho_2) \qquad (12)$$

At close ranges (but not close enough to invalidate the thin-layer approximations),

$$Z_{surface} \cong \rho_1/(\lambda^2 \Delta z_1) \qquad (13)$$

and one has the $1/\lambda^2$ behavior associated with a conductive layer overlying a more resistive section. One can determine only minimum parameters for the underlying section, however, until the form of the impedance changes. This occurs when

$$\lambda^2 \Delta z_1 \Delta z_2 \rho_2 \cong \rho_1$$

Thus, in order to estimate the resistivity-thickness product, one must extend the resistivity-measurement spreads out to distances comparable to

$$(\Delta z_1 \Delta z_2 \rho_2/\rho_1)^{1/2}$$

Since 100 km represents an upper limit for achievable resistivity-measurement spreads and since we are probably dealing with a $\Delta z_1 \Delta z_2$ value of approximately 100 $(km)^2$, one cannot expect to identify resistivity-thickness products for resistive zones in the crust unless the average conductivity of the top 10 km is less than 100 times the conductivity of the resistive zone. Thus, the resistivity technique will not be useful in identifying resistive crustal zones under sedimentary areas. Figure 1 illustrates this trend.

The logistic problems that limit the depth of investigation for resistivity measurements are not a factor in the magnetotelluric measurements, since here the depth of investigation is determined by the frequencies, and there is ample energy in the very low frequency magnetic-variation spectrum. Unfortunately, the method provides even less information concerning resistive, horizontal layers than the resistivity method. As we shall see, it can supplement the resistivity information, but it cannot replace it.

For this problem, we can go directly into

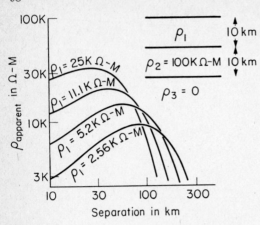

Fig. 1. The effect of surface resistivities on apparent-resistivity profiles for Schlumberger arrays.

transmission line equations from Maxwell's equations. For an E_x, H_y polarization using a $\rho^{-i\omega t}$ time dependence, and assuming horizontally uniform fields and ignoring displacement currents, Maxwell's equations become

$$dE_x/dz = i\mu\omega H_y \qquad (14a)$$

$$dH_y/dz = -\sigma E_x \qquad (14b)$$

For homogeneous zones, we have for the propagation constant

$$k = (i\mu\omega\sigma)^{1/2} \qquad (15a)$$

and using E_x/H_y as our impedance, we have for the characteristic impedance

$$K = (-i\mu\omega/\sigma)^{1/2} \qquad (15b)$$

The thin-layer approximation again holds for $k\Delta z \ll 1$, which now means the layer thickness must be much less than the skin depth of the wave in that layer.

That is, for $k\Delta z \ll 1$ we have from equation 8

$$Z = Z_T[1 - i(i\mu\omega\sigma_T)^{1/2} \Delta z]/(1 - iZ_T\sigma \Delta z) \qquad (16a)$$

where Z_T is the impedance at the bottom of layer and is equal to

$$(-i\mu\omega/\sigma_T)^{1/2}$$

and σ_T is the apparent conductivity of the bottom.

If the layer is resistive, i.e., $\sigma \ll \sigma_T$, then

$$|Z_T\sigma \, \Delta z| < |(i\mu\omega\sigma)^{1/2} \, \Delta z|$$

and the thin-layer approximation reduces to

$$Z \cong Z_T[1 - i(i\mu\omega\sigma_T)^{1/2} \, \Delta z] \qquad (16b)$$

If the layer is conductive, i.e., $\sigma \gg \sigma_T$, then

$$|(i\mu\omega\sigma_T)^{1/2} \, \Delta z| < |(i\mu\omega\sigma)^{1/2} \, \Delta z|$$

and the thin-layer approximation reduces to

$$Z \cong Z_T/(1 - iZ_T\sigma \, \Delta z) \qquad (16c)$$

For a conductive layer, a conductivity-thickness product is determined just as in the resistivity case, but for a resistive layer only the thickness is determined. It would appear that combined resistivity and magnetotelluric data would allow one to separately determine the thickness and the resistivity of the resistive zone, but again there are difficulties due to the near-surface conductivity.

If the lower crust and upper mantle have only modest conductivities, which is the usual situation, and if the near-surface conductivity is thick enough, then, to those frequencies that can penetrate through the surface layer, the lower crust and upper mantle will also appear to be part of a 'thin' resistive layer. This condition results whenever

$$\sigma_3^{1/2} \, \Delta z_3 \ll \sigma_1^{1/2} \, \Delta z_1$$

where the modestly conducting lower crustal zone is designated as zone 3.

A very similar analysis can be made for the magnetic variation method except one must include the horizontal wave number, since

$$H_z = -k_y E_x/\mu\omega$$

for an E_x, H_y polarization. As long as

$$k_y \ll (i\mu\omega\sigma)^{1/2}$$

the conclusions concerning the resolving power of the method are essentially the same as those for the magnetotelluric method.

Two-dimensional geometries. Figure 2 shows a summary of some of the long-line resistivity measurements made in various parts of the United States by the United States Geological Survey [*Keller et al.*, 1966]. The data are all indicative of a crust that is more resistive than the surface, but only the Maine-New Hampshire and the Adirondacks data can

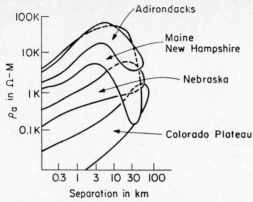

Fig. 2. Summary of some long-line resistivity measurements made by the United States Geological Survey.

be used to estimate resistivity-thickness values. The maximum value that would fit this data is about 10^8 ohm m km. Other deep-sounding electrical measurements have also given resistivity thickness products of this magnitude or smaller [*Van Zijl*, 1969; *Lahman and Vozoff*, 1965; *Migaux et al.*, 1960; *Cantwell et al.*, 1965; *Hauk*, 1960; *Schlumberger and Schlumberger*, 1932]. These resistivity-thickness product estimates are in the range of Brace's predicted values, which are based on the assumption of water saturated conditions [*Brace*, 1971]. If the electrical sounding interpretations reported are valid, one would have to conclude that no extensive zones of dry crustal rocks have been detected to date from surface measurements.

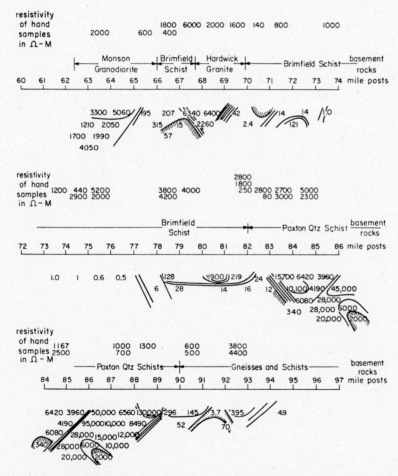

Fig. 3. Apparent-resistivity cross sections and surface geology along the Massachusetts turnpike. The apparent resistivities are given in ohm-meters.

Even when we realize that the near-surface resistivities restrict the areas in which surface measurements can be used to determine these crustal resistivities, these results do not present an optimistic picture in the search for extensive zones with resistivity values greater than 10^6 ohm-m.

The interpretations have been based on the assumption of layered mediums and it must therefore be assumed that no lateral variations exist. It is widely recognized that lateral variations do exist, and one must investigate their effects on the electrical sounding data. G. V. Keller (personal communication, 1970) has suggested that resistive zones in the crust may have lateral dimensions that are too small for them to be detectable from surface electrical measurements.

Variations of the near-surface resistivities also complicate the electrical sounding interpretations. Figure 3 shows dipole-dipole resistivity measurements made along the Massachusetts Turnpike [*Madden et al.*, 1962]. The apparent resistivities are plotted in the cross section below the measurement lines, and the resistivities of water-saturated hand samples collected along the line are shown above. The measurements covered a length of 37 miles between Palmer and Worcester in an igneous and metamorphic region west of Boston. The dipoles were one mile long, and the maximum separations obtained were 6 miles, or about 10 km. These measurements delineate quite well various boundaries in rock type and dramatize the wide range of resistivity values that can exist even in igneous and metamorphic areas. The very good conductor, the Brimfield schist, is a pyritic-graphitic schist that outcrops at only a few spots along the profile but that is interpreted to underlie most of the center section. Data of this kind clearly show the inadequacies of layered-medium interpretations in regions where lateral variations are pronounced.

TWO-DIMENSIONAL MODEL CALCULATIONS FOR RESISTIVITY MEASUREMENTS

In order to investigate the effects of both variable surface resistivities and the finite widths of possible crustal resistive zones one must use more complicated models.

When the features have a consistent strike direction that extends for distances a good deal longer than the depths of interest, one can use two-dimensional models for a better approximation to the actual resistivity distributions.

From the basic potential equations

$$\nabla \Phi = -\rho \mathbf{J} \qquad (17a)$$

where

$$\nabla \cdot \mathbf{J} = I \qquad (17b)$$

where I is the symbol for current sources.

If the strike direction is y,

$$\rho = \rho(x, z) \qquad (18)$$

one can transform out the y variable. Using a Fourier transform and assuming symmetry about the x axis, we define

$$\Phi = \int_0^\infty \Phi_\lambda(x, z) \cos (\lambda y) \, \partial \lambda \qquad (19a)$$

$$J_x = \int_0^\infty J_{\lambda x}(x, z) \cos (\lambda y) \, \partial \lambda \qquad (19b)$$

$$J_z = \int_0^\infty J_{\lambda z}(x, z) \cos (\lambda y) \, \partial \lambda \qquad (19c)$$

$$I = \int_0^\infty I_\lambda(x, z) \cos (\lambda y) \, \partial \lambda \qquad (19d)$$

From (17) one also obtains

$$J_y = \int_0^\infty (\lambda/\rho)\phi_\lambda \sin (\lambda y) \, \partial \lambda \qquad (19e)$$

and a set of two-dimensional equations

$$\partial \Phi_\lambda / \partial x = -\rho J_{\lambda x} \qquad (20a)$$

$$\partial \Phi_\lambda / \partial z = -\rho J_{\lambda z} \qquad (20b)$$

$$\partial J_{\lambda x}/\partial x + \partial J_{\lambda z}/\partial z = -(\lambda^2/\rho)\Phi_\lambda + I_\lambda \qquad (20c)$$

These equations can be approximated by a lumped-circuit rectangular network with conductances between nodes aligned in the x direction, given as

$$Y_x = (1/\rho) \, \Delta z/\Delta x \qquad (21a)$$

and conductances between nodes aligned in the z direction, given as

$$Y_z = (1/\rho) \, \Delta x/\Delta z \qquad (21b)$$

and conductances from node to ground given as

$$Y = (\lambda^2/\rho) \, \Delta x \, \Delta z$$

Solutions of these difference equations or network equations can be obtained by relaxation or by direct methods [*Madden and Thompson*, 1965] for different λ values, and the final solution can be obtained by an inverse Fourier transform.

A computer program that is based on this numerical technique was used to calculate two-dimensional models to test the effect of lateral variations in masking the resistivity signature of resistive crustal zones.

In order to extend the range of measurement distances modelled without using very large networks, three scales were used with unit spacings of 2, 7.5, and 30 km. The model represents a crustal-resistivity section that is similar to the predictions for a water-saturated crust in the New England heat flow environment [*Brace*, 1971] that surrounds a zone of more resistive rocks 150 km wide, whose right edge appears at the center of the section. The

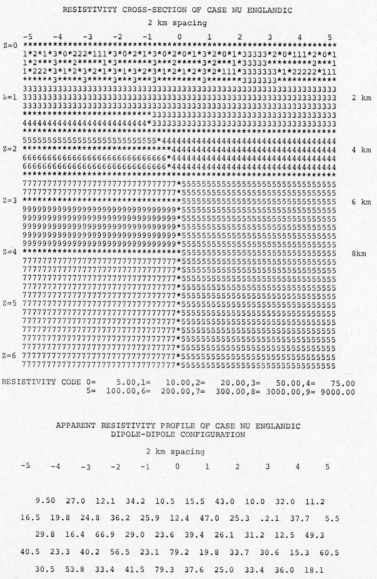

Fig. 4. Resistivity and apparent-resistivity cross sections for a two-dimensional crustal model. Resistivity values are given in kilo-ohm-meters. The unit spacing is 2 km.

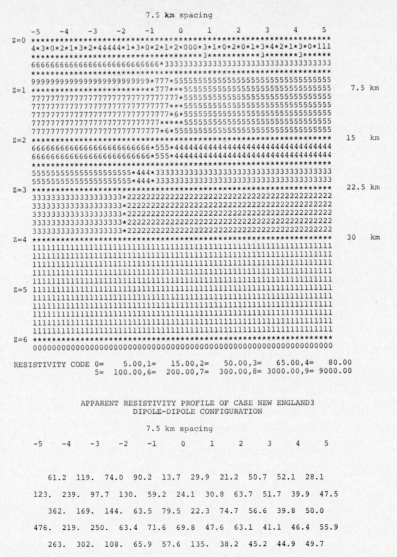

Fig. 5. Resistivity and apparent-resistivity cross sections for a two-dimensional crustal model. Resistivity values are given in kilo-ohm-meters. The unit spacing is 7.5 km.

background-resistivity profile has a resistivity thickness product of 1.5×10^6 ohm m km compared to 4×10^7 ohm m km for the more resistive zone. Variations of near-surface resistivities from 5000 to 50,000 ohm m are also included in the model. Figures 4, 5, and 6 show the resistivity and apparent-resistivity cross-sections of the model. The right-hand side of the resistivity cross-sections is the assumed background-resistivity section, and the resistive zone appears on the left. The apparent-resistivity values are plotted beneath the center of the dipole-dipole locations and at a distance below the dipole locations proportional to the dipole-dipole separation. The first row of numbers consists of apparent-resistivity values for dipoles whose centers are two units apart. The next row represents dipoles three units apart, et cetera. The dipoles and the separation vectors are all aligned across strike.

The near-surface resistivity variations complicate the picture by giving a wide scatter to the apparent-resistivity values, but the average trend is still discernible. This pattern is shown in Figure 7, in which all the data taken over the two zones are included. The data from over the more resistive zone are quite distinct from the background data at separations of 30 km, but at separations of 100 km or more the resistive zone disappears from the electrical view.

This effect occurs whenever any of the measurements overlap onto the more conductive background zones.

SUMMARY

These results show that surface electrical measurements have severe limitations in delineating resistive crustal zones. The best that can be done is to obtain a resistivity-thickness product estimate for such zones by means of re-

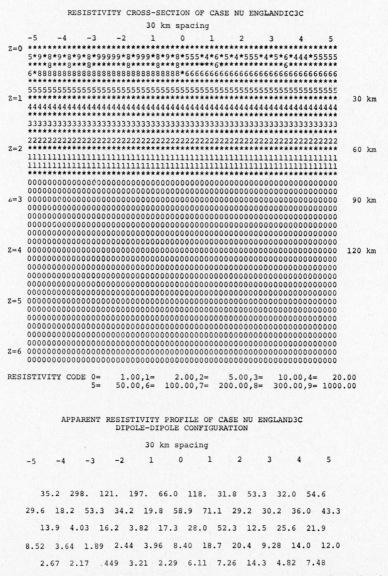

Fig. 6. Resistivity and apparent-resistivity cross-sections for a two-dimensional crustal model. Resistivity values are given in kilo-ohm-meters. The unit spacing is 30 km.

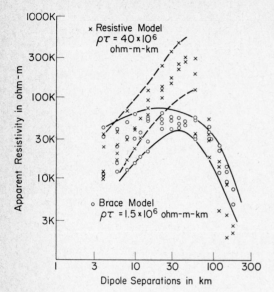

Fig. 7. Summary of theoretical apparent-resistivity profiles over a two-dimensional crustal model. The background resistivity represents a water-saturated crust in which is enclosed a more resistive zone of finite width.

sistivity measurements. Even this task can be accomplished only if the near-surface conductivity is not too great compared to that of the resistive zone and if the resistive zone has a great enough lateral extent. Some reported resistivity measurements have appeared to give a reasonable sampling of the resistive part of the crust, but in no cases have resistivity thickness products of greater than 10^6 ohm m km been reported. Since the temperature and pressure conditions most suitable for achieving high resistivities should persist for approximately 10 km in depth extent, these resistivity results would appear to indicate that free water is still present in the upper and middle regions of the crust.

Acknowledgments. The two-dimensional apparent-resistivity calculations were done by a computer program made available to me by Mandrel Industries, and I am grateful for this help. My knowledge and interest in crustal electrical properties has benefited from the opportunity to work on aspects of this problem with Dr. T. Cantwell, Dr. K. Vozoff, and Professor W. Brace.

I also wish to acknowledge my appreciation for the support I received for work in this area from the Office of Naval Research, through contract NR-371-401, and from the National Science Foundation.

REFERENCES

Brace, W., Resistivity of saturated crustal rocks to 40 kilometers based on laboratory measurements, in *The Structure and Physical Properties of the Earth's Crust, Geophys. Monogr. Ser.,* vol. 14, edited by J. G. Heacock, AGU, Washington, D.C., this volume, 1971.

Cantwell, T., P. Nelson, J. Webb, and A. S. Orange, Deep resistivity measurements in the Pacific Northwest, *J. Geophys. Res., 70,* 1931–1937, 1965.

Compagnie Generale de Geophysique, Abaques de sondage electrique, *Geophys. Prospect., 3,* suppl. 3, 1955.

Hauk, A. M., Deep structure resistivity measurements in Massachusetts, M.S. thesis, Dept. Geol. and Geophys., M.I.T., Cambridge, Mass., 1960.

Keller, G. V., L. A. Anderson, and J. I. Pritchard, Geological Survey investigation of the electrical properties of the crust and upper mantle, *Geophysics, 31,* 1078–1087, 1966.

Lahman, H. S., and K. Vozoff, Deep resistivity results from dc ground tests at Hoover Dam, *Sci. Rep. 6, Contract AF 19(628)-2351, Project 4600, Task 460008,* Geoscience Inc., Air Force Cambridge Res. Lab, Cambridge, Mass., 1965.

Madden, T., T. Cantwell, D. Greenwalt, A. Kelly and A. Regier, Progress report on geomagnetic studies and electrical conductivity in the earth's crust and upper mantle, Report of Project NR-371-401, Dept. Geol. and Geophys., M.I.T., Cambridge, Mass., 1962.

Madden, T., and W. Thompson, Low-frequency electromagnetic oscillations of the earth-ionosphere cavity, *Rev. Geophys. Space Phys., 3,* 211–254, 1965.

Migaux, L., J. L. Astier, and P. Revol, Un essai de determination experimental de la resistivite electrique des couches profondes de l'ecorce terrestre, *Ann. Geophys., 16,* 555–560, 1960.

Mooney, H., and G. Wetzel, *The Potential about a Point Probe in a Two-, Three-, and Four-Layered Earth,* Univ. of Minnesota Press, Minneapolis, 1957.

Schlumberger, C., and M. Schlumberger, Electrical studies of the earth's crust at great depths, *Geophys. Prospect.,* A.I.M.E. 134–140, 1932.

Stratton, J. A., *Electromagnetic Theory,* McGraw-Hill, New York, 1941.

Sunde, E. D., *Earth Conduction Effects in Transmission Systems,* D. Van Nostrand, New York, 1949.

Van Zijl, J.S.V., A deep schlumberger sounding to investigate the electrical structure of the crust and upper mantle in South Africa, *Geophysics, 34,* 450–462, 1969.

DISCUSSION

Healy: We are observing quite generally now a cutoff in the occurrence of earthquakes at a depth of about 10 km in the western United States. One explanation is that something happens which materially changes the physical properties of the crustal material at this depth, and this could be the absence of free water. What really happens if you heat a rock, increase the pressure, and leave the system open so water can escape?

Kaufman: What happens depends on whether the rock has good permeability or whether the pores are closed. I suppose the water pressure reaches lithostatic if it cannot get out because the permeability is too low.

Healy: You are talking about geological time now, so good permeability is six orders of magnitude different from what you would mean by good permeability in an oil field. On this time scale, even unfractured granite looks like a sieve. Could the water go into some mineral phase? At some depth you would have water in the gaseous phase in equilibrium with the mineral, and below that no free water at all.

Madden: I think that what would happen is that the water pressure would be reduced by the water going into a mineral phase until you reach equilibrium and the reaction stops. Now if you bring in more water you have a little more reaction, but through geologic time you may have brought in so little water, because of the low permeability, that you have never developed any real change in mineralogy. But all the time there is free water present at the equilibrium pressure.

Electrical Studies of the Crust and Upper Mantle

GEORGE V. KELLER

Department of Geophysics, Colorado School of Mines, Golden, Colorado   80401

Abstract. In recent years, there have been many efforts made to determine the profile of electrical conductivity through the crust and into the upper part of the mantle. A variety of techniques have been used, including direct-current resistivity soundings, geomagnetic deep soundings, magnetotelluric soundings, and electromagnetic soundings. As a result of the intricacies in the interpretation of the data obtained with each of the techniques, it is difficult to make generalizations about the variation in electrical conductivity in the crust from one geological province to another unless large volumes of data are considered. When such volumes are available, several consistencies in the behavior of electrical conductivity may be noted that appear to be in agreement with modern concepts of crustal mobility and plate tectonics. In tectonically active areas of the continents, most data indicate that conductivity increases to moderately high values, in the range from 0.01 to 0.1 mho/m, at depths between 50 and 100 km. In stable or nuclear continental areas, this increase is not so marked. In areas overlying subduction zones, the conductivity appears to be high through the crust and into the mantle. Where the crust is believed to be the most mobile, conductivity is highest. This increase in conductivity might be attributed to higher than normal temperatures, to the presence of higher than normal water contents, or to both factors. Either factor would also be expected to lower the viscosity of the rock in the upper mantle.

It has been established with some certainty on the basis of seismic investigations that the crust and upper part of the mantle can be separated into a sequence of layers with differing acoustic properties, separated by well developed interfaces that are parallel or subparallel to the earth's surface over considerable areas. The nature of these interfaces is still a matter of conjecture, and additional evidence about the reason for the change in acoustic properties at these boundaries is still being sought. Considerable effort has been expended in the past decade on the application of electrical surveying methods to mapping these same interfaces and to measuring the electrical properties of the rocks that comprise the various shells of the earth.

Laboratory studies have provided a basis for the belief that electrical sounding methods can be useful in studies of the crust and upper mantle [see *Parkhomenko*, 1967; *Keller*, 1963, 1966a]. The results of laboratory measurements of electrical conductivity of rock samples can be summarized briefly as follows:

1. The electrical conductivity of water-free crystalline rocks is profoundly affected by temperature. At temperatures of a few hundred degrees Kelvin, the conductivity of a dry rock is 10^{-8} mho/m, with a range of $\pm$ 2 orders of magnitude. Near the melting point, the conductivity is 10^{-3} mho/m, with a range of $\pm$ 2 orders of magnitude. At upper mantle temperatures, a change in temperature of 100° may change the conductivity of a dry rock by an order of magnitude.

2. The electrical conductivity of a dry rock is appreciably affected by the chemical and mineralogical composition of that rock. At high temperatures, conductivity is contributed by ions thermally excited from the crystal lattice. Rocks that contain a large number of small ions, such as magnesium, iron, aluminum, hydrogen, and hydroxyl, will be more conductive than rocks with a low content of such ions. As a result, light colored or felsic rocks tend to be the most highly resistant. Dark-colored or mafic rocks are more conductive by one or two orders of magnitude, and the most conductive

rocks are those that are rich in hydrated minerals and that contain the hydrogen and hydroxyl ions in their crystal structure.

3. Pressure has a relatively minor effect on the conductivity of a dry rock unless it induces a phase change. Compression of a rock by pressure may reduce the mobility of ions by as much as a factor of two at upper mantle depths, but this amount of change in conductivity is negligible in comparison with the changes caused by variations in temperature or composition. A change in state will cause more significant changes in conductivity, particularly if the change in state is one that increases the binding energy of the small ions. Such an increase would markedly reduce the number of ions formed by thermal excitation at any given temperature and would consequently reduce the conductivity.

4. Free water is a very important constituent in determining the conductivity of a rock, particularly at low temperatures and pressures. If the solid minerals in a rock are much less conductive than the free water filling the pores in that rock, the resistivity can be calculated approximately from the expression:

$$\rho = \rho_w S^{-n} \tag{1}$$

where ρ_w is the electrical resistivity of the water filling the pores, S is the volume fraction of water present in a rock, and n is an experimentally determined parameter having a value of about 1.6 for crystalline rocks. The resistivity of a rock may vary over wide limits with variation of the porosity and, thus, of the free-water content. At low temperatures near the earth's surface, even a part per thousand of water in a rock will dominate the conduction of electricity. At greater depths, where temperature renders the rock-forming minerals more conductive, higher water contents are required before they become the dominant factor in determining the electrical properties. At upper mantle depths, any reasonable water content in the rock would not contribute enough conduction to mask ionic conduction in the solid minerals.

In view of these generalizations, we can expect the conductivity profile through the crust to exhibit three major sections: first, a surficial zone in which the conductivity of a rock is determined by the free-water content; second, a deeper zone in which the conductivity passes

through a minimum because porosity is reduced by overburden pressure and because the associated decrease in conduction through the decrease in free-water content is not fully compensated for by the development of thermally excited ionic conduction in the solid minerals; and third, a region of generally increasing conductivity at still greater depths, with thermally excited conduction becoming progressively more important.

The changes in conductivity caused by temperature should be gradual. In addition to such gradual changes, abrupt changes in conductivity may occur at depths where there are changes in the chemistry, mineralogy, or water content of the rock. For example, one might expect an abrupt increase in resistivity at the base of a sequence of porous sedimentary rocks underlain by crystalline basement rocks. Another example would be the increase in conductivity one might expect at a change from rocks of granitic composition to rocks of basaltic composition, as is hypothesized to occur at the Conrad seismic boundary. Still another type of discrete boundary might be observed where the pressure and temperature are sufficiently great to cause a change in mineral composition of a rock without a change in chemistry. Such a phase change from less dense minerals to more dense minerals might occur at the M discontinuity. Assuming that such a change results in the small ions being bonded more tightly in the crystal structure, one would expect the dense phase to be less conductive than the less dense phase.

All these considerations indicate that determinations of the conductivity profile through the crust and upper mantle should be useful for determining the nature of the boundaries recognized in seismic studies and for establishing the ambient conditions in these outer shells of the earth.

In fact, in examining the literature, one finds that very extensive efforts are being made to determine this conductivity profile. Descriptions of this work are widely distributed through the literature, and it is the purpose of the present paper to review and collate the results that have been obtained with the use of various electrical surveying methods under widely different geological conditions.

Four specific surveying techniques have been used in crustal-scale resistivity surveys; direct-current sounding, magnetotelluric sounding, induction sounding, and geomagnetic deep sounding. I shall review only the results obtained with the first three, because I am unfamiliar with the geomagnetic deep sounding method. For more information on this method, see *Rikitake* [1966].

DIRECT-CURRENT SOUNDING

In direct-current sounding, current is driven into the ground between widely spaced current electrodes, and the resulting voltage drops that develop in the ground are measured and converted to apparent resistivities. Two types of electrode arrangement have been used in crustal-scale direct-current surveys: the Schlumberger array and the dipole array. With the Schlumberger array, the current electrode pair is gradually expanded along a traverse line while the voltage drop between a closely-spaced pair of measuring electrodes fixed at the center of the array is recorded. The results of surveys made with this method have been reported by *Kraev* [1952], *Migeaux et al.* [1960], *Keller* [1966a, b], and *van Zijl* [1969]. Figure 1 shows the resistivity sounding curve obtained by van Zijl in South Africa. The apparent-resistivity values are plotted as a function of half the distance between the current electrodes, as is the convention in presenting such data. This half-distance is loosely related to the depth to which boundaries may be recognized. In the very simple case of a single uniform layer resting on a uniform substratum, the apparent resistivity has deviated from the value for the surface layer by a significant amount by the time the spacing has been increased to equal the thickness of the layer. If there are several layers, it becomes more difficult to specify how large a spacing is required to detect a deep boundary; usually larger spacings are required when the surface material consists of several discrete layers.

With the Schlumberger method, very long current-carrying wires must be laid out, and, as a result, the method is rather cumbersome. Some investigators have made use of the dipole method, in order to avoid this requirement. With the dipole method, a current field is developed in the earth by driving current between two contacts placed relatively close together. For crustal scale surveys, this separa-

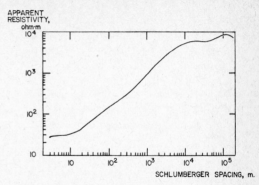

Fig. 1. Direct-current sounding curve published by *van Zijl* [1969] for a Schlumberger electrode array.

tion may be as great as 5 or 10 km. The voltage drop is then measured between pairs of electrodes at considerable distances from this dipole source, with the separation between measuring electrodes also being relatively short compared to the other dimensions of the array. Measurements can be made along radial traverses away from the source dipole, or the measurement points can be sited randomly at various distances. The depth to which boundaries can be recognized is associated with the distance of the receiver site from the source dipole. A dipole sounding curve obtained by the first procedure by *Antonov* [1969] is shown in Figure 2a; a set of sounding data obtained with random location of the receiver is shown in Figure 2b [*Keller et al.*, 1966].

Dipole data are characterized by a high degree of scatter in many cases, although such scatter is not so evident on data obtained with the Schlumberger array or with measurements made on radials about a dipole source. Scatter is caused by surface inhomogeneities in resistivity. Such inhomogeneities also affect the Schlumberger data, but the effect is not so apparent, because of the consistent way in which measurements are made.

Measurements that effectively comprise a direct-current sounding have also been obtained in a number of cases by measuring voltages caused by grounding of high-voltage direct-current (HVDC) transmission lines [*Lundholm*, 1946; *Cantwell et al.*, 1965; *Keller*, 1968b]. Inasmuch as such transmission lines are usually hundreds of kilometers long, measurements made about one end can be viewed as mapping

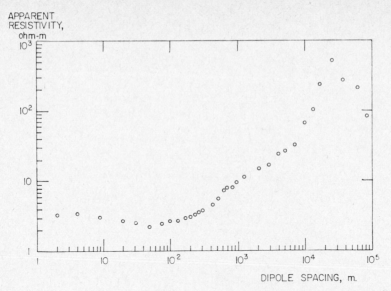

Fig. 2a. Dipole sounding curve with measurements made along a single radial from the dipole source [from *Antonov*, 1970].

the field from a single-pole current source and apparent resistivities can be computed accordingly. These apparent-resistivity values will form the same sounding curve that would be obtained with the Schlumberger array over a layered medium if the distance from the end of the line to the measurement site is taken to be equivalent to the Schlumberger spacing. With HVDC tests a great number of measurements of electric field strength can be made, and this multiplicity of data usually demonstrates the randomness in field strengths caused by surface changes in electrical properties. A plot of electric field strength as a function of distance from one end of the HVDC test line is shown in Figure 3 for measurements made in southern Nevada and nearby areas. Scatter is large, but the number of data available make it possible to establish a sounding curve with good reliability.

All direct-current sounding curves that have appeared in the literature have a strikingly similar form; for spacings of up to 50 km or more, apparent resistivity tends to increase with increasing distance from the end of the line, whereas at greater spacings, the apparent resistivity goes through a maximum. In no case has the sounding curve been followed very far down the descending branch at large spacings.

In discussing the interpretation of direct-current soundings, it is useful to define two parameters describing the electrical properties of a sequence of rocks, in addition to the resistivity; these are the longitudinal conductance and the transverse resistance. Longitudinal conductance is the conductivity integrated over some depth interval H:

$$S = \int_{z_1}^{z_2} \sigma(z) \ dz \qquad (2)$$

and the transverse resistance is the resistivity integrated over the same depth interval:

$$T = \int_{z_1}^{z_2} \rho(z) \ dz \qquad (3)$$

where $\sigma(z)$ and $\rho(z)$ are the conductivity and resistivity, respectively, as functions of depth over the interval H.

The resistive part of the crustal conductivity profile causes current flow to be restricted to the more conductive surface layers at short distances from the end of a source line. At greater distances, leakage through the resistant zone allows current to penetrate to the more conductive rocks at depth. When current is confined to the surface layer, the electric field decreases less rapidly with distance than it would in the case of a uniform earth, and the

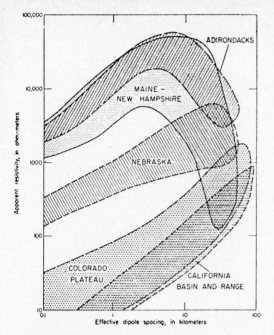

Fig. 2b. Summary of dipole resistivities for a series of soundings in which measurements were made randomly about the source dipole [from *Keller et al.*, 1966].

apparent resistivities computed from such data increase with distance from the source. At greater distances, leakage of current through the resistant zone becomes important, and the electric field will decrease more rapidly with distance than in the case of a uniform earth. The apparent resistivity will pass through a maximum at a distance that depends on the conductance of the surface layer and the transverse resistance of the resistive layer in a well defined manner. The distance at which half the current will have leaked through the resistant zone and at which the maximum apparent resistivity is observed will be:

$$R_{max} = (T_2 S_1)^{1/2} \qquad (4)$$

where T_2 is the transverse resistance of the resistant zone and S_1 is the conductance of the surface layers.

The distance at which the apparent-resistivity curve passes through a maximum becomes greater with an increase in either T_2 or S_1, whether or not the depth to the third layer varies. As a result one has the best chance of detecting the third zone in the crust when

surveys are made in areas of low surface conductance. Surface conductance may vary from less than 1 mho in an area of crystalline rocks to 5000 mhos or more in areas covered by thick sedimentary rocks.

The two parameters T_2 and S_1 can be determined easily from sounding curves by using relationships given by *Keller et al.* [1966]. A summary of values for these parameters from some of the published sounding curves is given in Table 1.

Data from areas with a higher surface conductance have been reported by *Keller et al.* [1966], but in such cases, the apparent resistivity curve does not pass through a maximum even for equivalent Schlumberger separations as great as 200 km.

One difficulty in interpreting direct-current sounding data of this type is that the curve shape is relatively insensitive to combined changes in the resistivity and thickness of the

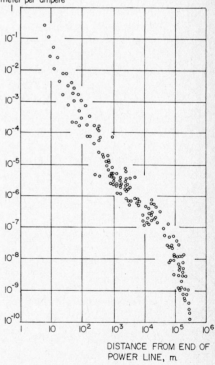

ELECTRIC FIELD INTENSITY, volts per meter per ampere

DISTANCE FROM END OF POWER LINE, m.

Fig. 3. Electric field measurements about a grounded high-voltage, direct-current transmission line in southern Nevada. Field strengths are given per ampere.

TABLE 1. Summary of Values for T_2 and S_1 derived from Direct-Current Sounding Data*

Location	S_1, mhos	T_2, ohm-m^2	Source
France	3.5	3.5×10^8	*Migaux et al.* [1960]
Near Leningrad	20.0	6.0×10^7	*Kraev* [1952]
Adirondacks, USA	0.1	3.6×10^8	*Keller et al.* [1966]
Maine-N. Hampshire	0.2	3.6×10^8	*Keller et al.* [1966]
Nebraska	3–10	$>1.0 \times 10^8$	*Keller et al.* [1966]
Southern Nevada	15–30	1.5×10^8	*Keller* [1968]
North Island, New Zealand	10–20	3.0×10^7	*Keller* [1969]
Near Novosibirsk, USSR	10–15	9.0×10^7	*Antonov* [1969]
South Africa	1.0	11.0×10^8	*van Zijl* [1969]

* Here T_2 is the transverse resistance of the resistant zone and S_1 is the conductance of the surface layers.

second layer, such that the parameter T_2 is not changed. The minimum resistivity for the second layer permitted by the data ranges from 10^4 to 10^5 ohm-m. The corresponding maximum thickness for the resistant portion of the crust is no more than 15 to 20 km. Little can be said about the electrical properties of the third zone from the direct-current sounding data obtained to date, except that the resistivity is much less than that of the second zone, and probably less than 1000 ohm-m.

Perhaps one of the most discouraging aspects of the direct-current sounding data obtained so far is the very large electrode spans required to penetrate only several tens of kilometers into the earth. Even under good conditions where the surface conductance is 10 mhos or less, electrode separations of more than 100 km are required to provide the first clue suggesting the existence of the third layer. Electrode separations of thousands of kilometers would be required in order to obtain penetration through the crust under sedimentary basins, where the surface conductance is measured in hundreds of thousands of mhos. With such large spans required, the direct-current method lacks horizontal resolution for detecting changes in the nature of the third layer.

MAGNETOTELLURIC SOUNDINGS

Most attempts to determine the electrical conductivity profile through the crust and upper mantle have been based on the use of the magnetotelluric method, described originally by *Cagniard* [1953]. This survey technique makes use of the natural electromagnetic field of the earth as a power source, at frequencies ranging

downward from 1 to 10^{-4} Hz, or even less. The technique requires that simultaneous observations be made of orthogonal electric and magnetic variations in the electromagnetic field. The electric field is detected simply by measuring the voltage drop between two electrodes aligned in the direction of the field component to be studied, whereas the magnetic field can be detected with either an induction coil of large area or a sensitive magnetometer. In early attempts to use the magnetotelluric method, it was assumed that orthogonal field components could be measured by orienting the electric and magnetic sensors geometrically at right angles. Such a procedure provided data that showed a very large degree of scatter when converted to apparent resistivity, using Cagniard's formula:

$$\rho_a = \frac{i}{\omega\mu} \frac{E^2}{H^2} \tag{5}$$

where ω is the frequency in radians per second, μ is the magnetic permeability, and E and H are the electric and magnetic field strengths, respectively, measured in mks units. A typical set of apparent resistivity values showing this scatter is given in Figure 4.

One explanation for this scatter appears to be 'anisotropy' in the electromagnetic field. A time-varying magnetic field incident on an anisotropic earth can be resolved into components oriented along the principal directions of the anisotropy ellipse. Each of these components will then see a different resistivity and will therefore induce currents of differing densities. Because the ratio of the two principal components of current density is not the same as the ratio

APPARENT RESISTIVITY,
ohm-m

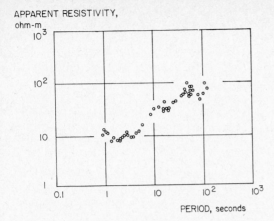

Fig. 4. Magnetotelluric sounding curve showing normal scatter of computed apparent resistivity values [from *Vozoff and Ellis*, 1966].

of the two principal components of magnetic field strength, the direction of the resultant current density is not at right angles to the direction of the magnetic field strength vector. Resistivity calculations made under the assumption that they are at right angles can be seriously in error, and the amount of error will vary with the polarization of the incident magnetic field.

In current magnetotelluric surveys, two components of both the electric field and the magnetic field are measured simultaneously and are correlated to define the orthogonality of the field components. In many cases, a vertical component of the magnetic field is also measured to assure that events are not used that have a significant vertical component. Procedures for analysis of magnetotelluric data in this manner have been described in detail by *Word et al.* [1971] and *Berdichevskiy* [1968]. A set of magnetotelluric sounding curves showing the importance of anisotropy is given in Figure 5. The difference in character between the two curves computed for the principal directions of anisotropy is striking, and it is apparent that the average curve resembles neither.

Many magnetotelluric surveys around the world have been described in the literature. However, because magnetotelluric data tend to be imprecise even when the best field techniques and data reduction procedures are used, it is difficult to assess the significance of differences between interpreted profiles within one survey

area, and it is even more difficult to assess the significance of differences found between different investigators. In reviewing the many interpreted sections that are available in the literature, it would seem that there are three gross types of results being reported (actually four, if surveys made over deep sedimentary basins were to be included):

1. *Sections in which the resistivity beneath the surface veneer remains high (greater than 300 to 1000 ohm-m) to depths beyond 200 km, with the exception of a thin, moderately conductive zone at depths of several tens of kilometers (see Figure 6).* Such an interpretation has been reported for several hundred magnetotelluric soundings in all: by *Kovtun et al.* [1968] for a single sounding made on the Island of Kerguelen in the Indian Ocean; by *Kopitenko et al.* [1967] for soundings made in the volcanic area of the Kamchatka Peninsula; by *Berdichevskiy et al.* [1969] for soundings in a portion of the area surveyed in Yakutia; by *Gokhberg et al.* [1968] for soundings in Turkemenia; by *Portnyagin* [1968] and *Pospeev et al.* [1969] for soundings made in eastern Siberia; by *Militzer and Porstendorfer* [1968] for soundings made in the south of the German Democratic Republic; by *Fournier and Metzger* [1969] for a sounding made in Senegal; by *Matveev et al.* [1969] for soundings made in

APPARENT RESISTIVITY,
ohm-m

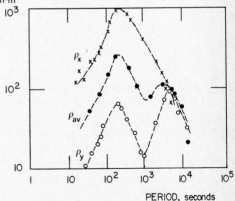
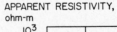

Fig. 5. Magnetotelluric sounding curves showing the effect of anisotropy [from *Kopitenko et al.*, 1967]. Curves for ρ_x and ρ_y are resistivities computed for principal directions, and the ρ_{av} curve is the geometric average of these two.

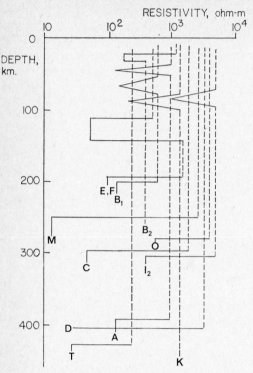

Fig. 6. Generalized summary of resistivity sections through the crust and upper mantle that show a reversal in the profiles and continued high resistivity to considerable depth. The alphabetical key indicates the source for each profile, as given in the caption for Figure 11. Broken lines represent minimum possible resistivities.

the Perm area of Russia; by *Kovtun and Chicherina* [1969] for soundings made in the central Russian depression; and by *Porstendorfer and Kühn* [1965] for measurements made in Zanzibar.

2. *Sections in which the resistivity remains high only to moderate depths (60 to 150 km).* Such interpretations (see Figure 7) have been reported by *Niblett and Sayn-Wittgenstein* [1960]; by *Vozoff and Ellis* [1964] and by *Srivastava et al.* [1963] for soundings in Alberta, Canada; by *Petr et al.* [1964] for a sounding in Czechoslovakia; by *Dowling* [1970] for soundings in Wisconsin; by *Enenshtein et al.* [1969] for a sounding north of the Aral Sea; by *Swift* [1967] for soundings made in the Basin and Range province of the southwestern United States; by *Plouff* [1966] for soundings made in the mountainous parts of the western United States; by *Mitchell et al.* [1971] for

measurements made in northern Texas; and by *Pospeev et al.* [1969] for soundings made over the Baikal rift zone.

3. *Sections in which the resistivity decreases to moderate values at shallow depths (less than 60 km).* Such interpretations (see Figure 8) have been reported by *Portnyagin* [1968] for a few soundings in the Irkutsk region of Russia; by *Gornostaev* [1967] for soundings over the Baikal rift zone; by *Gugunava* [1969] for soundings in the Georgian SSR; by *Militzer and Porstendorfer* [1968] for several soundings made in the northern part of the German Democratic Republic; and by *Hermance and Garland* [1968] for soundings in Iceland.

The sections of Figures 6, 7, and 8, compiled from these references are shown in highly schematic fashion. There is no unique association between the type of section and the structural setting, though there is an apparent tendency for the third type of profile, in which the resistivity decreases at shallow depths, to be found in volcanic areas. Indeed, investigators working in these areas have suggested that the low resistivity is caused by partial melting of

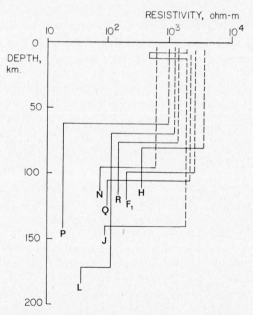

Fig. 7. Generalized summary of resistivity sections though the crust and upper mantle for moderate depths to conductive rock (60 to 150 km). The key indicates the source for each profile and is given with Figure 11. Broken lines represent minimum possible resistivities.

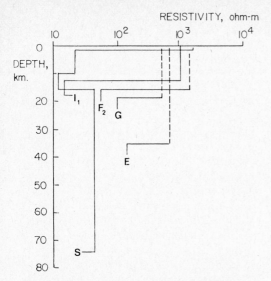

Fig. 8. Generalized summary of resistivity sections through the crust and upper mantle for shallow depths to conductive rock (less than 60 km). The key indicates the source for each profile, as given with Figure 11. Broken lines represent minimum possible resistivities.

the rock at depths of a few tens of kilometers [*Hermance and Garland*, 1968; *Kopitenko et al.*, 1967; *Vanyan et al.*, 1970]. At the other extreme, many of the soundings showing high resistivity to several hundred kilometers depth have been made in stable platform areas of the Eurasian continent. The soundings showing depths of the order of 100 km to the bottom of the resistant part of the section have been made largely in the western United States, an area that may be considered tectonically active.

It is interesting to note that measurements made on the coast of Africa and on Kerguelen Island in the Indian Ocean [*Kovtun and Chichirina*, 1969; *Fournier and Metzger*, 1969; *Porstendorfer and Kühn*, 1965] show high resistivity to considerable depths. It may be that the high conductivity of the nearby oceans in these cases has altered the shape of the magnetotelluric curve from the shape it would have assumed if the ocean were not present. The uncertainty introduced into the interpretation of magnetotelluric soundings by the effect of lateral changes in electrical properties of the near-surface materials is a major shortcoming of the technique. Alternative methods that make use of very deep electromagnetic sounding may

offset this shortcoming, but it is difficult to provide the amount of power required to investigate to depths of hundreds of kilometers.

ELECTROMAGNETIC SOUNDING

Electromagnetic methods using controlled sources for generating an electromagnetic field have not yet been much used in crustal-scale studies, but they appear to offer a means for obtaining more definitive resistivity sections than those obtainable with other electrical surveying techniques. The principal difficulty is in obtaining a strong enough electromagnetic field to permit investigations to mantle depths.

An electromagnetic field can be generated by passing a time-varying current through a grounded length of wire or through an ungrounded loop of wire. The time-varying current may be primarily of a single frequency in the 'frequency-domain' sounding methods or it may consist of a step current containing a wide spectrum of frequencies in the time-domain sounding methods. Because the frequencies required for penetration into the crust are very low, the step-current method seems the more practical.

The electromagnetic fields developed from an ungrounded loop and from a short grounded wire are not identical in behavior. The field from an ungrounded loop is much more localized than the field from a grounded wire, so that it becomes progressively more difficult to provide the power necessary to make soundings with a loop source as attempts are made to reach greater depths. *Keller* [1968a] has computed the currents required with a loop source and with a grounded wire source to provide penetration to significant depths. He finds that a loop source is probably limited to use for depths of less than 5 km.

Vanyan et al. [1967] has investigated both the theory and practice of electromagnetic sounding with a short grounded wire source for the case in which the separation between the source and the receiver is at least five times as great as the depth to which resistivity is to be measured. This simplifies the analysis of the problem markedly, but the requirement that the source-receiver separation be this large is an unnecessary restriction on the use of deep electromagnetic methods. More recently, studies have indicated that it is practical to measure the resistivity to a depth comparable to the separation between the source and receiver or even to greater depths under ideal conditions [G. V. Keller, unpublished manuscript, 1971; *Harthill*, 1969; *Silva*, 1969; *Kaufman and Morozova*, 1970]. The basic method can be seen from the theoretical curves for electromagnetic coupling over a two-layer earth shown in Figure 9. Figure 9 shows two sets of curves for the electromagnetic coupling between a short grounded wire source and a vertical-axis loop receiver. *Vanyan et al.* [1967] has shown that an apparent resistivity ρ_a can be computed from the early part of the transient coupling in this system when a step current is driven through the source:

$$\rho_a = \frac{2\pi R^4 V}{3ILA}$$

where R is the separation between the center of the grounded wire source and the center of the receiving loop, V is the voltage detected in the loop, I is the amplitude of the current step in the source wire, L is the length of the source wire, and A is the area of the receiving loop. It is assumed here that both L and the radius of the receiver loop are small compared to R and that the loop is located on the equatorial axis of the transmitter wire.

The variation of apparent resistivity with time following the start of the transient coupling is shown by the set of curves to the left in Figure 9; each curve represents a single value for the ratio of layer thickness to the separation between the source and receiver. If the layer thickness is small compared to the separation, the apparent resistivity tends to show the presence of the medium below the layer accurately. If the layer thickness is more than a third of the spacing, the effect of the lower medium becomes obscured.

G. V. Keller (unpublished manuscript, 1971),

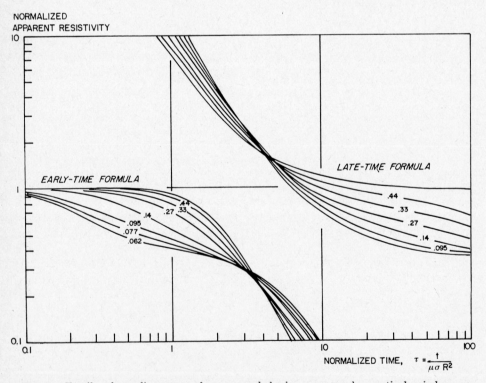

Fig. 9. Family of coupling curves for a grounded wire source and a vertical axis loop receiver for the case in which a layer with unit resistivity rests on a substratum with a resistivity of ⅓. The curve parameter is the ratio of layer thickness to spacing between source and receiver.

Jacobson [1969], and *Silva* [1969] have shown that an apparent resistivity can also be calculated from the same transient coupling by using the expression:

$$\rho_a = \left(\frac{A\,ILR\mu^{5/2}}{40\pi^{3/2}t^{5/2}V} \right)^{2/3}$$

where t is the time following the start of the transient coupling. The apparent-resistivity curves computed with this formula are shown to the right in Figure 9. It is important to note that these curves yield the resistivity of the underlying medium no matter how great the depth is in comparison to the spacing between the transmitter and receiver, provided that the transient signal can be followed for a sufficiently long time.

G. V. Keller (unpublished manuscript, 1971) has described an extensive electromagnetic survey in the volcanic region of the North Island of New Zealand, using this approach. An example of an experimental sounding curve is shown in Figure 10 for comparison with the theoretical curves as shown in Figure 9. The study indicated that crustal resistivities are relatively low along the line of volcanic activity on the North Island, with resistivity being no more than a few hundred ohm-meters in the crust.

Long current-carrying wires are even more convenient for deep electromagnetic soundings than are short wires. A wire can be considered to be infinitely long if its length is more than three times the offset distance at which the field is measured or more than three times the depth to be probed, whichever is greater. It is quite impractical to lay out such a long line solely for the purpose of crustal-scale resistivity surveys, but on occasion it may be possible to take advantage of power distribution lines to make such surveys. In recent years there has been growing interest in the transmission of high-voltage direct current, and part of the engineering studies associated with such systems involve transmission of direct current along such systems with return through grounded electrodes. Transient magnetic fields have been recorded from two such experimental programs, one in Manitoba, Canada [*Jacobson*, 1969; *Harthill*, 1969] and the other in the Sierra Nevada Mountains, California [N. Harthill and G. V. Keller, unpublished manuscript, 1971]. In the first case, the data were interpreted to indicate a uniform resistivity of approximately 2000 ohm-m to depths greater than 20 km. In the second case, the data were interpreted to indicate a resistivity of approximately 10,000 ohm-m to depths of 52 to 55 km beneath the

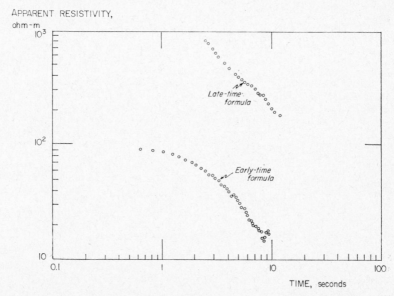

Fig. 10. Time-domain electromagnetic sounding curve from New Zealand. The source, a grounded wire, and the receiver, a vertical-axis loop, were separated by 53.2 km.

Sierra batholith and resistivities of approximately 2000 ohm-m at greater depths.

EVALUATION

The data on the electrical conductivity at crustal and mantle depths that have been reviewed in the preceding sections might be considered to support modern concepts of plate tectonics and crustal mobility. The locations at which various types of deep electrical soundings have been made and that have been described in the literature are shown on the map in Figure 11. It is apparent that these surveys cover a wide geographical area of the continents, though few measurements have been reported in the World Ocean.

The profile of resistivity through the crust and the upper mantle in continental areas ap-

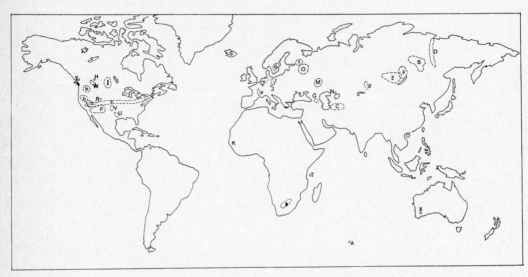

Fig. 11.   Locations of deep electrical studies reviewed in this paper.

Magnetotelluric studies

A   is Kerguelen [*Kovtun et al.*, 1968].
B] is Yakutia [*Berdichevskiy et al.*, 1969].
C   is Turkmenia [*Gokhberg et al.*, 1968].
D   is Yakutia [*Vanyan et al.*, 1970].
E   is Baikal area [*Portnyagin*, 1968; *Gornostaev*, 1967; *Pospeev et al.*, 1969].
F   is South Siberian platform [*Gornostaev*, 1967].
G   is Georgia [*Gugunava*, 1969].
H   is Manitoba [*Niblett and Sayn-Wittgenstein*, 1960].
I   is German Democratic Republic [*Militzer and Porstendorfer*, 1968].
J   is Czechoslovakia [*Petr et al.*, 1964].
K   is Senegal [*Fournier and Metzger*, 1969].
L   is Wisconsin [*Dowling*, 1970].
M   is Perm district [*Matveev et al.*, 1969]
N   is Aral Sea [*Enenshtein et al.*, 1969].

O   is Central Russia [*Kovtun and Chicherina* 1969].
P   is southwestern United States [*Swift*, 1967].
Q   is German Federal Republic [*Vozoff and Swift*, 1968].
R   is southwestern United States [*Plouff*, 1966].
S   is Iceland [*Hermance and Garland*, 1968].
T   is Zanzibar [*Porstendorfer and Kühn*, 1965].
U   is Texas [*Word et al.*, 1971].
V   is Texas [*Mitchell et al.*, 1971].
W   is Alberta [*Srivastava et al.*, 1963; *Vozoff and Ellis*, 1966].
X   is Australia [*Everett and Hyndman*, 1967].
Y   is British Columbia [*Caner and Auld*, 1968].
Z   is Massachusetts [*Madden and Cantwell*, 1960].

Controlled-source surveys (direct-current and electromagnetic)

a   is United States [*Keller et al.*, 1966].
b   is Western United States [*Keller*, 1968b].
c   is eastern United States [*Cantwell et al.*, 1964].
d   is Sweden [*Lundholm*, 1946].
e   is France [*Migaux et al.*, 1960].
f   is near Leningrad [*Kraev*, 1952].

g   is eastern Siberia [*Antonov*, 1969].
h   is South Africa [*van Zijl*, 1969].
i   is New Zealand [*Keller*, 1971].
j   is Manitoba (N. Harthill and G. V. Keller, unpublished manuscript, 1971).

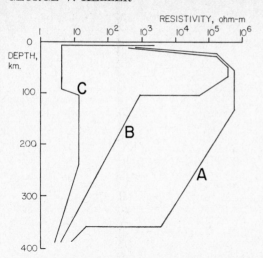

Fig. 12. Highly idealized resistivity profiles through the crust and upper mantle. Profile A pertains to a stable continental nucleus, profile B to a mobile crustal plate, and profile C to a volcanic rift area.

pears to follow one of three general forms, as shown generalized in Figure 12. In all areas, resistivity decreases to values in the range from 10 to 100 ohm-m at depth in the lower crust or upper mantle. Such a decrease in resistivity is predictable on the basis of sample measurements that have shown that the resistivity of dry rocks decreases to the range of 10 to 100 ohm-m at temperatures just below the melting point for surface ambient pressures. One interpretation of the variation in the depth at which such resistivities are observed is that the 1000° to 1100° C isotherm varies in depth from 10 to 20 km in volcanic rift zones to several hundred kilometers in the nuclear continental areas.

In the nuclear continental areas, where re-

sistivity remains high or moderately high to several hundred kilometers depth, a zone of low resistivity is commonly observed in the lower part of the crust or in the upper part of the mantle, at depths of less than 60 km. It is difficult to believe that this reversal in the resistivity profile is accompanied by a reversal in the geothermal gradient, and, inasmuch as water is the component of a rock most likely to cause marked changes in resistivity, it is reasonable to guess that this conductive zone is one in which the water content is relatively high.

If a decrease in resistivity is associated with temperatures near the melting point of a rock, that decrease is presumably accompanied by a decrease in the viscosity of the rock. There may also be a decrease in viscosity associated with a resistivity change caused by the presence of water because, in many rock types, the water causes a lowering of the melting point. Therefore, if plates of crustal and upper mantle rocks are free to move, the low-viscosity substratum on which they ride should be detectable with resistivity measurements. Considering this, the three types of resistivity profile shown in Figure 12 might be associated with three types of area in a mobile crust, as indicated in Figure 13. The low resistivities associated with rift areas may be characteristic of subduction zones, where the evolution of water from a downgoing slab of crust causes partial melting of the rock over the slab and generally lowers the resistivity of all the overlying rock. The low resistivities at intermediate depths may be typical of mobile plates of crust, where a plate approximately 100 km thick moves on a zone of rock that is near its melting point. The electrical resistivity profiles in the stable continental areas may be indicative of crustal plates that are not mobile,

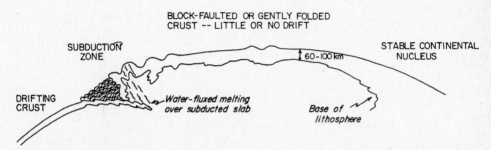

Fig. 13. Hypothetical association of resistivity profiles with regions in a mobile crust.

because the conductivity is not markedly increased until depths of several hundred kilometers are reached.

This association of the results reported for deep electrical soundings is highly speculative but should provide some reason for further more detailed studies of the electrical resistivity profile through the crust and upper part of the mantle.

Acknowledgment. The development of the time-domain electromagnetic sounding method was carried out with the support of the National Science Foundation under grant GA-1029.

REFERENCES

Antonov, Yu. N., On deep soundings with direct current in western Siberia (in Russian), *Geol. Geofiz., 8,* 98, 1969.

Berdichevskiy, M. N., *Electrical Prospecting with the Magneto-Telluric Profiling Method* (in Russian), 255 pp, Nedra, Moscow, 1968.

Berdichevskiy, M. N., V. P. Borisova, V. P. Bubnov, L. L. Vanyan, I. S. Fel'dman, and I. A. Yakovlev, Anomalous conductivity of the earth's crust in Yakutia (in Russian), *Fiz. Zemlya, 10,* 43, 1969.

Bondarenko, A. T., and D. I. Savrosov, On the electrical conductivity of eclogite and kimberlite stocks in Yakutia at high temperatures with respect to upper mantle structure (in Russian), *Geol. Geofiz., 5,* 72–80, 1969.

Cagniard, L., Basic theory of the magneto-telluric method of geophysical prospecting, *Geophysics, 18,* 605, 1953.

Caner, B., and D. R. Auld, Magneto-telluric determination of the upper mantle conductivity structure at Victoria, British Columbia, *Can. J. Earth Sci., 5* (5), 1209, 1968.

Cantwell, T., J. N. Galbraith, Jr., and P. Nelson, Deep resistivity results from New York and Virginia, *J. Geophys. Res., 69* (20), 4367, 1964.

Cantwell, T., P. Nelson, J. Webb, and A. S. Orange, Deep resistivity measurements in the Pacific northwest, *J. Geophys. Res., 70* (8), 1931, 1965.

Dowling, F. L., Magneto-telluric measurements across the Wisconsin Arch, *J. Geophys. Res., 75* (14), 2683, 1970.

Enenshtein, B. S., E. L. Krul', N. V. Lipskaya, O. A. Skugarevskaya, and M. A. Ivanov, Description of simultaneous studies with frequency sounding and magneto-telluric sounding, in *Magneto-Telluric Methods for Studying the Structure of the Earth's Crust and Upper Mantle* (in Russian), p. 225, Nauka, Moscow, 1969.

Everett, J. E., and R. D. Hyndman, Magneto-telluric observations in southwestern Australia, *Phys. Earth Planet. Interiors, 1,* 49, 1967.

Fournier, H. G., and J. Metzger, Sondage mag-neto-tellurique directional a la station magnetique de M'Bour (Senegal) *C. R. H. Acad. Sci., Ser. D, 269,* 297, 1969.

Frischknecht, F. C., Fields about an oscillating magnetic dipole, *Quart. J. Colo. Sch. Mines, 62* (1), 1–326, 1967.

Gokhberg, M. B., V. G. Dubrovskiy, and K. Nepesov, Results of the use of magnetic storms for deep sounding in Turkmenia (in Russian), *Fiz. Zemlya, 12,* 97, 1968.

Gornostaev, V. P., Some supplementary evidence on deep structure of the Pre-Baikal from electrical prospecting data (in Russian), *Geol. Geofiz., 11,* 98, 1967.

Gugunava, G. E., On determining some properties of the 'B' layer with the magneto-telluric sounding method (in Russian), *Geol. Geofiz., 10,* 96, 1969.

Harthill, N., Deep electromagnetic sounding, geological considerations, D. Sc. thesis T-1257, Colorado School of Mines, Golden, Colorado, 1969.

Hermance, J. F., and G. D. Garland, Deep electrical structure under Iceland, *J. Geophys. Res., 73* (12), 3797, 1968.

Jacobson, J. J., Deep electromagnetic sounding technique, D. Sc. thesis T-1252, Colorado School of Mines, Golden, Colorado, 1969.

Kaufman, A. A., and G. M. Morozova, Theoretical basis for transient field sounding in the near zone (in Russian), 124 pp., Siberian Dept. of Nauka, Novosibirsk, 1970.

Keller, G. V., Electrical properties in the deep crust, IEEE Trans. *Antennas Propagat., 11,* (3), 344–357, 1963.

Keller, G. V., Electrical properties of rocks and minerals, in *Handbook of Physical Constants,* edited by S. P. Clark, Jr., *Geol. Soc. Amer. Mem. 97,* 546–577, 1966a.

Keller, G. V., Dipole method for deep resistivity studies, *Geophysics, 31* (6), 1088–1104, 1966b.

Keller, G. V., Electrical prospecting for oil, *Quart. J. Colo. Sch. Mines, 63* (2), 268 pp., 1968a.

Keller, G. V., Statistical study of electrical fields from earth-return tests in western states compared with natural electric fields, *IEEE Trans. Power App. Syst., 87* (4), 1050, 1968b.

Keller, G. V., L. A. Anderson, and J. I. Pritchard, Geological Survey investigations of the electrical properties of the crust and upper mantle, *Geophysics, 31,* 6, 1078, 1966.

Kopitenko, Yu. A., E. S. Gorshkov, T. A. Gorshkova, I. S. Fel'dman, and T. A. Fel'dman, Magneto-telluric sounding near Klyuchi in the Kamchatka territory (in Russian), *Fiz. Zemlya, 9,* 66, 1967.

Kovtun, A. A., and N. D. Chicherina, Results of magneto-telluric investigations in the central Russian depression, in *Magneto-Telluric Methods for Studying the Structure of the Earth's Crust and Upper Mantle* (in Russian), 195, Nauka, Moscow, 1969.

Kovtun, A. A., O. M. Raspopov, V. A. Troitskaya, B. N. Kazak, and J. Loran, Magneto-telluric investigation in the region of Port of France

(Kerguelen) (in Russian), *Fiz. Zemlya*, 7, 89, 1968.

Kraev, A. P., *Osnovi Geoelektriki*, 445 pp., Gostoptekhizdat, Moscow, 1952.

Lundholm, R., The experimental sending of D. C. through the earth in Sweden, paper 134, in *Proc. Conf. Int. des Grands Reseaux Electriques a Haute Tension*, 1946.

Madden, T. R., and T. Cantwell, Preliminary report on crustal magneto-telluric measurements, *J. Geophys. Res.*, 65 (12), 4204, 1960.

Magnitskiy, V. A., and V. N. Zharkov, Low-velocity layers in the upper mantle, in *The Earth's Crust and Upper Mantle, Geophys. Monogr. Ser.*, vol. 13, edited by P. J. Hart, pp. 664–675, AGU, Washington, D. C., 1969.

Matveev, B. K., M. N. Yudin, and V. A. Ponosov, Some results of trials of magneto-telluric sounding in the Perm district, in *Magneto-Telluric Methods for Studying the Structure of the Earth's Crust and Upper Mantle* (in Russian), p. 220, Nauka, Moscow, 1969.

Migaux, L., J.-L. Astier, and P. Revol, Essai de determination experimental de la resistivite electrique des couches profoundes de l'écorce terrestre, *C. R. Sci. Acad. Sci.*, 251, 567, 1960.

Militzer, H., and G. Porstendorfer, Stand und Entwicklungs-stendenzen geophysikalischer Untersuchungen fur die Tiefenerkundung, *Zh. Angew. Geol.*, 14 (12), 617, 1968.

Mitchell, B. J., M. Landisman, and S. Mueller, Electrical and seismic properties of the earth's crust in the southwestern Great Plains, in *The Structure and Physical Properties of the Earth's Crust, Geophys. Monogr. Ser.*, vol. 14, edited by J. C. Heacock, AGU, Washington, D. C., this volume, 1971.

Niblett, E. R., and C. Sayn-Wittgenstein, Variation of electrical conductivity with depth by the magneto-telluric method, *Geophysics*, 25 (5), 998, 1960.

Noritomi, K., Migration of charged carrier in the case of electric conduction of rocks, *Sci. Rep. Tohoku Univ. Ser. 5*, 9 (2), 120–127, 1958.

Noritomi, K., The electrical conductivity of rocks and the determination of the electrical conductivity of the earth's interior, *Akita Univ. Mining Coll. J., Ser., A.*, 1 (1), 29–59, 1961.

Parkhomenko, E. I., *Electrical Properties of Rocks*, 314 pp., Plenum, New York, 1967.

Petr, V., J. Pecova, and O. Praus, A study of the electric conductivity of the earth's mantle by magneto-telluric measurement at Srovarova (Czechoslovakia), *Pr. Geofys. Cesk. Akad. Ved.*, 208, 407, 1964.

Plouff, D., Magneto-telluric soundings in the southwestern United States, *Geophysics*, 31 (6), 1145, 1966.

Porstendorfer, G., and P. Kühn, Works on the International Upper Mantle Project on Zanzibar/Tanzania, *Monatsber. Deut. Akad. Wiss. Berlin*, 7 (10/11), 3 pp., 1965.

Portnyagin, M. A., On the anomalous electrical conductivity at the crust-mantle boundary in the Irkutsk Amphitheatre (in Russian), *Geol. Geofiz.*, 12, 93, 1968.

Pospeev, V. I., V. I. Mikhalevskiy, and V. P. Gornostaev, Results of the use of the magneto-telluric method in regions of eastern Siberia and the Far East, in *Magneto-Telluric Methods for Studying the Structure of the Earth's Crust and Upper Mantle* (in Russian), p. 139, Nauka, Moscow, 1969.

Rikitake, T., *Electromagnetism and the Earth's Interior*, 308 pp., Elsevier, Amsterdam, 1966.

Silva, L. R., Two-layer master curves for electromagnetic sounding, M. Sc. thesis T-1250, 120 pp., Colorado School of Mines, Golden Colorado, 1969.

Srivastava, S. P., J. L. Douglass, and S. H. Ward, The application of magneto-telluric and telluric methods in central Alberta, *Geophysics*, 28, 998, 1963.

Swift, C. M., A magneto-telluric investigation of an electrical conductivity anomaly in the southwestern United States, Ph.D. thesis, Dept. Geophys. Geol., M.I.T., Cambridge, Mass., 1967.

Vanyan, L. L., L. Z. Bobrovnikov, V. L. Loshenitzina, V. M. Davidov, G. M. Morozova, A. N. Kuznetsov, A. I. Shtimmer, and E. I. Terekhin, *Electromagnetic Depth Soundings*, 312 pp., Consultants Bureau, New York, 1967.

Vanyan, L. L., N. A. Zabolotnaya, E. P. Kharin, and G. I. Shtekh, Cause of the anomaly in geomagnetic variations in Yakutia (in Russian), *Fiz. Zemlya*, 6, 96, 1970.

van Zijl, J. S. V., A deep Schlumberger sounding to investigate the electrical structure of the crust and upper mantle in South Africa, *Geophysics*, 34 (3), 450, 1969.

Vozoff, K., and R. M. Ellis, Magneto-telluric measurements in southern Alberta, *Geophysics*, 31 (6), 1153, 1966.

Vozoff, K., and C. M. Swift, Jr., Magneto-telluric measurements in the North German Basin, *Geophys. Prospect.*, 26 (4), 454, 1968.

Word, D. R., H. W. Smith, and F. X. Bostick, Jr., Crustal investigations by the magnetotelluric tensor impedance method, in *The Structure and Physical Properties of the Earth's Crust, Geophys. Monogr. Ser.*, vol. 14, edited by J. C. Heacock, AGU, Washington, D. C., this volume, 1971.

DISCUSSION

Professor Keller was not present during the following discussion; his comments below were written after he received a transcript of the discussion.

Bostick: I take exception to the depth to which Keller carries his interpretations using finite source techniques. From my prospecting experience I would say you need a receiver-transmitter separation of five to ten times the depth to which you are looking, in order to get reliable results. This means we need a separation of 100 km to determine properties at 20-km depth; you can see the difficulty of having to use such large separations before we begin to get involved in the problem we are trying to solve. The magnetotelluric method is sensitive only to conductors but does have the very nice feature of long wavelength, enabling us to penetrate below the near-surface multidimensional features.

Ward: In my study we carried the calculations for tens of kilometers depth out to hundreds of kilometers separation. In one example we did try stopping at 10- and 30-km separation and found there was insufficient resolution, so we are supporting your statement quantitatively with a specific model. As a rule of thumb, when looking for inhomogeneities, not layered structures, we explore no deeper than half the separation when using a two-coil system.

Also, it is misleading to treat all the magnetotelluric data throughout the world as if it is of equal quality. It is not of equal quality and some of the interpretations are not valid.

Keller (Reply to the comments of Bostick and Ward on the depth to which conductivity can be determined with a controlled-source electromagnetic sounding technique):

The depth of investigation for any electrical sounding technique is difficult to define in a satisfactory manner, but for purposes of comparison, consideration of the interpretation in the case of a two-layer medium is useful. For the horizontal-wire—vertical-axis loop system, assuming a step-current excitation of the transmitter, a value of apparent resistivity for the early part of the transient coupling may be computed using formulas given by *Jacobson* [1969] or *Silva* [1969]. Silva has compiled apparent-resistivity curves for the two-layer

case we wish to consider by doing a Fourier transformation of the frequency-domain tables of *Frischknecht* [1967].

Time-domain curves for these two types of apparent resistivity in this manner were shown in Figure 9 of my paper. It may readily be seen from this illustration that in principle at least, the presence of the substratum may be detected with roughly equivalent ease at any depth beneath the transmitter-receiver system, and there is no limit to the depth of investigation for the system.

In practice, a finite dynamic range in the receiving equipment does limit the depth to which conductivity interpretations may be made. The voltage recorded during the late part of the transient decreases approximately as the inverse 5/2 power of time and so it rapidly drops to unmeasurable values. The voltage level at which the transient coupling becomes unmeasurable with the required precision sets a limit to the depth to which conductivity can be determined.

Detection of the substratum in a two-layer case is possible with the simplest analog recording system, one having a 40-db dynamic range, for a depth of burial of about 0.4 times the spacing between source and receiver. If the dynamic range is increased by signal processing, the substratum may be detected at greater depths. With 72 db of dynamic range, the presence of the second layer could be detected for a depth of burial somewhat greater than the spacing between the source and the receiver.

These considerations are based on the requirement that a signal received from the transmitter be strong enough to enable the achievement of the required signal-to-noise ratio. The source moment for a horizontal grounded-wire source to provide this signal strength may be estimated from the data presented in Figure D1, a summary of experience gathered in research projects carried out with the time-domain sounding methods at the Colorado School of Mines [*Jacobson*, 1969; G. V. Keller, unpublished data, 1971]. This plot shows the source-receiver separations for which 24-db signal to noise ratios have been obtained, on the average, as a function of source moment (the product of wire length and current step

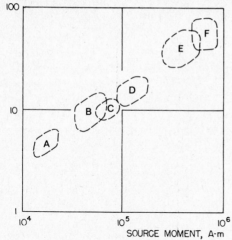

DISTANCE, km

SOURCE MOMENT, A·m

Fig. D1. Summary of experience with the amount of source moment required to provide a 24-db signal-to-noise ratio at the receiver in a time-domain electromagnetic coupling system using a grounded wire as a source and a vertical-axis loop as a receiver. Here, A is the Sierra Nevadas of California [*Jacobson*, 1969]; B is the Volcanic province in North Island, New Zealand; C is the Kilauea volcano, in Hawaii; D is the volcanic belt in Nicaragua; E is the volcanic province in North Island, New Zealand; and F is eastern Colorado.

amplitude). It may be seen from these data that controlled-source electromagnetic soundings to the depth of the low velocity layer in the mantle are quite feasible.

Porath: On the basis of geomagnetic deep sounding data, you cannot have such a resistive section as Keller showed in the continental sheared regions—you must have higher conductivities at depth. In order to explain the attenuation of the normal vertical field, you have to go down to at least 10 ohm-m at 400 kms. All the experimental data on high pressure and temperature properties of silicates also indicate much larger conductivity at shallower depth than shown in Keller's models.

Keller (Reply to Porath concerning rock conductivity at depth):

In my opinion, the presence of highly conductive rocks in the crust and outer mantle requires more effort to explain than does the presence of resistive rocks. Generally, resistivities above a few hundred to a thousand ohm-m

are not easily differentiated with field techniques in use today; conductive rocks whose presence in the crust and mantle may be recognized have resistivities ranging downward from a hundred to a few tens of ohm-m and, in rare cases, to less than 10 ohm-m.

There are numerous data in the literature on the electrical properties of rocks at temperatures below the melting point [*Magnitskiy and Zharkov,* 1969; *Parkhomenko,* 1967; *Bondarenko and Savrosov,* 1969; *Keller,* 1963; *Noritomi,* 1958, 1961; *Keller,* 1966a]. In the outer mantle, it is generally accepted that the temperature approaches the melting point at depths corresponding to the seismic low velocity layer and that then, at greater depths, the melting point increases more rapidly with depth than does the temperature. If there were no changes in the state or the composition of the rock, this behavior would result in there being a maximum in conductivity at the depth of the low velocity zone.

In general, we might expect the following types of behavior for the conductivity profile in the earth:

1. In areas of low to moderate heat flow, the temperature may come no closer than 100° to 200° to the melting temperature, with the minimum conductivity in the outer mantle being less than 0.001 mho/m. This would probably not be detectable with the electrical sounding techniques in use today. At some greater depth, the structure of the rock may change to the point where electronic conduction becomes important. However, it is reasonable that resistivities in excess of 1000 ohm-m may exist to a depth of 700 km or more. If the ambient temperature remains at a nearly constant fraction of the melting point, the conductivity may remain nearly constant over a considerable range in depth.

2. In zones of high heat flow, the temperature may reach the melting point over some depth interval, and a zone with a conductivity of 0.1 to 1 mho/m will be present, bounded above and below by rocks with resistivities in the range from 10 to 100 ohm-m. This zone would be detectable with induction electromagnetic surveys.

3. In zones with intermediate heat flow, the conductivity may approach a value of the order of 0.01 mho/m at the depth of the low velocity

layer and be detectable with induction measurements.

Landisman: Very few groups (other than the Austin and, recently, the Wisconsin group) have used sufficiently high frequencies (up to several Hz) to get resolution from magnetotelluric observations for the shallower portions of the crystalline crust. Papers often report data that begin at ten and even thirty seconds, and these observations really do not contain much information about the present problem.

Madden: I think you might say that none of the electromagnetic methods will ever tell you how resistive the resistive layers are. You have to use galvanic methods to answer those questions. Unless the electromagnetic measurement is a transmission measurement, it will only tell you the thickness of the layer, and I am sure that will be true of the measurements Keller referred to.

Ward: You can force the current to go normal to the interfaces by using horizontal dipoles and get a little better sensitivity. We have done calculations for both horizontal and vertical axis sources and the former is sensitive to conductivity either increasing or decreasing downwards, whereas the latter is sensitive only to conductivity increasing with depth.

Keller (Reply to comments of Landisman, Madden, and Ward):

The comments concerning the difficulty of determining the resistivity of a resistive layer immersed in an otherwise conductive medium are quite pertinent, and I would like to pursue the question in some detail. Consideration of any sounding method based on the use of an electromagnetic field (including the 'dc' or galvanic method) from a grounded wire is considerably simplified in the case of a strictly layered earth if Maxwell's equations are rewritten in terms of two vector potential components, one parallel to the layer boundaries and the other normal to these boundaries [*Vanyan et al.*, 1967; *Keller*, 1968]. Then, the parameters describing the layering (the resistivity and the thickness of each bed) enter the problem when the boundary conditions are applied to these two vector potentials. Application of the boundary conditions generates two auxiliary functions, sometimes called Stefanescu functions. One of these, R_1, depends only on horizontal components of current flow in the earth, and the other, R_2, depends only on vertical components of current flow.

The physical coupling between any source and receiver can be expressed in terms of these two Stefanescu functions, R_1 and R_2, which contain all the information on the properties of the resistivity section. The nature of the functions depends on the geometry of the source and receiver. Therefore, the capacity of any method for recognizing the resistivity of a highly resistive bed immersed in otherwise conductive rocks can be studied by analyzing the Stefanescu functions rather than the more complicated expressions for the coupling between specific sources and receivers.

It happens that R_1 (when $m = 0$, which corresponds to a very large source-receiver separation) is the square of the well-known apparent-resistivity function for the magnetotelluric method. For the three-layer case $N = 3$ and $m = 0$, R_1 reduces to:

$$R_1(N = 3)_{m=0} = \frac{A_1 - B_1 e^{-2\gamma_1 h_1}}{A_1 + B_1 e^{-2\gamma_1 h_1}}$$

where both A_1 and B_1 are functions of σ_1, σ_2, σ_3, γ_2, and h_2.

$$\gamma_i{}^2 = i\omega\mu\sigma_i$$

Thus in the case of a resistant layer included between two more conductive layers, R_1 depends only on the conductivities of the first and third layers and the thicknesses of the first and second layers. The conductivity of the second layer has a negligible effect on the accuracy of the approximation used. In this approximation, the expressions are expanded in series form; second- and higher-order terms are dropped, and the first term, $(\sigma_2/\sigma_1)^{1/2}$ is also dropped, where σ_i is the conductivity of the ith layer.

In expanding the Stefanescu function as a series in powers of γ_i, the ratio of the size of the first term dropped to the size of the last term kept is $(\sigma_2/\sigma_1)^{1/2}$ (where σ_1 and σ_2 are the conductivities of the first and second layers, respectively). Therefore, the accuracy of the approximation is determined by the ratio of the two conductivities. If an accuracy of 5 per cent for the truncated series expression is desired,

$$(\sigma_2/\sigma_1)^{1/2} < 0.05$$

and the conductivity contrast should be 400 to 1 or greater.

The second Stefanescu function, R_2, behaves asymptotically if a large separation between the source and the receiver is assumed. At low frequencies, the resulting approximate expression is a function of the second-layer resistivity regardless of its value. For values of $\gamma_1 h_1$ significantly different from 0, R_2, the second Stefanescu function then becomes:

$$R_2(N = 3)_{m=0} \frac{1 + e^{-2\gamma_1 h_1}}{1 - e^{-2\gamma_1 h_1}}$$

which has the same form as for the two-layer case and thus does not depend on the resistivity of the third layer. If

$$\gamma_1 h_1 \to 0$$

then

$$R_2(N = 3)_{m=0} = \frac{\sigma_2/\sigma_1}{1 - \dfrac{\gamma_3}{\omega h_2}}$$

An alternate asymptotic form of the second Stefanescu function is obtained when frequency is taken to zero. This applies to dc soundings where space is changed rather than frequency to control the depth of investigation. In evaluating the R_2 function in this limit, the second-order terms in σ_2/σ_1 and σ_2/σ_3 are discarded [*Keller*, 1966] to obtain:

$$R_2(N = 3)_{\omega=0}$$

$$= \frac{\rho_1 + mT_2 - (\rho_1 - mT_2)e^{-2mh_1}}{\rho_1 + mT_2 + (\rho_1 - mT_2)e^{-2mh_1}}$$

Although σ_2 and h_2 have both disappeared explicitly, they contribute to the transverse resistance, T_2 (h_2/σ_2).

In the second case, the argument is similar. In expanding the second Stefanescu function as a series in powers of γ_i, the ratio of the size of the first term dropped to the size of the last term kept is (σ_2/σ_1). In this case, if an accuracy of 5 per cent for the truncated series expression is desired, we have the condition:

$$(\sigma_2/\sigma_1) < 0.05$$

and the conductivity contrast should be 20 to 1 or greater. In this respect, the dc method offers less promise than the ac methods based on measurement of the first Stefanescu function, where a contrast of 400:1 might be resolved.

In my opinion, the following generalizations can be made about the problem of determining the resistivity of a resistant layer in the crust:

1. If the resistant layer is overlain by a thin, resistant veneer of sedimentary rock or weathered crystalline rock, electromagnetic methods based on the measurement of the first Stefanescu function are more likely to provide a value for the resistivity of the second layer than the dc method is, because the electromagnetic methods do not suffer from nonuniqueness until relatively high contrasts in resistivity between the first and second layers are reached.

2. If the resistant layer is overlain by more normal, conductive rocks, it may be possible to determine the resistivity of the resistant zone by complete determinations of the first and second Stefanescu functions, using a combination of electromagnetic methods and dc sounding methods. However, because the two types of methods are differently affected by lateral resistivity changes, it is probable that a large number of each type of sounding would have to be done in a single geological province before the results would be convincing.

3. Perhaps the most promise is offered if the low-frequency part of the second Stefanescu function can be determined with an electromagnetic method. Then, the question of equivalence in the second layer does not arise. In electromagnetic systems employing a vertical axis coil for either the transmitter or the receiver or both, the coupling is a function of only the first Stefanescu function [*Vanyan et al.*, 1967]. With horizontal-axis coils or grounded wires, the coupling is a function of both the first and second Stefanescu functions; the way they are combined depends on the geometry of the system. It is possible, for example, to choose orientations and positions for two grounded wires so that the coupling is a function of only the second Stefanescu function, as is desired here. Whether such measurements are practical in the field remains to be determined.

A Review of the Evidence on Low-Resistivity Layers in the Earth's Crust

H. PORATH[1]

Geosciences Division, University of Texas at Dallas, Dallas, Texas   75230[2]

Abstract. It has been suggested in a number of publications that data from geomagnetic deep sounding, magnetotelluric, and electrical resistivity studies of several regions of the earth indicate the presence of conductive layers in the earth's crystalline crust. This review shows that, in all instances studied, an alternate solution can be presented or that the geological setting of a region in which measurements were made does not allow a unique interpretation of observations. Anomalous magnetic-variation fields observed in geomagnetic deep sounding studies in the western United States are shown to be related to lateral variations in the electrical conductivity of the upper mantle and do not require a conductive lower crust. In geological provinces with an extensive, conductive, sedimentary sequence, magnetic variation anomalies result from the concentration and channeling by sedimentary structures of large-scale current systems induced either in the oceans or in the continental crust. Apparent-resistivity curves derived from magnetotelluric studies are affected by the anomalous currents in the vicinity of sedimentary conductive structures. It is shown that these lateral changes in the resistivity of the upper crust rather than an increase of conductivity at depth can cause the observed features in the apparent resistivity curves. This conclusion is supported by examples of apparent-resistivity curves in regions of known, intense, anomalous currents and by resistivity curves computed for stations near two-dimensional structures which under certain assumptions can model resistivity changes in the upper crust. Deep sounding, using some form of direct-current electrode method appears to be the approach best suited for detailed investigations of the electrical structure of the earth's crust, since the source field has a known configuration and is limited compared to the global dimensions of fields used in magnetotelluric studies.

Evidence on the electrical conductivity structure of the earth's crust and upper mantle comes from electrical resistivity, magnetotelluric, and geomagnetic deep sounding studies. Results from these studies indicate that the earth's crust and upper mantle can, in general, be divided into three regions. The surface zone, which is composed of sediments or weathered rocks, forms a moderate to good conductor, owing to the presence of significant amounts of electrolytic solutions in the pore spaces of rocks. The magnetic variation anomalies associated with this zone have been discussed by *Porath and Dziewonski* [1971*b*]. In the intermediate zone, where rocks become less porous, owing to higher pressures, and contain very little water, the resistivity of the earth's crust increases to at least several thousand ohm meters [*Cantwell*, 1960; *Keller et al.*, 1966]. At some depth in the upper mantle, the resistivity decreases again because of the temperature effect on semiconducting silicates [*Tozer*, 1959].

Several reports have recently appeared in the literature in which the presence of a conductive layer in the intermediate resistive zone of the earth's crust is suggested. It is the purpose of this paper to examine the evidence for conductive layers in the earth's crust, particularly in the light of recent advances in the understanding of the distribution of the global current systems which provide the source field for the geomagnetic and magnetotelluric deep sounding methods [*Porath and Dziewonski*, 1971*b*].

EVIDENCE ON CONDUCTIVE LAYERS IN CRYSTALLINE CRUST FROM GEOMAGNETIC DEEP SOUNDING STUDIES

Lateral variations in the electrical conductivity of the earth's crust and upper mantle

---

[1] Deceased.

[2] Formerly the Southwest Center for Advanced Studies, Dallas, Texas 75230.

cause anomalies in the horizontal and vertical components of geomagnetic variations generated by current systems in the ionosphere and magnetosphere. Mapping of the anomalous variation fields with arrays of three-component magnetometers is termed 'geomagnetic deep sounding.'

An appreciable internal magnetic field is induced in the presence of a transient source due to a considerable increase of the conductivity at depths of a few hundred kilometers [Lahiri and Price, 1939]. The vertical component Z of this induced internal field opposes the external field, whereas the external and internal horizontal fields reinforce each other [Rikitake, 1966, p. 129]. Magnetic-variation fields in the period-range from several minutes to a few hours are, therefore, almost horizontal at a distance from the source-field region. Lateral inhomogeneities in the conductivity structure distort this normal variation field and result in anomalous vertical and horizontal magnetic fields. Anomalous fields are most obvious in the vertical component, where they can be observed in the presence of only a small normal field. The scale lengths of the source fields range from about 5000 km for magnetic disturbances in the auroral zone [Oldenburg, 1969] to global dimensions for the daily variation fields. The currents induced in the conductive strata of the earth will be of corresponding dimensions.

The most intense current systems are induced in the highly conducting sea water of the oceans for variations of periods between a few minutes and a few hours. Some of the currents induced in the oceans leak into the moderately conductive upper continental crust in the regions where the resistivity contrast between sediments and sea water is small. Examples are the north German basin, in Europe, and the coastal plains bordering the Gulf of Mexico, in North America, both of which contain thick sedimentary sequences with conductivities close to those of sea water [Porath and Dziewonski, 1971a, b].

For short-period variations, currents induced in the conductive upper crust of continents become important. These currents are concentrated in conductive sedimentary structures such as deep basins. The inhomogeneous distribution of induced currents in the upper crust results in numerous anomalies in geomagnetic variations for periods of up to several hours [Porath and Dziewonski, 1971a, b].

In general, even if a conductivity structure is nearly two-dimensional, it is not possible to approximate the observed anomalous fields associated with crustal conductivity variations by the electromagnetic response of two-dimensional models. The problem is three-dimensional, since only part of the current is induced near the conductivity anomaly.

Currents induced in the upper mantle are affected by lateral changes in conductivity due to temperature variations. The most comprehensive study of mantle conductivity anomalies has been conducted in the western United States, where it can be shown that enhanced electrical conductivity correlates with regions of high heat flow [Gough and Porath, 1970] and a well-developed seismic low velocity zone in the upper mantle [Porath and Gough, 1971; Porath, 1971].

Caner et al. [1967] and Caner [1970] have suggested, however, that the geomagnetic-variation anomalies in the western United States are caused by a conductive lower crust and not by conductivity variations in the upper mantle. This conclusion was based on numerical-model studies of the attenuation of vertical normal fields for a sequence of plane layers. A conductive lower crust has also been proposed by Hyndman and Hyndman [1968] on the grounds of petrological considerations. The resistivity model proposed by Caner [1970] for the crust and upper mantle in tectonic regions of western North America is shown in Figure 1, section a. His model for the Great Plains province, a moderately resistive upper crust over a more conductive lower crust and upper mantle, is shown in Figure 1, section b. For western North America (Figure 1, section a) Caner proposes a low-resistivity layer of 5–15 Ω-m for the lower crust and a thickness of 20 to 40 km.

Porath and Gough [1971] attempted to approximate the anomalous horizontal and vertical fields observed in the western United States by using the techniques of Madden and Swift [1969] and Wright [1969] and the model of Caner [1970]. Anomalous fields computed for this model depend strongly on period, whereas the observed fields change very little in the period range from 30 to 90 minutes. Phase differences of 30° or more should be observed between anomalous and normal fields, provided the assumptions of the two-dimensional model calculations are correct. Inspection of the magneto-

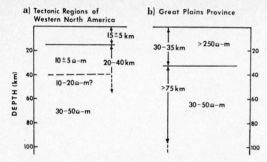

Fig. 1. Conductivity models for tectonic regions of (a) western North America and (b) the Great Plains province, as suggested by *Caner et al.* [1969], from geomagnetic deep sounding and magnetotelluric studies [after *Caner et al.*, 1969].

grams from stations in regions of pronounced anomalies indicate, however, that anomalous vertical fields are almost in-phase with the inducing east-west fields [*Porath and Gough*, 1971]. Variation anomalies related to models of conductive structures in the upper mantle, on the other hand, show the observed dependence of amplitude on variation period and only small phase differences between anomalous and normal fields [*Porath and Gough*, 1971].

In addition to anomalous vertical and horizontal fields, conductivity inhomogeneities give rise to differences in the attenuation of the normal Z field which consists of the sum of an external vertical field and the induced internal field of opposite sign. Regions of anomalously high conductivity are, therefore, characterized by an attenuated vertical field. Amplitudes of normal Z as a function of period can be approximated by plane layered earth models, provided the scale length of the inducing field can be estimated [*Schmucker*, 1970].

A strong attenuation of the vertical field in the tectonic regions of western North America is evident in magnetograms of a magnetic disturbance from stations in northwestern United States and southwest Canada [from *Camfield et al.*, 1971] shown in Figure 2. In the western regions, amplitudes of normal Z are attenuated by about a factor of two as compared to amplitudes for stations in the Great Plains province [*Camfield et al.*, 1971]. The ratios of $Z(X^2 + Y^2)^{1/2}$ observed for these two provinces are approximated by fields associated with plane layered conductivity models, if we assume that the

inducing field has dimensions of several thousand kilometers [*Porath et al.*, 1971]. The observed attenuation factors of the normal field can be satisfactorily explained by decreasing the depth of the mantle conductor in the western region (Table 1). In the model, a 2-km thickness of conductive sediments ($\rho = 5$ Ω-m) is assumed for the Great Plains province, and a 15-km moderately conducting upper crust is assumed for the tectonic western region, in accordance with the resistivity soundings of *Cantwell et al.* [1965]. The corresponding values in the normal fields calculated for the model by *Caner* [1970] are appreciably below those observed (see Table 1).

Caner's interpretation can be attributed to two incorrect assumptions. First, *Caner et al.* [1967] assume that regional variations in the vertical-field amplitudes represent variations in the attenuation of the normal Z field, whereas they are related to the anomalous fields associated with upper mantle and sedimentary conductivity anomalies of these regions [*Porath and Gough*, 1971; *Porath and Dziewonski*, 1971a]. Second, *Caner et al.* [1967] assume in their model calculations scale lengths of the inducing fields of several hundred kilometers, whereas magnetic disturbance fields have dimensions of at least several thousand kilometers [*Swift*, 1967; *Oldenburg*, 1969].

Based on the more recent studies mentioned in the preceding paragraph, variation anomalies are caused either by concentration and channeling of induced currents in conductive sedimentary structures or by lateral variations in upper mantle conductivities. Geomagnetic deep sounding studies provide no evidence for a conductive layer in the lower crust under tectonic regions in the western United States.

EVIDENCE ON CONDUCTIVE LAYERS IN THE CRYSTALLINE CRUST FROM MAGNETOTELLURIC STUDIES

Geomagnetic deep sounding studies in regions of nearly uniform incident fields can detect only lateral changes in conductivity; a conductive layer of uniform depth and thickness will not be recognized. Magnetotelluric studies, on the other hand, are best suited for regions where the conductivity structure of the earth can be approximated by plane layers. Both methods use the same source fields, which are generated in the

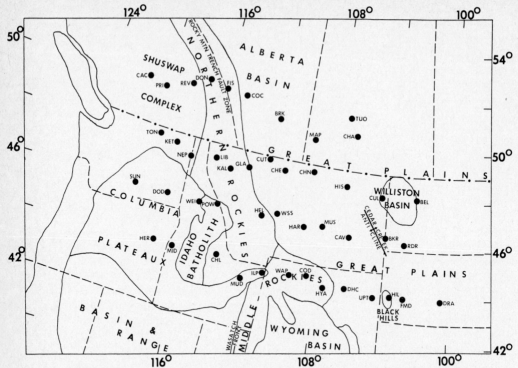

Fig. 2a. Simplified tectonic map of northwestern United States and southwestern Canada. Magnetograms for the stations for the two northern lines are shown in Figure 2b.

12 AUGUST 1969

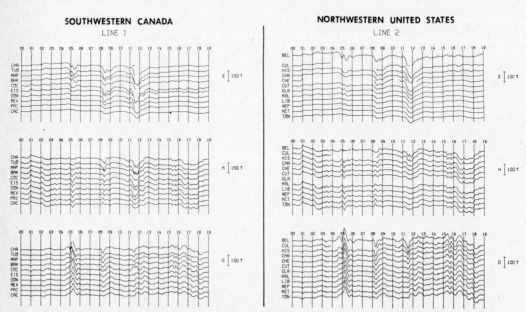

Fig. 2b.   Magnetograms for a magnetic disturbance of August 12, 1969, showing attenuation of the vertical field at stations west of the northern Rockies front ranges. Stations are plotted from east to west; Line 1 is in southwestern Canada and Line 2 is in the northwestern United States [after *Camfield et al.*, 1971].

TABLE 1a. Resistivity Models for the Northwestern United States and Southwest Canada

Crustal Model*				Upper Mantle Model†			
West		East		West		East	
Layer Thickness, km	Resistivity, Ω-m	Layer Thickness, km	Resistivity, Ω-m	Layer Thickness, km	Resistivity, Ω-m	Layer Thickness, km	Resistivity, Ω-m
15	1000	35	1000	15	200	2	5
30	10	∞	40	155	1000	348	1000
∞	40			∞	5	∞	5

* From *Caner* [1970]; reciprocal wavelength $K = 10^{-8}$ cm^{-1}.
† From *Porath et al.* [1971]; $K = 10^{-8}$ cm^{-1}.

ionosphere and magnetosphere and are usually of global dimensions.

The magnetotelluric deep sounding method is based on the observations of orthogonal components of the horizontal electric (E) and magnetic (H) variation fields. The apparent resistivity E/H and the phase angle between E and H are plotted as functions of frequency. The resulting curves are interpreted by matching them with curves obtained from numerical solutions of plane layered conductivity models [*Cagniard,* 1953]. The incident field is usually assumed to be uniform, but model curves can also be computed for sources of finite dimensions [*Srivastava,* 1965].

The most important assumption in magnetotelluric studies is, however, the uniformity of the induced currents. It has been shown in the preceding section and in *Porath and Dziewonski* [1971b] that the distributions of currents used in geomagnetic deep sounding and magnetotelluric studies are very complex, owing to the fact that the conductivity of the earth's upper crust is highly nonuniform. Currents induced in the oceans or in the upper continental crust are channeled into conductive sedimentary structures or into water channels between two resistive continental blocks.

Apparent-resistivity curves derived from magnetotelluric measurements made in the vicinity of anomalous current systems should no longer be interpreted by the plane layered approximation. Two examples of apparent-resistivity curves in the neighborhood of current concentrations in sea channels running between two more resistive continental blocks are shown by *Porath and Dziewonski* [1971b, Figure 2]. Figure 11a shows results from a magnetotelluric station on the Island of Rügen in the Baltic Sea [*Porstendorfer,* 1965], and Figure 11b shows results from a station on Ellesmere Island in the Canadian Arctic. The latter station was situated about 100 km from the Robeson Channel [*Whitham and Andersen,* 1965], which concentrates currents induced in the Arctic Sea and North Atlantic to give rise to the Alert anomaly in geomagnetic

TABLE 1b. Attenuation of Normal Z (Z/H ratio) for the Northwestern United States and Southwest Canada*

	Crustal Model				Upper Mantle Model			
	West		East		West		East	
Period, min	Observed Attenuation	Theoretical Attenuation	Observed Attenuation	Theoretical Attenuation	Observed Attenuation	Theoretical Attenuation	Observed Attenuation	Theoretical Attenuation
47.6		0.09		0.14	0.18	0.20	0.32	0.31
102.4		0.13		0.19	0.24	0.22	0.34	0.35

* Theoretical values of attenuation are calculated for the models presented in Table 1a.

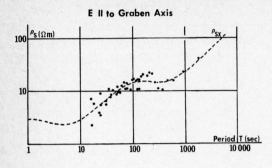

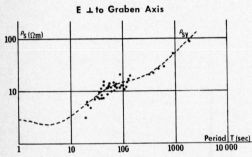

Fig. 3. Apparent-resistivity curves for a station in the Rhine graben [*Losecke, 1970*].

variations [*Porath and Dziewonski, 1971b*], and to an additional inland anomaly [*Niblett and Whitham, 1970*].

The effect of an anomalous conductor on magnetotelluric apparent-resistivity curves depends on the distance between the magnetotelluric station and the conductor and on the resistivity as a function of depth in the vicinity of the station. Anomalous features in apparent-resistivity curves shift to longer periods and become less pronounced as the distance between the anomalous conductor and the station increases. Lateral effects are particularly serious if the magnetotelluric station is located over a highly resistive crust, for then even short-period variations have considerable skin depths associated with them.

It is difficult to calculate numerically the influence of anomalous current systems on apparent-resistivity curves. Two-dimensional methods developed by *Madden and Swift* [1969], *Wright* [1969], *Jones and Price* [1970], and *Patrick and Bostick* [1969] are usually not applicable, since conductive structures channel currents that are induced elsewhere. Two-dimensional models used to approximate the amplitude of a geomagnetic-variation anomaly overestimate, therefore, the

conductivity of the structure producing the anomalous fields.

An example of resistivity curves near a two-dimensional conductive structure [after *Wright*, 1970] is shown in Figure 12 of *Porath and Dziewonski* [1971b]. Apparent-resistivity curves for electric and magnetic field vectors polarized along the structure are 'anisotropic' up to the longest periods and show small inflections between 50 and 100 sec. *Wright* [1970] notes that failure of the apparent-resistivity curves to converge at long periods is not an indication of an inhomogeneity at depth but can be explained by near-surface distortions of telluric currents.

Figure 3 shows apparent-resistivity curves recorded in the Rhine graben [*Losecke*, 1970] which are similar to those of Alert and Rügen. Current concentrations in the conducting sediments of the graben result in an enhanced east-west horizontal field over the center of the graben and a reversal of the vertical field between the western and eastern edges of the graben [*Winter*, 1967] and may be responsible for the observed inflections in the apparent-resistivity curves. In addition to concentrations of currents induced in the continental crust, currents induced in the Atlantic Ocean may be leaking across the conductive sediments of the Paris Basin [*Fournier*, 1969] into the Rhine graben sediments.

The apparent-resistivity curves from the island of Rügen and the Rhine graben were interpreted as indicative of a conductive layer in the lower crust [*Porstendorfer*, 1965; *Losecke*, 1970]. In these studies, the effects of the distribution of the induced current systems were neglected. Where the upper crustal layer is resistive, minimums occur in the apparent-resistivity curves over a range of frequencies at stations in the vicinity of anomalous current systems. This was observed by *Caner and Auld* [1968] at Victoria on Vancouver Island in British Columbia (Figure 4) where the apparent-resistivity curves show a minimum for a period of about 100 sec. This effect was attributed originally by Caner and Auld to a thin layer of lower resistivity in the upper mantle. Later, *Caner et al.* [1969] concluded that current concentrations in the nearby ocean offered a more likely interpretation of the results. The ocean effect is enhanced by channeling of the induced ocean currents through the Juan de Fuca Straits to within a close distance of the magnetotelluric station at Victoria.

Caner et al. [1969] proposed, however, from magnetotelluric measurements in Alberta and British Columbia, a conductive lower crust under the mountain and plateau regions of western Canada. The conductivity model suggested is shown in Figure 1, whose validity was discussed in the second section from the standpoint of geomagnetic deep studies. Apparent-resistivity curves from *Vozoff and Ellis* [1966] and *Caner et al.* [1969] are shown in Figure 5, together with the locations of the magnetotelluric stations. The apparent resistivities show a large scatter with considerable anisotropies, particularly for the stations at Grand Forks and Fernie. *Caner et al.* [1969] suggest, from the 'smoothed bands of mean ± standard deviations,' a zone of reduced resistivity in the lower crust beneath the western region. Within the error limits, however, the simplest interpretation is a smooth decrease of apparent resistivity toward longer periods for stations in the western region and almost constant resistivities over the period range of observations for the eastern stations Pincher and Fernie.

Caner et al. [1969] assume that magnetotelluric results from the stations Fernie and Pincher typify the electrical structure beneath the Great Plains province. This conclusion is tenuous, because as *Camfield et al.* [1971] show, using variometer data, these stations are in a zone of magnetic-variation anomalies that are probably related to changes in the conductivity of the upper mantle beneath the northern Rocky Mountains. In addition, they may be influenced by a variation anomaly due to the conductive sediments in the Alberta basin. Vulcan, a magnetotelluric station operated by *Vozoff and Ellis* [1966], is cited by *Caner et al.* [1969] in support of their model.

Vozoff and Ellis [1966] present results for the stations at Brooks and Beiseker (Figure 5a), which were located farther away from the zone of variation anomalies. The results for Brooks are shown in Figure 6. Their interpretation shows a conductive surface zone, an intermediate resistive zone, and an increase in conductivity at a depth of the order of 100 km in the upper mantle. Comparing the magnetotelluric results for Beiseker and Brooks and those for Penticton and Grand Forks indicates a more conductive upper mantle under the western region, in accordance with the geomagnetic deep sounding

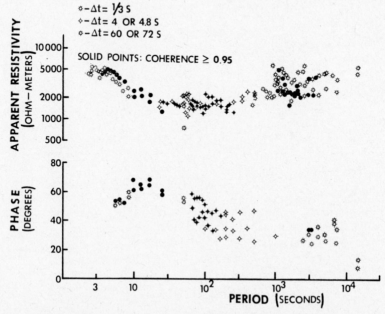

Fig. 4. Apparent resistivities and phase angles as a function of period for Victoria Observatory on Vancouver Island, Canada. *t* denotes the interval of digitization [after *Caner and Auld,* 1968].

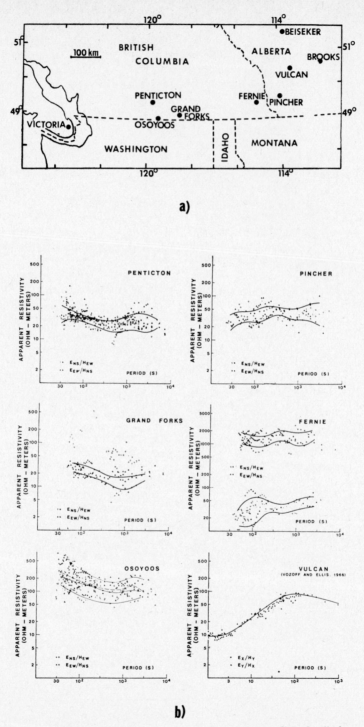

Fig. 5. Section *a* shows magnetotelluric stations in southwest Canada (index map); section *b* shows apparent resistivities versus period and is after *Vozoff and Ellis* [1966] and *Caner et al.* [1969].

results by *Camfield et al.* [1971], shown in Figure 2.

Other magnetotelluric evidence on conductive layers in the resistive zone of the earth's lower crust comes from the Great Plains province of the United States [*Dowling, 1970; Mitchell and Landisman, 1970; Word et al., 1970*] and from Siberia [*Berdichevskiy et al., 1969*].

Dowling operated sixteen magnetotelluric sites across the Wisconsin arch. Apparent-resistivity curves of five of those sites are shown in Figure 7 [*Dowling, 1970*]. They are strongly anisotropic over the entire period range. For three of the five sites, a distinct minimum appears in the apparent resistivity at periods between 40 and 120 sec; this pattern has been interpreted by Dowling as an indication that a moderately conductive basic lower crust underlies a resistive granitic crust.

However, the existence of lateral inhomogeneities in the distribution of crustal conductivities offers a more likely interpretation of these features of the apparent-resistivity curves. Figure 8 shows a simplified tectonic map of the United States with the principal sedimentary features in the Great Plains province and the eastern United States. It is seen that Dowling's magnetotelluric stations are in the vicinity of one

of the largest sedimentary basins in the United States, the Michigan basin, which measures several hundred kilometers in diameter and has a maximum depth of about 4 km. Apparent-resistivity curves show features that would be expected in the neighborhood of an anomalous conductor. Minimums in the apparent-resistivity curves shift to longer periods and become less pronounced as the distance from the sedimentary basin increases between sites 1 and 4. The sediments in the Michigan Basin, as in other Paleozoic sedimentary basins in the United States, have relatively high conductivities and represent a large resistivity contrast with the basement rocks of the Wisconsin arch. Currents induced in the upper crust will, therefore, concentrate in the basin. For long-period variations, additional currents induced in the oceans may reach the Michigan basin via the extremely conductive sediments of the coastal plains of the Gulf of Mexico, the Mississippi embayment, and the Illinois basin. Since the upper crust in Wisconsin is resistive, apparent-resistivity curves in the vicinity of anomalous currents in the Michigan Basin resemble those obtained by *Caner and Auld* [1968] on Vancouver Island.

Mitchell and Landisman [1971] have interpreted apparent-resistivity curves from two

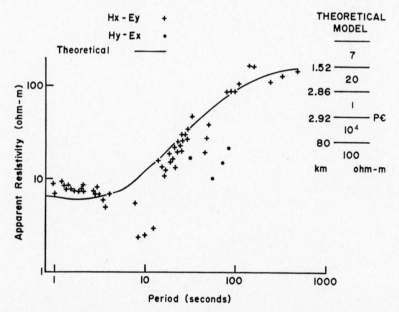

Fig. 6. Apparent resistivities and theoretical model for a magnetotelluric station at Brooks in Alberta, Canada [after *Vozoff and Ellis*, 1966].

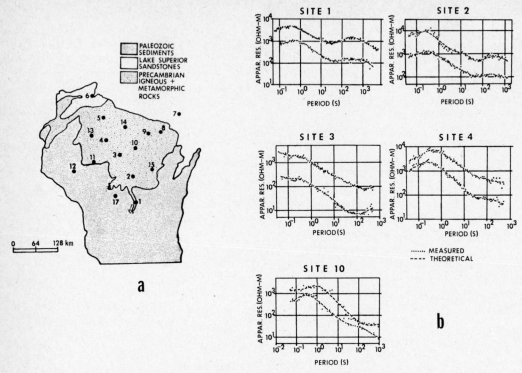

Fig. 7. Section *a* shows magnetotelluric stations in Wisconsin; section *b* shows apparent-resistivity curves for sites 1–4 and site 10 [after *Dowling*, 1970].

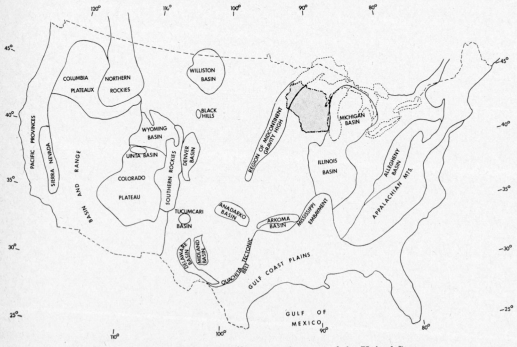

Fig. 8. Major tectonic and sedimentary features of the United States.

magnetotelluric profiles obtained by Mobil Oil Company that cover an area 30 km by 50 km in the southwestern Great Plains province. Their plane layered model includes a thin, anisotropic layer between depths of 21 and 23 km. The introduction of this layer is necessary to explain the anisotropy of the apparent-resistivity curves at longer periods by a plane layered model.

It is possible to present an alternate interpretation of these observations on the basis of the geological setting of the profiles. Two sedimentary structures, the Midland basin and the Delaware basin, are located to the southwest of the profiles and contain sections of sediments of resistivities as low as 5 Ω-m [Word et al., 1965]. Currents induced in the upper crust of the southwestern plains will be concentrated in these basins. In addition, current systems induced in the oceans may be leaking into the continental crust of the Great Plains via the conductive coastal plains sediments [Porath and Dziewonski, 1971b].

Apparent-resistivity curves similar to those reported by Mitchell and Landisman [1971] are obtained near lateral-conductivity inhomogeneities, which can be approximated by two-dimensional model structures. In Figure 9, the apparent-resistivity curves are shown at a distance of 120 km from a sedimentary structure that has roughly the dimensions of the Delaware and Midland basins for the electric field polarized parallel ($E_{\parallel}$) and perpendicular ($E_{\perp}$) to the model structure. In the model, the resistivity of the sediments is not reduced below 180 Ω-m, since lower resistivities, though more realistic, would require a smaller grid size for the numerical computations [Wright, 1969] and result in the expenditure of a prohibitive amount of computer time. The divergence between apparent-resistivity curves ($E_{\perp}$ and $E_{\parallel}$) for the model increases with period, and the resistivity curves do not converge again at long periods. Anisotropies in apparent-resistivity curves at long periods are, therefore, not necessarily an indication of an anisotropic layer at depth but are, more simply, an expression of conductivity inhomogeneities in the upper crust. Because of the size of the conductive structures, similar resistivity curves are obtained over a range of distances from the conductor. This explains the comparatively uniform features in the apparent-resistivity curves observed by Mitchell and Landisman [1971] over several tens of kilometers.

Sedimentary resistivities are usually several times less than those used in the model. Crustal conductivity inhomogeneities of much smaller dimensions than those in the model can, therefore, produce features in the apparent-resistivity curves similar to those observed by Mitchell and Landisman [1971], since currents induced elsewhere are channeled into such conductive structures.

A thin conductive layer in the earth's crust has been reported by Word et al. [1970, 1971]

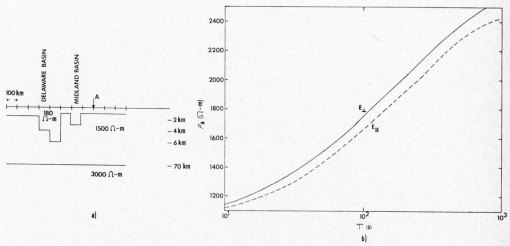

Fig. 9. Two-dimensional conductivity model and apparent-resistivity curves for station A 120 km from the right edge of the structure.

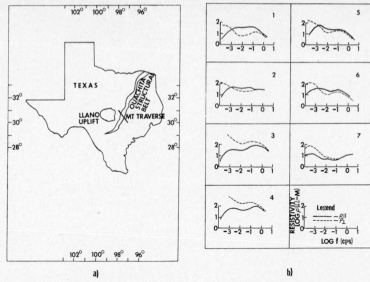

a)   b)

Fig. 10.   Apparent resistivities for stations across the Ouachita tectonic belt along the profile
shown in Figure 9a [after *Word et al.*, 1970].

for several stations along a 100-km magnetotel-luric traverse crossing the Ouachita tectonic belt near the Llano uplift in Texas. In their inter-pretation, the conductive layer is located at the interface between sediments and granitic base-ment rocks; this location has been attributed by the authors to high conductivities associated with the weathered and water-saturated base-ment rocks in the interface zone. Their results are, therefore, not an example of a conductive zone in the earth's crystalline crust.

Tensor-impendance curves determined by

Word et al. show a considerable anisotropy over most of the frequency band of observations. Rather pronounced minimums are observed in the period range from 50 to 100 sec (Figure 10). Two-dimensional conductivity inhomogeneities usually distort apparent resistivity curves more severely when the electric field vector is po-larized normal to the long axis of the structure [*Patrick and Bostick*, 1969]. *Word et al.* [1970] have, therefore, interpreted apparent resistivities corresponding to $E_{\parallel}$ by using plane layered mod-els. However, it has been shown by *Wright*

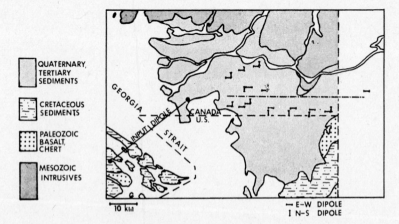

Fig. 11a.   Simplified geological map of the Fraser Valley showing the location of the current
dipole and the voltage dipoles.

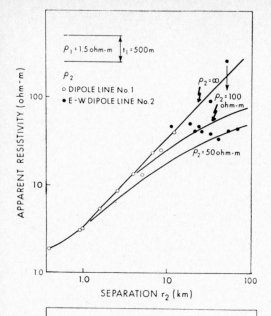

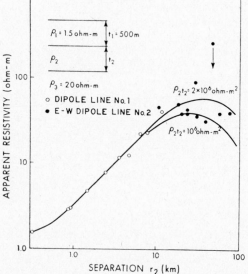

Fig. 11b. Apparent resistivities together with two- and three-layer model curves [after *Samson*, 1969].

[1970] for two-dimensional models that this approximation is valid only for very long periods.

The one-dimensional model interpretation included a conductive layer of 1 Ω-m resistivity and a thickness of several hundred meters on top of a resistive basement. It has been shown by *Porath and Dziewonski* [1971b], however, that the Ouachita tectonic belt is a region of

intense anomalous currents induced in the ocean and channeled into the sedimentary cover of the Great Plains province. The anomalous features in the apparent-resistivity curves, particularly at long periods, may, therefore, be explained by the effects of lateral inhomogeneities.

Magnetotelluric apparent-resistivity curves resembling some of those found by Dowling in Wisconsin have also been observed by *Berdichevskiy et al.* [1969] in the Lena River valley in Siberia. Again, the resistivity curves exhibit the strong anisotropy indicative of lateral conductivity inhomogeneities. The magnetotelluric stations are in a region of extreme changes in sedimentary thickness. In addition, currents induced in the Arctic Ocean may be channeled by the conductive fluvial sediments of the Lena River valley.

It can be concluded, therefore, that magnetotelluric measurements interpreted in terms of conductive layers in the earth's crystalline crust can be more simply explained by lateral variations in the conductivity of the upper crust. It has been shown that magnetotelluric results can be attributed to a heterogeneous and, in places, highly conductive upper crust and a resistive crystalline crust. The basic difficulty in the interpretation of magnetotelluric measurements they are suitable for detailed investigations of the conductivity structure of the crystalline lineating depths to resistive basement rocks and large-scale regional differences in the electrical structure of the earth, it is doubtful whether such as oceans and conductive sediments, result in highly nonuniform induced currents. Although magnetotelluric studies may be useful in de- results from the assumption that the globally current electrodes; separations of more than induced current system is uniform. Large-scale conductivity inhomogeneities of the upper crust, crust.

EVIDENCE ON CONDUCTIVE LAYERS IN THE CRYSTALLINE CRUST FROM DIRECT-CURRENT SURVEYS

Direct-current (dc) electrical deep sounding studies and similar methods based on a controlled source field are best suited to study the conductivity structure of the crust. Lateral-conductivity inhomogeneities still affect dc apparent-resistivity curves, especially those determined by the dipole-dipole method [*Keller*,

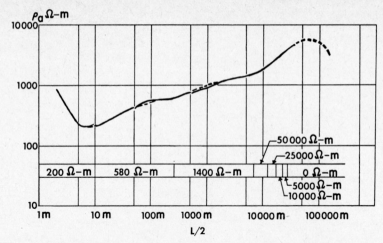

Fig. 12. Experimental (continuous lines) and theoretical (broken line) apparent-resistivity curves for a deep Schlumberger sounding in the Mortagne massif, western France [after *Migaux et al.*, 1960]. *L* is the electrode spacing.

1966]. However, in contrast to geomagnetic and magnetotelluric deep sounding, apparent resistivities from dc deep soundings depend only on the distribution of the local electric fields and not on currents generated elsewhere.

A study of the conductivity properties of the earth's lower crust by dc electrical soundings has been attempted in only a limited number of cases. The effective depth of penetration in these methods depends on the separation between the several kilometers provide considerable technical difficulties. Results from four deep soundings by dc methods will be reviewed; two of these soundings indicate a resistivity inversion in the crust.

The apparent resistivities shown in Figure 11 were obtained by *Samson* [1969] from dipole-dipole soundings in the Fraser valley in British Columbia, southwest Canada. The observations indicate that for large separations between the current dipole and the measurement dipole, apparent resistivities tend to level off. This pattern has been interpreted by Samson as evidence for a decrease in resistivity at depths between 7 and 8 km.

It is well known that dipole-dipole deep soundings are highly sensitive to lateral conductivity inhomogeneities [*Keller*, 1966]. This effect appears to account for the variations in the apparent-resistivity curves measured in the Fraser valley. The dipole array is expanded along a line roughly parallel to the contact be-

tween conducting sediments in the Fraser valley and resistive Mesozoic intrusives to the north. It has been shown by *Keller* [1966, Fig. 12] that in such a situation the lateral resistivity change between sediments and igneous rocks in the surface layer may be confused with a conducting layer at depth.

Schlumberger deep soundings with electrode spacings of up to 70 km in a granite massif, made in western France by *Migaux et al.* [1960], have been reported as indicating a sharp drop in the resistivity of the lower crust [*Keller et al.*, 1966]. This conclusion is not based on real data but, rather, on a hypothetical resistivity model for the lower crust which was used by *Migaux*

TABLE 2. Summary of Interpretation of Deep Schlumberger Sounding on the Namaqualand Complex, South Africa [after *van Zijl*, 1969]

Layer No.	Resistivity (ρ), Ω m	Thickness (h), km	Max. Depth, km
1	30	0.012	
			0.012
2	255	0.168	
			0.18
3	15,000	3.5	
			3.7
4	3000	17	
			21
5	37,000	30	
			51
6	0	∞	

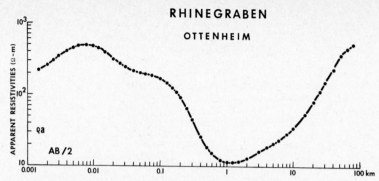

Fig. 13. Apparent resistivities for a deep Schlumberger sounding in the Rhine graben [after *Blohm and Flathe*, 1970]; *AB* is the electrode spacing.

et al. [1960] to extrapolate the apparent-resistivity curves beyond the range of observations (see Figure 12). Observed apparent resistivities increase less steeply for the last part of the curve; this pattern can be attributed [*van Zijl*, 1969] to the fact that for large spacings the electrodes are in the vicinity of more conductive sediments which surround the comparatively small granite massif.

Electrode spacings of up to 270 km were used by *van Zijl* [1969] in Schlumberger deep soundings in South Africa at a site carefully selected to avoid lateral effects. The soundings were made over a granite-gneiss complex which is several hundred kilometers long and 150 km wide. A summary of van Zijl's interpretation is shown in Table 2. Beneath two thin layers of moderate resistivity due to weathered rocks, a three-layered highly resistive crust is observed with the 17-km thick middle layer of lowered resistivity.

Van Zijl suggests that the difference in resistivity between the upper two resistive layers is due to a difference in age between them. A highly resistive layer of the younger granite-gneiss complex possibly overlies an ancient crust, which may have retained more water than the highly metamorphic top layer. This is a reasonable interpretation, for all three crystalline layers fall within the range of resistivities which is normal for the intermediate zone of the crust.

A correlation between zones of low seismic velocity in the crust and enhanced electrical conductivities has been suggested by *Berdichevskiy et al.* [1969] and by *Mitchell and Landisman* [1971]. However, the fourth dc deep sounding survey to be discussed here seems to

deny this correlation. This survey was conducted by *Blohm and Flathe* [1970] in the Rhine graben in southwest Germany. Seismic data suggest velocity inversions in the crust [*Mueller et al.*, 1967], unlike the resistivity results.

A Schlumberger electrode arrangement was also used in this experiment, and apparent resistivities were determined for spacings of up to 150 km between electrodes. The profile followed roughly the strike of the Rhine graben. The apparent-resistivity curve observed for the potential electrodes in the Rhine graben is shown in Figure 13. After a minimum observed for electrode separations of 2 km, which has been attributed by *Blohm and Flathe* [1970] to the effect of Rhine graben sediments, the apparent-resistivity curves increase smoothly up to the largest electrode separations with no evidence of resistivity inversions at depth.

SUMMARY

It has been shown that in all the reported cases of conductive layers in the crystalline crust there exists an alternate explanation related to the effects of lateral conductivity inhomogeneities. These effects are particularly significant in magnetotelluric studies because of the complex distribution of the globally induced current system. It appears that the methods based on a controlled field source are the most promising for a detailed investigation of the conductivity structure of the earth's crust. Direct current electrical resistivity soundings represent a method of this type that is presently available. Other electromagnetic techniques should be improved and developed.

Acknowledgments. I am grateful to Dr. A. L. Hales for advice and encouragement during the progress of this work. Dr. A. Dziewonski suggested several improvements in the presentation. Dr. J. A. Wright generously supplied the computer program used for two-dimensional model calculations. Mr. D. Mosier provided assistance with programming, and Miss P. Patrick helped in preparing the drawings.

It is a particular pleasure to acknowledge the generosity of the Sun Oil Company which made available to me free computer time. Financial support was provided by the National Science Foundation grant GA-18265 and by the Office of Naval Research under contract N00014-67-A-0310-0005 (formerly NONR-4455601).

REFERENCES

Berdichevskiy, M. N., V. P. Borisova, V. P. Bubnov, L. L. Van'yan, I. S. Feldman, and I. A. Yakovlev, Anomaly of the electrical conductivity of the earth's crust in Yakutiya, *Izv. Nauk SSSR, Phys. Solid Earth, 10*, 633–637, 1969.

Blohm, E. K., and H. Flathe, Geoelectric deep sounding in the Rhine graben, paper presented at Int. Rift Symp., Karlsruhe, *Int. Upper Mantle Proj. Rep. 27*, 239–241, 1970.

Cagniard, L., Basic theory of the magnetotelluric method of geophysical prospecting, *Geophysics, 18*, 605–635, 1953.

Camfield, P. A., D. I. Gough, and H. Porath, Magnetometer array studies in the northwestern United States and Southwest Canada, *Geophys. J. Roy. Astron. Soc., 22*, 201–221, 1971.

Caner, B., Electrical conductivity structure in western Canada and petrological interpretation, *J. Geomagn. Geoelec.*, in press, 1970.

Caner, B., and D. R. Auld, Magnetotelluric determinations of upper mantle conductivity structure at Victoria, British Columbia, *Can. J. Earth Sci.*, 1209–1220, 1968.

Caner, B., W. H. Cannon, and C. E. Livingstone, Geomagnetic depth sounding and upper mantle structure in the Cordillera region of western North America, *J. Geophys. Res., 72*, 6335–6351, 1967.

Caner, B., P. A. Camfield, F. Andersen, and E. R. Niblett, A large-scale magnetotelluric survey in western Canada, *Can. J. Earth Sci., 6*, 1245–1261, 1969.

Cantwell, T., Detection and analysis of low frequency electromagnetic signals, Ph.D. thesis, Dep. Geology and Geophysics, MIT, Cambridge, Mass., 1960.

Cantwell, T., P. Nelson, J. Webb, and A. S. Orange, Deep resistivity measurements in the Pacific Northwest, *J. Geophys. Res., 70*, 1931–1937, 1965.

Dowling, F. L., Magnetotelluric measurements across the Wisconsin arch, *J. Geophys. Res., 75*, 2683–2698, 1970.

Fournier, H. G., Essai d'une coupe magnetotellurique du manteau superieur selon le rayon sud-est du bassin de Paris, (Abstract) *IASPEI-IAGA Assembly Program*, p. 201, Madrid, 1969.

Gough, D. I., and H. Porath, Long-lived thermal structure under the Southern Rocky Mountains, *Nature, 226*, 837–839, 1970.

Hyndman, R. D., and D. W. Hyndman, Water saturation and high electrical conductivity in the lower continental crust, *Earth Planet. Sci. Lett., 4*, 427–432, 1968.

Jones, F. W., and A. T. Price, The perturbation of alternating geomagnetic fields by conductivity anomalies, *Geophys. J. Roy. Astron. Soc., 20*, 317–334, 1970.

Keller, G. V., Dipole method for deep resistivity studies, *Geophysics, 31*, 1088–1104, 1966.

Keller, G. V., L. A. Anderson, and J. I. Pritchard, Geological survey investigations of the electrical properties of the crust and upper mantle, *Geophysics, 31*, 1078–1087, 1966.

Lahiri, B. N., and A. T. Price, Electromagnetic indication in non-uniform conductors and the determination of the conductivity of the Earth from terrestrial magnetic variations, *Phil. Trans. Roy. Soc. London A, 237*, 509–540, 1939.

Losecke, W., Ergebnisse magnetotellurischer Messungen bei Speyer, Proc. Int. Rift Symposium, Karlsruhe, *Int. Upper Mantle Prof. Rep. 27*, 242–243, 1970.

Madden, T. R., and C. M. Swift, Magnetotelluric studies of the electrical conductivity structure of the crust and upper mantle, in *The Earth's Crust and Upper Mantle, Geophys. Monogr. Ser.*, vol. 13, edited by P. J. Hart, pp. 469–479, AGU, Washington, D. C., 1969.

Migaux, L., J. L. Astier, and Ph. Revol, Un essai de détermination expérimentale de la résistivité electrique de couches profondes de l'écorce terrestre, *Ann. Geophys., 16*, 555–560, 1960.

Mitchell, B., and M. Landisman, Electrical and seismic properties of the Earth's crust in the southwestern Great Plains, *Geophysics, 36*, 363–381, 1971.

Mueller, S., E. Peterschmitt, K. Fuchs, and J. Ansorge, The Rift structure of the crust and upper mantle beneath the Rhinegraben, in The Rhinegraben Progress Report, *Abh. Geol. Landesanst., 6*, 109–113, 1967.

Niblett, E. R., and K. Whitham, Multi-disciplinary studies of geomagnetic variations anomalies in the Canadian Arctic, *J. Geomagn. Geoelec., 22*, 99–111, 1970.

Oldenburg, D. W., Separation of magnetic substorm fields for mantle conductivity studies in the western United States, M.S. thesis, University of Alberta, Edmonton, Canada, 1969.

Patrick, F. W., and F. X. Bostick, Magnetotelluric modeling techniques, *Tech. Rep. 59*, Electron. Geophys. Res. Lab., Univ. of Texas, Austin, 1969.

Porath, H., Magnetic variation anomalies and seismic low velocity zone in the western United States, *J. Geophys. Res., 76*, 2643–2648, 1971.

Porath, H., and A. Dziewonski, Crustal electrical

conductivity anomalies in the Great Plains Province of the United States, *Geophysics, 36,* 382–395, 1971a.

Porath, H., and A. Dziewonski, Crustal resistivity anomalies from geomagnetic deep sounding studies, submitted to *Rev. Geophys. Space Phys.,* 1971b.

Porath, H., and D. I. Gough, Mantle conductive structures in the western United States from magnetometer array studies, *Geophys. J. Res. Roy. Astron. Soc., 22,* 261–275, 1971.

Porath, H., D. I. Gough, and P. A. Camfield, Conductive structures in the northwestern U.S. and southwest Canada, submitted to *Geophys. J. Res. Roy. Astron. Soc.,* 1971.

Porstendorfer, G., Methodische und apparative Entwicklung magnetotellurischer Verfahren mit Anwendung auf die Tiefenerkundung im Bereich der norddeutschen Leitfähigkeitsanomalie, *Deut. Akad. Wiss. Berlin, Heft 3,* in German, p. 76, 1965.

Rikitake, T., Electromagnetism and the Earth's interior, 308 pp., Elsevier, Amsterdam, 1966.

Samson, J. C., Deep resistivity measurements in the Fraser Valley, British Columbia, *Can. J. Earth Sci., 6,* 1129–1136, 1969.

Schmucker, U., Anomalies of geomagnetic variations in the southwestern United States, *Monograph,* Scripps Inst. Oceanography, 1970.

Srivastava, S. P., Method of interpretation of magnetotelluric data when source field is considered. *J. Geophys. Res., 70,* 945–954, 1965.

Swift, C. M., A magnetotelluric investigation of an electrical conductivity anomaly in the southwestern U.S., Ph.D. thesis, Dep. Geology and Geophysics, MIT, Cambridge, Mass., 1967.

Tozer, D. C., The electrical properties of the earth's interior, pp. 414–436, in *Physics and Chemistry of the Earth,* vol. 3, edited by L. H. Ahrens, F. Press, K. Rankama, and S. K. Runcorn, Pergamon, New York, 1959.

Van Zijl, J. S. V., A deep Schlumberger sounding to investigate the electrical structure of the crust and upper mantle in South Africa, *Geophysics, 34,* 450–462, 1969.

Vozoff, K., and R. M. Ellis, Magnetotelluric measurements in southern Alberta, *Geophysics, 31,* 1153–1157, 1966.

Whitham, K., and F. Andersen, Magnetotelluric experiments in northern Ellesmere Island, *Geophys. J. Roy. Astron. Soc., 3,* 317–345, 1965.

Winter, R., Geomagnetic deep sounding at the Rhinegraben, in The Rhinegraben Progress Report, *Abh. Geol. Landesanst., 6,* 127–130, 1967.

Word, D. R., G. Hopkins, and F. W. Patrick, Subsurface resistivity sounding in the Delaware Basin, *Rep. 137,* Electron. Eng. Res. Lab., Univ. of Texas, 22 pp., Dallas, 1965.

Word, D. R., H. W. Smith, and F. X. Bostick, An investigation of the magnetotelluric tensor impedance method, *Tech. Rep. 82,* Electron. Geophys. Res. Lab., Univ. of Texas, Dallas, 1970.

Word, D. R., H. W. Smith, and F. X. Bostick, Crustal investigations by the magnetotelluric tensor impedance method, in *The Structure and Physical Properties of the Earth's Crust, Geophys. Monogr. Ser.,* vol. 14, edited by J. G. Heacock, this volume, AGU, Washington, D. C., 1971.

Wright, J. A., The magnetotelluric and geomagnetic response of two-dimensional structures, GAMMA 7, *Inst. Geophys. Metaalinst. Techn.,* Univ. Braunschweig, Germany, p. 102, 1969.

Wright, J. A., Anisotropic apparent resistivities arising from nonhomogeneous, two-dimensional structures, *Can. J. Earth Sci., 7,* 527–531, 1970.

DISCUSSION

Landisman: We really have to go into the question of a wet versus a dry crust. The record sections from active tectonic provinces such as the western United States, western Colorado, and the Rockies showed extremely large amplitudes in the vicinity of the critical point compared with those we have just seen for the mid-continent. The two sets of data from Colorado had very similar velocity profiles. This may be taken as evidence for an inversion just above the refracting horizon. It could be an indication of higher temperature. However, it is unlikely that melting has occurred at these shallow depths, and I suggest that water is responsible. There is a remarkable correlation between earthquake epicenters and hot springs in North America, and in Japan, at Matsushiro, the earthquake foci moved up from depths of 10 km and the ground was observed to swell. When the foci became shallower, there was an outpouring of 10^7 m^3 of hot saline water, and at that point the activity moved away from the area.

H. W. Smith: These lateral inhomogeneities can create a lot of disturbance in the interpretation of magnetotelluric data. We wanted to show that, if you really have two-dimensionality, the E-parallel case gives you a good account of resistivity versus depth. If we invert the H-parallel case for the particular slide you showed, it would show extremely high resistivity to great depths. This is why we are going to have to be very careful on this interpretation. I am afraid some of the other magnetotelluric data you reviewed were not very good.

Porath: I agree. I do not want to rule out the possibility of a conducting layer in the crystalline crust, just to present the available evidence considering the geological setting of the measuring stations.

Crustal Investigations by the Magnetotelluric Tensor Impedance Method

D. R. WORD, H. W. SMITH, AND F. X. BOSTICK, JR.

Electrical Engineering Department, The University of Texas at Austin
Austin, Texas 78712

Abstract. A magnetotelluric (MT) sounding method that involves a tensor impedance relationship between the surface electric and magnetic micropulsation fields of the earth has been developed. The role of the tensor impedance, including the vertical magnetic field relationship, is discussed. MT soundings made at 7 sites along a sixty-mile traverse east of Austin, Texas, in the frequency band 10^{-4} to 10 Hz are analyzed. The traverse bears northwest-southeast and crosses the Ouachita fold belt system on the flanks of the Llano uplift. The sounding covers the approximate depth range of 0.1 to 100 km. An earth model is derived from the sounding and includes a vertical cross section of the estimated conductivity distribution and estimates of the strike directions of the various formations represented by the cross section. The model is compared with available geological information, and an attempt is made to explain some of the observed three-dimensional properties of the impedance function. The model is in good agreement with expected geology to depths for which surface and borehole data can be reasonably extrapolated. The sounding responds to materials ranging from Cretaceous sediments through the upper mantle and provides reasonable definition of a high-resistivity region which appears to be the Precambrian granite basement. Along the traverse, the top of this basement occurs at depths ranging from 3 to 10 km. In the upper mantle, a resistivity decrease to about 1 Ω-m or less is found to occur at depths of about 60 to 100 km. An apparent northwesterly rise in the conductive substrate is found. This and other evidence seems to suggest an uplift in the conductive mantle in association with the Llano uplift.

Southeast of the Llano uplift in central Texas, the Precambrian granite dips sharply toward the Gulf coast beneath a thickening overburden of sedimentary rocks. The surface of the granite basement becomes indeterminable within a few tens of miles from its outcrop in Llano County. An MT survey along a traverse extending from near a point of outcrop southeastward toward the coast was performed for two purposes.

The first purpose was to extend the known depths to the granite basement beyond the boundaries within which it was defined by borehole penetrations. It was reasoned that the Precambrian material would, on the average, be more resistive than the sedimentary rocks, so that a sharp contrast in the electrical resistivity would occur at the base of the sediments. Such an interface, as well as other significant conductivity features both within and below the sediments, should be defined by the MT survey.

The second purpose of the study was to apply and evaluate the tensor-impedance method

of MT analysis to data acquired over reasonably well known inhomogeneous geologic structures. The Ouachita fold belt and the overlying sediments provide such structures in the sedimentary rocks beneath the survey path. The Ouachita system is a major Paleozoic orogenic belt of severely folded and faulted Paleozoic rocks that has several borehole penetrations in the frontal zone near the Llano uplift. The inhomogeneity of the lithology in this system leads one to suspect corresponding inhomogeneities in the electrical conductivity. The Ouachita system is overlain by Cretaceous and younger sediments within which the lithological units are fairly planar and dip southeastward. Electric well logs indicate a corresponding tendency toward stratification of the electrical conductivity.

In his original discussion of the magnetotelluric method of subsurface resistivity sounding, *Cagniard* [1953] assumed that the conductive earth is horizontally stratified and is illuminated by plane electromagnetic waves propagating vertically downward. The frequent failure of this model to explain the measured data and to produce repeatable results has brought about two main criticisms. The first of these was directed at the source assumption, and the second was directed at the one-dimensional earth model.

Regarding the source, *Wait* [1954], *Price* [1962], and *Srivastava* [1965] discussed the effects of horizontal variations in the primary fields due to limited spatial dimensions of the source. They pointed out that field variations must be small over a lateral distance comparable to a skin depth in the conductive earth at the frequency of concern. The degree of source effect is still in question and possibly is considerable at times and in some situations, especially at high latitudes where the source amplitude for some frequency bands is usually much larger and peaked near the auroral zone [see, for example, *McNish*, 1964; *Heirtzler*, 1964], where large lateral gradients might become significant. However, empirical measurements and coherence analyses of micropulsation signals at widely separated stations [*Bloomquist et al.*, 1967; *Orange and Bostick*, 1965; *Swift*, 1967], as well as theoretical source studies [*Prince et al.*, 1964], lend credence to the assumption of large source dimensions for low and middle latitudes, often of continental extent or greater. Based on this evidence, the effects of finite source dimensions were ignored in this study. This omission is undoubtedly responsible for some of the mild scatter in the processed MT data.

Of more direct concern here is the question of a multidimensional earth model. The scalar impedance described by Cagniard for his one-dimensional earth model is inadequate for the interpretation of lateral inhomogeneities or anisotropies in the conductive earth. The result of these spatial irregularities is that the electric and magnetic fields at the earth's surface are no longer orthogonal. *Cantwell* [1960] suggested that the tangential fields can be related in general by a two-dimensional tensor quantity that relates each electric-field component to the total tangential magnetic field. *Bostick and Smith* [1962] also recognized the tensor nature of the surface impedance and demonstrated the behavior with axis rotation; they pointed out means for determining the apparent strike of an earth anomaly from the principal direction of the tensor.

A considerable amount of experimental MT evidence has been reported since the Cagniard theory was first proposed [*Bostick and Smith*, 1961; *Bryunelli*, 1964; *Cantwell and Madden*, 1960; *Fournier*, 1963; *Hopkins and Smith*, 1966; *Parkinson*, 1962; *Pokityanski*, 1961; *Pospeyev*, 1965; *Spitznogle*, 1966; *Srivastava et al.*, 1963; *Swift*, 1967; *Vozoff et al.*, 1963, 1964]; however, many of the questions and problems involved, particularly with interpretive methods, still remain. To date, there has been reported only limited experimental evidence using the tensor impedance method.

Interpretation of the MT tensor impedance, once it has been measured, needs further study. Analytical solutions for three-dimensional geometries are almost non-existent and unobtainable with present methods. Analog solutions are feasible, but they are usually cumbersome and are seldom attempted. Consequently, interpretation has been limited to two-dimensional geometries. Approximate analytical solutions have been obtained for a few simple two-dimensional models such as the vertical plane contact or the dike [*d'Erceville and Kunetz*, 1962; *Rankin*, 1962; *Weaver*, 1963] and the small-amplitude sinusoidal horizontal interface [*Mann*, 1964]. Numerical solutions can be generated for almost any two-dimensional model [*Patrick*, 1969], but there is an obvious need for improvement in interpretive techniques for three-dimensional geometries. Three dimensionality in earth structure is readily identified in the measured impedance [*Swift*, 1967; *Sims*, 1969], but thus far the resulting data have only been rejected and not interpreted.

The vertical magnetic field is of considerable importance. Its linear relationship with tangential fields has been known for some time [*Bryunelli*, 1964; *Parkinson*, 1962] as well as its potential for determining strike direction in the anomaly, especially in connection with the seacoast effect [*Weaver*, 1963]. However, it has not been used to any extent in interpreting MT soundings.

THEORY

For the analytical model used here in the interpretation of the MT data, the conductive earth is assumed to occupy the half space $z \geq 0$, as shown in Figure 1*b*. Unless otherwise noted, the positive sense of each of the orthogonal lateral coordinates x and y is north and east, respectively. The half space $z < 0$ is considered to be free space. Electromagnetic excitation is assumed to be provided by plane waves propagating vertically downward; this assumption eliminates the vertical components of the electric and magnetic fields from the primary waves. Secondary fields caused by reflections from lateral variations in the conductivity of the earth are the only source of the vertical components of the electromagnetic fields.

The MT analysis involves the estimation of various relationships between the Fourier resolved electromagnetic field components measured at the earth's surface. The vertical and two orthogonal horizontal components of the magnetic field plus two orthogonal horizontal components of the electric field constitute the set of field components measured at each site for this study. The vertical field in the air space above the earth was not measured, because it is strongly affected by static atmospheric electricity at the relatively low MT frequencies, nor was an attempt made to measure the vertical field just within the surface of the conductive earth. At the low frequencies involved in the MT study, displacement currents within the conductive earth may be neglected in favor of the conduction currents. Consequently, there is very little vertical component of current density and thus of vertical electric field within the conductive earth just beneath the surface. Since all the measured components of the electromagnetic field are continuous at the earth's surface, it is necessary to measure them only near the interface on either side.

The frequency domain relations between the Fourier resolved tangential components of the surface electromagnetic fields are written as

$$[E] = [Z][H] \qquad (1)$$

where

$$[E] = \begin{bmatrix} E_x \\ E_y \end{bmatrix}$$

and

$$[H] = \begin{bmatrix} H_x \\ H_y \end{bmatrix}$$

are the column vectors representing the tangen-

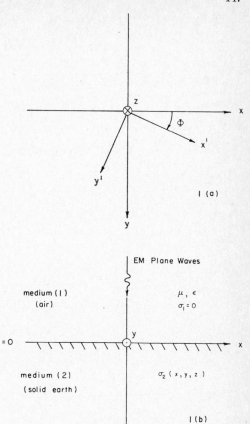

Fig. 1. Coordinate system and earth model.

tial electric and magnetic field vectors, respectively, and

$$[Z] = \begin{bmatrix} Z_{xx} & Z_{xy} \\ Z_{yx} & Z_{yy} \end{bmatrix}$$

is the rank two tensor impedance.

It is assumed that, if (1) were derived for some arbitrary earth model, the impedance elements would be a function of position, the source field configuration and, of course, the frequency. For an example of the source dependence, see *Wait* [1954], in which the nature of a scalar impedance estimate that is derived for a homogeneous isotropic earth model with a complex source distribution is examined. It is further assumed here that, since the source fields are geomagnetic micropulsations and since the spatial configuration of these fields varies with time, the dependence of the impedance elements on the source is reflected as a time variation of the elements themselves. The experimental estimates of the impedance elements obtained along the central Texas traverse do vary from one data sample to another, but this variation is relatively small except at the very lowest recorded frequencies. It is assumed that this temporal stability of the estimates indicates that they are only a slight function of the configuration of the source fields having the spatial extent of the geomagnetic micropulsations. This is not to say that the total micropulsation fields are uniform over any

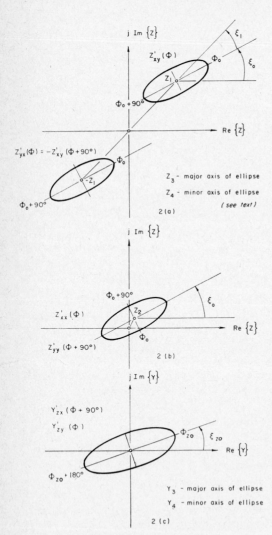

Fig. 2. Loci of impedance and admittance elements in the complex plane with respect to rotation angle Φ.

large area. On the contrary, the total fields, including both the relatively uniform source or incident fields and the possibly complex reflected fields, may have rather large spatial derivatives producing the functional dependence of the impedance estimates on position. For the remainder of this paper, the dependence of the impedance elements on the character of the source is suppressed.

Subject to the same considerations concerning the source, the vertical component of the magnetic field, H_z, can be related to the tangential electric field components with the expression

$$[H_z] = [Y_z][E] \qquad (2)$$

where

$$[Y_z] = [Y_{zx} Y_{zy}]$$

is a tensor admittance function.

Both $[Z]$ and $[Y_z]$ are estimated from the measured field data and are used for the purpose of constructing an interpretational model. All six elements of the combined tensors $[Z]$ and $[Y_z]$ can be estimated from two independent time sequences of the recorded field components or, effectively, from two different polarizations of the source field [Sims, 1969].

Rotation of coordinates. If $[\eta]$ is the coordinate rotation matrix for a vector in the xy plane, then in the primed coordinates of Figure 1a, rotated clockwise by the angle Φ,

$$[Z'] = [\eta][Z][\eta]^{-1} \qquad (3)$$

The loci of the $[Z'(\Phi)]$ elements in the complex impedance plane as a function of rotation angle Φ are generally elliptical, as recognized by Sims [1969], who suggests the graphical representation of Figure 2, showing the impedance behavior with rotation. Here, the ellipses are the loci of the impedance element phasors. All ellipses for the $[Z']$ element have the same dimensions and orientation. The ellipses for Z_{xy}', Z_{yx}', and Z_{xx}' are centered at $\pm Z_1$ and Z_2, respectively, as shown in Figures 2a and 2b. Z_1 and Z_2 are invariant with axis rotation. Z_3 is the value of the major axis, and Z_4 is the minor axis for each ellipse. The impedance loci have a rotation period of 180° in Φ. With Φ_0 as defined in Figure 2, the major and minor axis intercepts correspond to the angles $\Phi = \Phi_0 \pm n\,45°$, where $n = 0, 1, 2, \cdots$.

Complex plane loci for $[Y_z'(\Phi)]$ are also elliptical, in general, as illustrated in Figure 2c. For a given frequency, the ellipse is centered on the origin, and the angle between the major axis of the ellipse and the real axis is designated as ξ_{z0}. Y_3 and Y_4 are the values of the major and minor axes, respectively. The rotation period for these loci is 360° in Φ, and the major and minor axes correspond to $\Phi = (\Phi_{z0} \pm n\,90°)$ $n = 0, 1, 2, \cdots$, where Φ_{z0} is as defined in the figure.

One-dimensional earth. When the earth conductivity is a function of z only, the surface impedance is representable as a complex scalar $Z_s = Z_1$ that is invariant with Φ, and the elements of $[Z']$ become

$$Z_{xx}' = Z_{yy}' = 0 \qquad (4a)$$

$$Z_{xy}' = -Z_{yx}' = Z_s \qquad (4b)$$

for all Φ. The rotation loci for $[Z']$ elements are thus point ellipses in the complex plane centered at the origin and $\pm Z_s$.

There is no vertical magnetic field, $H_z = 0$, so that the rotation loci for elements of $[Y_z']$ are point ellipses centered at the origin.

Two- and three-dimensional earth. Arbitrary conductivity structures usually produce elements of $[Z']$ and $[Y_z']$ whose rotational loci are characterized by the open ellipses shown in Figure 2. Suppose, however, that there exists a vertical plane of symmetry through the measuring point (i.e., a plane containing the z axis, about which the structures on either side are mirror images of each other); this also includes the possibility of a two-dimensional earth as a special case. As the coordinate $x'y'$ axes are rotated, the tensor $[Z']$ must diagonalize whenever the x' axis or the y' axis lies within the plane of symmetry (i.e., for 90° increments of Φ) [*Word et al.*, 1970] such that

$$Z_{xx}'(\Phi) = Z_{yy}'(\Phi) = 0 \quad \Phi = \Phi_0 \pm n90°$$

$$Z_{xy}'(\Phi) = -Z_{yx}'(\Phi) \qquad \Phi \neq \Phi_0 \pm n90° \qquad (5)$$

where $n = 0, 1, 2, \cdots$. This causes the rotation loci for Z_{xx}' and Z_{yy}' to be a straight line ellipse centered on the origin. The loci for Z_{xy}' and Z_{yx}' are straight line ellipses also and are centered at $\pm Z_1$.

The vertical magnetic field H_z may or may not exist for the symmetrical earth structure.

In any case, each element of $[Y_z']$ must vanish for some value of $\Phi = \Phi_1$ and for $\Phi_1 + 180°$. This requires that the loci be in general straight-line ellipses centered on the origin. For a two-dimensional structure, the x' axis is aligned with the plane of symmetry for $\Phi = \Phi_{z0}$ and defines the characteristic dip direction of the structure. Note that for a two-dimensional or symmetrical structure the angle Φ_{z0} is a function of geometry only and is thus a useful diagnostic parameter.

In summary, the various conditions relating to axis rotation are the following.

For earth conductivity $\sigma_2(x, y, z)$ arbitrary, rotation loci must be primarily elliptical, as pictured in Figures 2a, 2b, and 2c.

For earth conductivity $\sigma_2(x, y, z)$ symmetrical about some vertical plane of symmetry, first, $[Z'(\Phi)]$ must and can only diagonalize for 90° increments in Φ. Rotation loci are line ellipses centered at the origin and $\pm Z_1$. Second, $[Z'(\Phi)]$ must diagonalize when the x' axis or y' axis is within the plane of symmetry. Third, $[Y_z']$ elements each must and can only vanish at 180° increments in Φ. Rotation loci are line ellipses centered at the origin. Fourth, an element of $[Y_z']$ must vanish whenever the x' axis or the y' axis lies within the plane of symmetry.

For earth conductivity $\sigma_2(x, y, z) = \sigma_2(z)$, rotation loci for $[Z']$, $[Y_z']$ are all point ellipses centered at the origin and $\pm Z_1$.

Three-dimensionality indicators. Precise interpretation of the two-dimensional tensor impedance $[Z]$ is limited at best to two-dimensional or simpler geometry. Some parameters can be defined that indicate departure from two-dimensionality; they thus serve as guidelines for interpretation.

The 'skew' index as defined by *Swift* [1967] is given by

$$\alpha \equiv \frac{(Z_{xx}' + Z_{yy}')}{(Z_{xy}' - Z_{yx}')} = \frac{Z_2}{Z_1} \qquad (6)$$

The index α is invariant with Φ. It may be noted that α is the ratio of the displacements of the centers from the origin of the rotational ellipses for the diagonal and cross tensor impedance elements. The condition $|\alpha| = 0$ is a necessary but not a sufficient condition for two dimensionality or symmetrical three dimensionality, since it requires only that the Z_{xx}', Z_{yy}' loci be centered on the origin.

Another indication of three dimensionality is the eccentricity of the loci ellipse for $[Z'(\Phi)]$. Consider the parameter defined as

$$\beta(\Phi) = \frac{(Z_{xx}' - Z_{yy}')}{(Z_{xy}' + Z_{yx}')} \qquad (7)$$

If the rotation loci are line ellipses, then the numerator must vanish for some $\Phi = \Phi_0$ and, thus, $\beta(\Phi_0) = 0$ for a two-dimensional earth. It is easily shown by substitution of (3) that $(Z_{xx}' - Z_{yy}')$ and $(Z_{xy}' + Z_{yx}')$ of Figure 2 must be in time phase in order for $\beta(\Phi_0)$ to vanish. Therefore, $\beta(\Phi)$ is real for a two-dimensional earth and complex in general for a three-dimensional earth. The 'ellipticity index' [Sims, 1969], defined as

$$\beta_0 \equiv \beta(\Phi_0) = Z_4/Z_3 \qquad (8)$$

is the ratio of minor to major axes of the impedance ellipse. A nonzero value for $|\beta_0|$ is indicative of three dimensionality.

The condition

$$|\alpha| = 0 \qquad |\beta_0| = 0 \qquad (9)$$

is a necessary and sufficient condition for two dimensionality that is used here to mean that $[Z'(\Phi)]$ is antisymmetric and will diagonalize for some value of Φ.

As the three-dimensional effect becomes small, it is possible for the axes of the rotation ellipse to become vanishingly small compared to the elements of $[Z']$, and this must be considered in judging the degree of three dimensionality. It might be desirable to weight the importance of β_0 as

$$\beta_{00} \equiv \beta_0 \frac{(Z_{xy}' + Z_{yx}')}{(Z_{xy}' - Z_{yx}')}\bigg|_{(\Phi = \Phi_0)}$$

$$= \frac{(Z_{xx}' - Z_{yy}')}{(Z_{xy}' - Z_{yx}')}\bigg|_{(\Phi = \Phi_0)} \qquad (10)$$

Ellipticity indices can also be defined for $[Y_z']$ in a similar manner. If

$$\beta_{z0} \equiv \beta_z(\Phi_{z0}) = \frac{Y_{zx}'}{Y_{zy}'}\bigg|_{(\Phi = \Phi_{z0})} \qquad (11)$$

$|\beta_{z0}|$ is zero for an apparent two dimensionality but is not necessarily nonzero for an apparent three dimensionality.

Interpretation technique. MT interpretation is based primarily on the point relationships:

$$[E] = [Z][H] \qquad (12a)$$

$$[H_z] = [Y_z][E] \qquad (12b)$$

which ideally would be known as a function of x and y as well as frequency.

The most practical approach to MT interpretation is to first synthesize a 'best' one-dimensional conductivity distribution $\sigma(z)$ from the tensor impedance $[Z]$ at each data site. A set of $\sigma(z)$ models for a grid of data sites can then be contoured or otherwise assembled to produce an initial estimate of the three-dimensional structure. Two- and three-dimensional model solutions are of little value at the outset, at least on a trial-and-error basis, but they serve more appropriately as a means of verification and refinement of the initial interpretation. Further refinement of the earth model can hopefully be accomplished by considering the coordinate rotation properties of $[Z]$ and $[Y_z]$ for each data site in view of the initial estimate of the structure with attempts then to achieve agreement.

Principal-axis estimates. The dip axis and principal directions of $[Z]$ are found from the coordinate rotation properties of the tensors in (12). The dip-axis estimates are defined by the x' axis for $\Phi = \Phi_{z0}$, obtained from

$$\Phi_{z0} \rightarrow |Y_{zy}'(\Phi)|_{\max} \qquad (13)$$

The higher impedance principal direction of $[Z]$ is defined by the x' axis for $\Phi = \Phi_0$, obtained from

$$\Phi_0 \rightarrow |Z_{xy}'(\Phi) + Z_{yx}'(\Phi)|_{\max} \qquad (14)$$

The rotation angles are usually functions of frequency.

For symmetrical geometry, the x' axis defines the vertical plane of symmetry ($x'z$ plane) and the dip axis for the angle Φ_{z0}. The maximum-impedance axis at angle Φ_0 is either along or normal to the dip axis, depending on the conductivities involved.

For nonsymmetrical geometry, the angle Φ_{z0} defined as in (13) responds to an average or pseudo dip axis of the material to which the electromagnetic field measurements are sensitive at a given frequency. The frequency function

$\Phi_{zo}(\omega)$ then tends to describe variations in pseudo dip axis with depth.

Frequency function for determining a one-dimensional model. If for x'-axis rotation angle Φ_0 the diagonal elements of $[Z']$ are normally small, the orthogonal E and H components are essentially related by

$$E_x' \cong Z_{xy}'H_y' \qquad (15a)$$

$$E_y' \cong Z_{yx}'H_x' \qquad (15b)$$

The impedances in (15) will be called $Z_{\parallel}$ and $Z_{\perp}$ for E-field components that are approximately parallel ($E_{\parallel}$) and perpendicular ($E_{\perp}$) to the estimated strike direction. The strike direction as used here is defined as the normal to the vertical plane of symmetry. The dip axis lies within the plane of symmetry. For a symmetrical structure, equations 15 become precise equalities and $Z_{\parallel}$ and $Z_{\perp}$ become precisely the impedances for E polarizations parallel and normal to the strike, respectively.

Consider a two-dimensional situation. If homogeneous isotropic material surrounds the origin or measuring point, $Z_{\parallel}$ and $Z_{\perp}$ become equal in magnitude in the high-frequency limit. As frequency is decreased $Z_{\parallel}$ and $Z_{\perp}$ are equal until a lateral inhomogeneity or anisotropy is sensed. The two functions respond differently to an anomaly and will thus begin to split apart. The behavior of these impedances for still lower frequencies must be considered in synthesizing a $\sigma(z)$ model.

$Z_{\perp}$ tends at first to be less sensitive to material off the z axis. This effect is crudely explained by the fact that, for a secondary (reflected) wave front incident on the origin at some angle θ to the z axis, the influence of the secondary electric field lying within the plane of incidence is proportional to $\cos \theta$. Consequently, $Z_{\perp}$ tends to have a sensitivity lobe that is directed along the z axis, a property that makes $Z_{\perp}$ a good locator of a first horizon in conductivity of the structure and probably a good delineator of lateral changes along the dip axis. The influence of a secondary E field normal to the plane of incidence does not have a geometrical factor that depends on θ; thus, $Z_{\parallel}$ is more sensitive to material off the z axis and tends to respond to some lateral averaging of the conductivity.

As frequency is decreased such that penetration depths extend beyond the nearest anomaly, $E_{\perp}$ and $Z_{\perp}$ remain permanently affected by the nearest anomaly for all lower frequencies; $E_{\parallel}$ and $Z_{\parallel}$ are continuous functions along the dip direction and become laterally constant in the low-frequency limit as though the shallower material were homogeneous. $Z_{\parallel}$ tends to become independent of shallow material in the two-dimensional structure for lower frequency as it would for a one-dimensional structure. It is thus able to represent the effect of deeper material in the structure without accumulative errors due to shallower anomalies. A one-dimensional interpretation of $Z_{\parallel}(\omega)$ would therefore seem to define a $\sigma(z)$ more nearly approximating the actual conductivity distribution along the z axis. *Patrick* [1969] discusses this point and demonstrates the phenomenon by fitting one-dimensional models to both $Z_{\parallel}$ and $Z_{\perp}$, which were computed for two-dimensional structures at various positions along the dip axis. A line of $\sigma(z)$ models produced from $Z_{\parallel}$ seems to reproduce a reasonable smoothed version of the original structure. $Z_{\perp}$ models are more in error beyond the first horizon of the structure and reflect lingering effects of a shallow anomaly on low-frequency values of $Z_{\perp}$; this pattern causes a constant bias if the deeper structure becomes one-dimensional again.

For three-dimensional structures, the arguments set forth above tend to apply as long as departure from two dimensionality is not too severe. Both E-field polarizations $E_{\parallel}$ and $E_{\perp}$ will usually retain some distorting effects of shallower anomalies, but $E_{\parallel}$ will probably be less distorted. In any case, corrections for such effects can be attempted only after some estimate of the structure has been obtained. Severe departure from two dimensionality can cause considerable error in the estimation of a $\sigma(z)$ model, and such a condition can be anticipated when three-dimensionality indicators of $[Z]$ become large. Real geological structures, however, can often be approximated by an assemblage of two-dimensional or symmetrical structures whose effects are sometimes separable (at least qualitatively) in the resulting three-dimensional impedance tensor [*Word et al.*, 1970].

One-dimensional model synthesis. Suppose an impedance function $Z_a(\omega)$, say $Z_{\parallel}$, has been chosen from $[Z]$ at a given measurement site such that Z_a behaves adequately as the surface impedance of a one-dimensional or Cagniard model $\sigma_m(z)$, approximating the conductivity below the site. The next step then is to perform the inversion process:

$$Z_a(\omega) \to \sigma_m(z) \qquad (16)$$

If Z_a is known for all ω, unique inversion is made possible by the skin-effect phenomenon [*Bailey*, 1970]. In fact, the range on ω need only be such that effective penetration depths encompass the z range of interest in most cases.

The sensitivity of Z_a to a particular portion of $\sigma_m(z)$ and, consequently, the relative importance of that portion of the model as obtained in the inversion process are of considerable interest. The sensitivity is usually a complicated function of $\sigma_m(z)$ and of measurement noise and is difficult, if not impossible, to formulate analytically. Qualitatively, a particular region of the model is reasonably well defined if it produces flux exclusion (i.e., attenuation of the electromagnetic fields in the z direction is rapid compared to the thickness of the region) over some frequency range. It is of some use to implement this rule as follows. Let $\sigma_m(z)$ be of the discrete form in Figure 3, let δ_n be the skin depth in σ_n, and let ω_n denote the frequency for which layer n would become effective in Z_a (i.e., at ω_n the E field at the surface of layer n is about $1/e$ of its value at $z = 0$). Confidence in layer n of the model can be based on the following qualitative rule:

$$d_n > k\delta_n(\omega_n) \to \text{layer } n \text{ well defined} \qquad (17a)$$

$$d_n < k\delta_n(\omega_n) \to \text{layer } n \text{ not well defined} \qquad (17b)$$

That k depends on measurement noise is implicit in this rule; $k = 0.75$ to 1.0, determined empirically by the authors, can be used as a rough guideline for present measurement capabilities. Layer n meets condition 17b when either d_n or σ_n is small. The layer may still affect the value of Z_a, although one or both of its parameters is not well defined by the inversion.

A closed-form solution for the inversion problem in (16) is not presently available. All methods now used involve iteration of the forward problem to achieve a fit to the frequency function Z_a. Such methods are well covered in the literature [*Cagniard*, 1953; *Wait*, 1962; *Price*, 1962; *Hopkins and Smith*, 1966; *Swift*, 1967; *Patrick*, 1969]. For the layered model in Figure 3, the solution for Z_a can be found from the easily obtainable expression:

$$Z_{n-1} = F(Z_n{}'\ \sigma_{n-1}{}'\ d_{n-1}) \qquad (18)$$

by iterating up the z axis, on n, beginning with the bottom layer impedance Z_N. Manual fit of the data requires a considerable amount of skill in manipulating the model parameters. *Hopkins and Smith*, [1966] have devised an orderly procedure to accomplish this. *Wu* [1958] and *Patrick* [1969] have written computer programs to achieve automatically a minimum mean square fit to the impedance function by minimizing the function

$$\Psi = \sum_\omega \frac{1}{|Z_a|^2} |Z_a - Z_m|^2 \qquad (19)$$

with respect to all model parameters. Z_m is the model impedance. Variables to which Ψ is insensitive can be made constant to eliminate instabilities in the process. The program written by *Patrick* [1969] was used on the results presented in a later section and has proven to be a very useful tool.

It is convenient to define an apparent-resistivity function for use in place of Z_a as

$$\rho_a(\omega) = \frac{1}{\omega\mu} |Z_a(\omega)|^2 \qquad (20)$$

The function ρ_a provides better insight into the resistivity behavior and for a homogeneous model

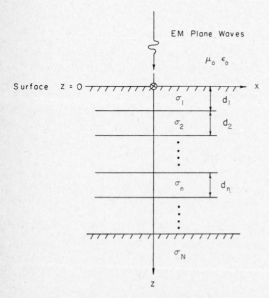

Fig. 3.   Layered conductivity model.

becomes the intrinsic resistivity of the medium. In the one-dimensional model, Z_a behaves as a minimum phase function, and the phase adds no information not contained in ρ_a. Minimum phase behavior seems very apparent in all multidimensional impedance functions measured by the authors, although it is doubtful that such could be expected in every case. The phase of Z_a should certainly be considered in the interpretation.

For a cursory interpretation of an apparent-resistivity function, standard sets of two- or three-layer apparent-resistivity curves [Cagniard, 1953; Yungul, 1961] can be used for comparison. Also, approximate inversion methods [Sims, 1969; Niblett and Sayn-Wittgenstein, 1960] have been devised that assume purely exponential decay of the E field in each layer. Such methods are good for mild conductivity contrasts but tend to ignore highly resistive layers.

The initial estimate of subsurface structure formed by an assemblage of one-dimensional models can possibly be refined by considering the behavior of [Z] for individual sites. For example, the estimated dip-axis direction at a particular frequency should be in agreement with the average dip axis of the corresponding portion of the model. Such a comparison provides a good cross check on the results and suggests possible steps for refinement.

A z-ω correspondence is desirable for interpretation of frequency functions such as $\Phi_{z0}(\omega)$. There is no one-to-one correspondence between depth and frequency for a given conductivity function $\sigma_m(z)$; however, an approximate relationship can be established on the basis of a composite skin depth $\bar{\delta}_n(\omega)$ or effective $1/e$ penetration depth into $\sigma_m(z)$ for frequency ω. Let $\bar{\delta}_n$ be defined by

$$\bar{\delta}_n(\omega) = z \qquad |E(z)| = 1/e|E(0)| \qquad (21)$$

The function $\bar{\delta}_n(\omega)$ is monotonic for most probable situations, although it is not precisely so in general (e.g., it would probably not be unique for thick air space between two conductors).

A computer program was written to evaluate $\bar{\delta}_n(\omega)$ and is applied to the results in this study. This depth frequency criterion is perhaps best suited for H_z phenomena because of a tendency for z-ω selectivity in $H_z(\omega)$ [Word et al., 1970]. Surface H_z is entirely a secondary field, for the

source assumed throughout this study, and its existence is an immediate indicator of lateral inhomogeneity. H_z is produced primarily by $H_\perp$ (precisely for two-dimensional cases); consequently, a normalized H_z function defined as

$$H_{zn} \equiv H_z/H_\perp \qquad (22)$$

should be time invariant for a given structure. A map of H_{zn} on an x-$\bar{\delta}_n$ or x-ω cross section tends to locate the position of an anomaly in that plane. Since the secondary wave front producing H_z must propagate from the anomaly, the phase of H_{zn} is advanced or retarded as the measuring point is moved, respectively, toward or away from the anomaly. Insofar as H_{zn} can be associated with a particular anomaly, the 180° directional ambiguity in the dip-axis estimate can be eliminated. In some cases the absolute value of H_{zn} phase seems to indicate whether the anomaly is more conductive or less conductive than the surrounding mediums.

MAGNETOTELLURIC SOUNDING IN CENTRAL TEXAS

Micropulsation measurements were made at seven sites along a 60-mile traverse east of Austin, Texas. Data were taken in the frequency band 0.0001 to 10 Hz, which corresponds to sounding depths ranging from about 0.1 km to over 100 km. Figure 4 shows the traverse on a regional scale along a line bearing approximately N 40°W. Site locations are shown in Figures 5 through 7 on a larger horizontal scale. The site numbers shown were assigned for reference purposes; also, the sites bear the names of nearby landmarks. These names are abbreviated in some figures and they are written out in Table 1. Texas counties are outlined in thin continuous lines in Figures 5 and 6; the MT traverse extends from northern Williamson County through Travis County to southeastern Bastrop County.

Geological background. The MT traverse lies on the flanks of the Llano uplift and crosses the outer portion of the Ouachita fold belt system at the edge of the Texas craton. The Llano Uplift is an exposed crystalline granitic uplift of the Precambrian basement and is associated with the Ouachita system [Flawn, 1964; Woods, 1956], extending from northern Mexico some 1300 miles to southwestern Alabama (see Figures 4–7).

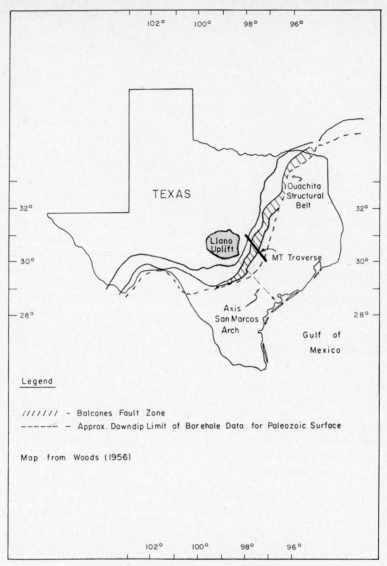

Fig. 4.   Magnetotelluric sounding traverse.

The younger sediments overlying the Ouachita system strike about N 20°E near the MT traverse and dip generally southeastward toward the Gulf coast. The Ouachita geosyncline is composed primarily of two structural salients, the east Texas and Rio Grande embayments [*Flawn*, 1964], respectively, separated by the San Marcos arch, whose axis extends from the Llano Uplift outcrop southeast to the coast [*Fowler*, 1956; *Halbouty*, 1966], as shown in Figure 4.

Subsurface geology in the region of the MT traverse is reasonably well known to borehole depths. Borehole penetration to Paleozoic material extends roughly southeast to the broken line in Figure 4. *Flawn et al.* [1961], using nearly all available geophysical evidence, derived a detailed map of the subsurface structure of the Llano uplift and the Ouachita system and provided the information for the contour map of the Paleozoic surface presented in Figure 5. The map in Figure 6 has been redrawn from *Kenney* [1967]. The Precambrian surface contours shown in Figure 6 are based largely on geophysical data and boreholes not reaching the basement. This surface is not well determined over most of the region shown. Southeast of the zone of overthrust faulting in the Premesozoic, the basement surface depth is beyond reliable estimation; contours are drawn on the Paleozoic or Premesozoic surface south of the fault zone.

Figure 8 is a geological section along the MT traverse; it is derived from various sources including data from maps of the Ouachita structural belt by *Flawn et al.* [1961]. This section shows the various lithologic units along the traverse on a logarithmic depth scale. Since electromagnetic fields tend to attenuate exponentially in the conducting earth, the logarithmic depth scale is a natural one for MT sounding. Paleozoic and Precambrian surfaces were obtained as described above from the contour maps by *Flawn* [1961] and *Kenney* [1967]. The Cretaceous and younger sediments are reasonably well known from borehole penetrations. Contacts between

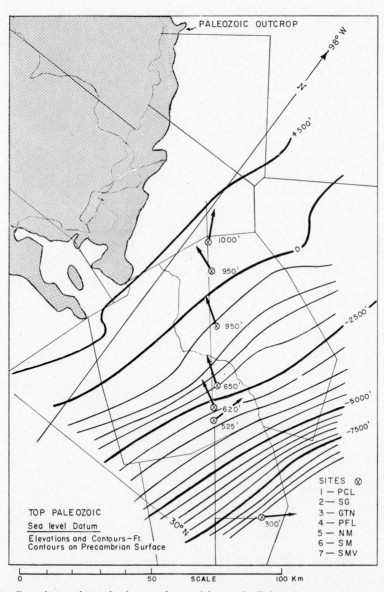

Fig. 5. Rotation angle Φ_{zo} for frequencies sensitive to the Paleozoic surface and shallower.

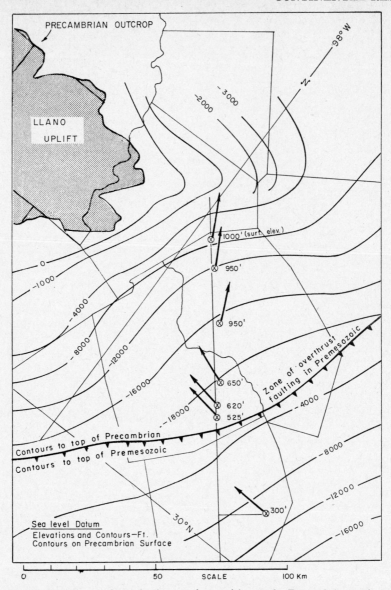

Fig. 6.   Rotation angle Φ_{z0} for frequencies sensitive to the Precambrian surface.

the various sedimentary groups seem to be fairly planar, except for occasional normal faulting, and they dip southeastward with a slope of about 0.01. Accurate estimates of contact depths were obtainable near sites 5 and 7. The interfaces, broken lines in the cross section, were drawn with constant slope from known points of surface outcrop through the estimated intersections at sites 5 and 7. Any fault displacements present have been ignored by this process, but

their effect is probably not pronounced, since a good fit with the MT data was obtained with the plane interface assumption. Resistivity logs were available for boreholes designated W1 through W4 at the locations shown in Figure 8. Their resistivity values correlate reasonably well with the broad range of values shown.

Data.   Field measurements were made of two components of the electric field (E_x, E_y) and three components of the magnetic field (H_x, H_y,

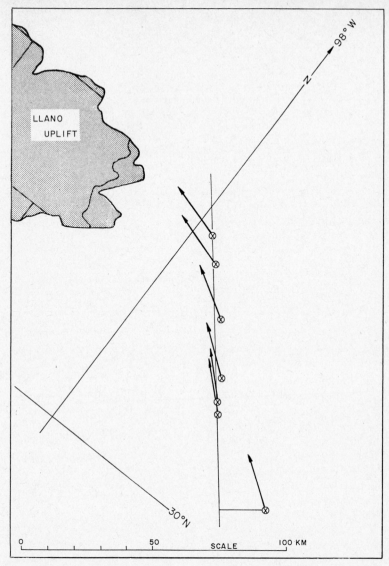

Fig. 7. Rotation angle Φ_{x0} for frequencies sensitive to the conductive substratum.

TABLE 1. Site Designations and Locations

Site No.	Name	Abbrev.	Location	Elev. ft.
1	Purcell	PCL	30°47.88′N, 97°55.48′W	1000
2	San Gabriel	SG	30°42.45′N, 97°52.45′W	950
3	Georgetown	GTN	30°35.25′N, 97°45.88′W	950
4	Pflugerville	PFL	30°27.15′N, 97°35.53′W	650
5	N. Manor	NM	30°22.45′N, 97°31.83′W	620
6	S. Manor	SM	30°20.40′N, 97°32.17′W	525
7	Smithville	SMV	30°02.63′N, 97°08.25′W	300

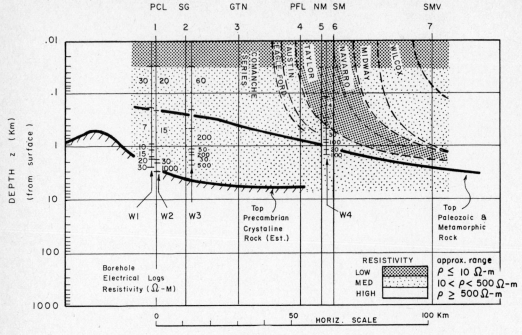

Fig. 8. Geological section along MT traverse with electric log resistivities.

H_z) at each of the seven sites by using the MT instrumentation system developed at the Geomagnetic Laboratory of The University of Texas at Austin [*Word et al.*, 1970]. The frequency range from about 0.0001 to 10 Hz was covered in three overlapping bands (low, medium, high). These were recorded with system responses adjusted to pre-whiten approximately the spectra within each range. An effort was made to obtain at least three good data sequences for each of the three frequency bands; up to six or eight were usually obtained for the medium and high bands. Record lengths averaged about 20 hours for the low band, 2 hours for the medium band, and 15 minutes for the high band.

Data analysis. All data sets were analyzed on a CDC 6600 computer using the techniques described by *Sims* [1969] and *Word* [1970]. Included were the computations of power density spectra, cross spectra, coherency, and polarization from which the following quantities were obtained: apparent resistivity and phase angle for the principal axes of $[Z']$; rotation angles Φ_0 and Φ_{z0} for the principal x' axes of $[Z']$ and $[Y_z']$, respectively; the skew $|\alpha|$, the ellipticity

β_0, for the measured impedance $[Z]$, and the ellipticity β_{z0} for the measured admittance $[Y_z]$.

RESULTS

A criterion for accepting data points for apparent resistivity and phase angle was based mainly on the phasor coherency as defined by *Sims* [1969]. Only data points having values greater than 0.8 were used in the interpretation. Figure 9 shows a plot of the apparent resistivity and phase versus frequency for the principal elements of $[Z'(\Phi_0)]$ at the North Manor site. The elements $\rho_{\parallel}$ and $\rho_{\perp}$ correspond to $E_{\parallel}$ and $E_{\perp}$; the continuous curves in this figure show the model fit obtained for the $\rho_{\parallel}$ case. Similar results were obtained for the other six sites and all curves are graphed on small scales in Figure 10 for convenience in examining them collectively.

A contour plot of $\rho_{\parallel}$ in a horizontal distance versus frequency plane along the MT traverse is shown in Figure 11. This is actually a plot of the best fit theoretical $\rho_{\parallel}$ function for the model to be discussed subsequently; however, the theoretical function, appearing as the continuous line in Figures 9 and 10, serves as a reasonable smoothed average for the experimental points.

One-dimensional profiles. A measured impedance function $Z_a(\omega)$ was chosen from each set of results as typified by Figure 9. One-dimensional layered models of the form of Figure 3 were obtained for all sites by two methods. The first method involved a computer visualization routine whereby a model fit was accomplished by successive manual adjustments of the model parameters; the second method used an automatic computer program written by *Patrick* [1969], which minimizes the mean square difference between the measured and model impedance function with respect to all model parameters.

Plots of $\rho(z)$ for layered models satisfying the apparent-resistivity function $\rho_a(\omega)$ are given in Figure 12. The solid line curves represent the profiles to be used in the two-dimensional section and the broken curves are alternate models; the relatively well defined parameters are drawn as thick solid lines. Certain parameters that were not well defined by the inversion were adjusted to probable values still in agreement with the sounding results: a surface layer of $\rho_1 = 5$ Ω-m was used at each site; thin conductive layers were set to 1 Ω-m; thin resistive layers were set

to 100 Ω-m or 10^4 Ω-m; and at site 1, ρ_2 and d_4 were taken from borehole log W1. The above assumptions were used when required by lack of enough high-frequency data or by insensitivity of the inversion to the value of the parameter in question.

The surface layer thickness d_1 was adjusted to a maximum permissible value when it was not determined by $\rho_1 = 5$ Ω-m. For sites 3 and 4, d_1 is determined by ρ_1; the broken curve is for $\rho_1 = 12$ Ω-m and is a good estimate of the upper limit on ρ_1 and d_1. For highly resistive layers $\rho_n \to \infty$, the layer thickness d_n is determined by the inversion if $d_n \gg \delta_{n+1}(\omega_n)$, where δ_{n+1} is the effective skin depth in the medium below layer n and ω_n is the frequency at which layer n would influence the surface impedance Z_a (see (17)). For small d_n, the layer is effectively transparent. The 10^4 Ω-m layers for sites 3 and 4 are possibly not actually present. The dimensions shown represent upper limits for which these layers are transparent. Admittedly, transparent resistive layers could be placed anywhere in the profile in this manner, and this always represents a potential source of error

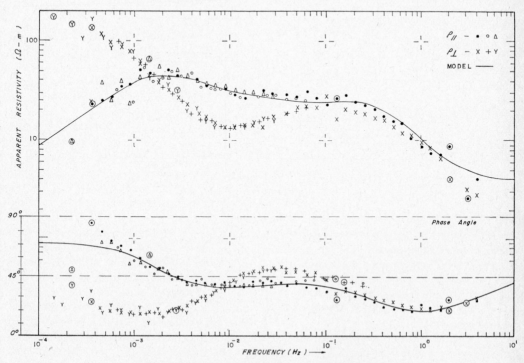

Fig. 9. Site 5 North Manor apparent resistivity.

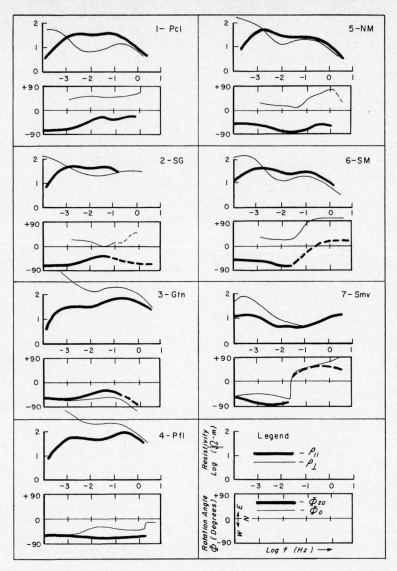

Fig. 10. Summary of apparent resistivity and rotation angle results (see lower right box for legend).

in depth resolution. Nevertheless, when there is reason to expect the presence of a thin resistive region, a useful estimate of the maximum possible thickness can be obtained as above.

For electrically thin and highly conductive layers, only the product $\sigma_n d_n$ is determined by the inversion process; as a result, it is required that a condition such as the setting of thin conductive layers to 1.0 Ω-m be made. Limits could be placed on the values of σ_n and d_n individually,

since the layer would be known electrically thin for the frequencies at which it is effective.

Two-dimensional section. The one-dimensional profile estimates were assembled to produce the vertical cross section shown in Figure 13. Vertical lines represent the data site locations, and the layered model parameters ρ_n and d_n are indicated at each site: the large numerals are layer resistivity in Ω-m and the short cross bars represent interfaces. The discrete functions

$\rho(z)$ for all sites were correlated along the traverse and contoured in 5 resistivity ranges, with interpolation between sites, to form a smoothed estimate of the two-dimensional resistivity distribution. A crude estimate of the horizontal averaging interval along the cross section at a depth z is about $2z$ or less, if a given $\sigma(z)$ is considered influenced by horizontally neighboring material up to a distance z from the measuring site. At a depth of 100 km, the distribution is averaged over the entire traverse at each site.

The relative importance of various features of the model can be gathered from Figure 12. Generally, the shallower portion of the cross section to the first horizon in resistivity is not well defined, owing to lack of adequate high-frequency data; frequencies of up to 10^2 and 10^3 Hz would be needed to obtain good definition for $0.01 < z < 0.1$ km. The high-resistivity bed, marked 10^4 Ω-m, is questionable at sites 3 and 4, and the thickness shown in that region represents an upper limit.

Comparison of model with geology. Comparison of the MT model with the expected geological structure described earlier shows good agreement for most major features of the conductivity distribution, including estimated dip axes. A comparison of the MT profile in Figure 13 with the expected geological section in Figure 8 is facilitated by the superposed sections in Figure 14.

The first horizon in resistivity seen by the sounding is the Austin limestone and lower Cretaceous formation, at sites 1 to 6 and the Wilcox sandstone, at site 7. There is particularly good correlation between the Navarro and Taylor shale beds and the associated conductive streak on the model, which is clearly defined by the sounding at sites 6 and 7.

The Paleozoic surface presents no discernible contrast over the lower Cretaceous and therefore is not defined by the sounding, nor are the various Ouachita facies of the folded Paleozoic material defined, although longer signal averaging in the sounding might detect this material. The lower resistivity regions of the Paleozoic and Cretaceous toward either end of the traverse are possibly associated with different formation facies. The interpolated boundaries between the regions are of course open to question between data sites; however, the geological and electrical boundaries are not expected to conform necessarily. At site 1, near $z = 1$ km, the 8 Ω-m and underlying 100 Ω-m materials prob-

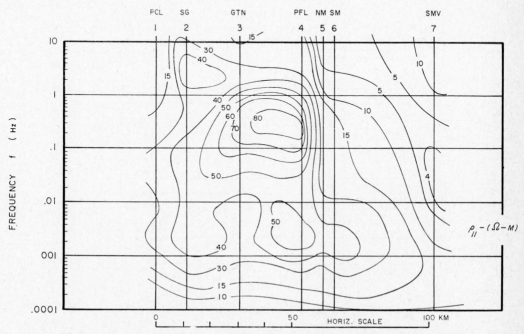

Fig. 11. Apparent-resistivity versus frequency cross section for E parallel to strike.

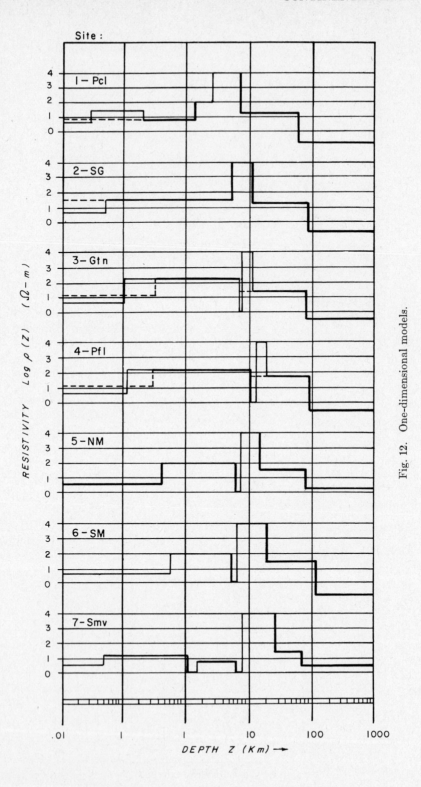

Fig. 12. One-dimensional models.

ably correspond, respectively, to the Upper and Lower Paleozoic rocks of foreland facies.

The resistive bed, 500–∞ Ω-m, is thought to correspond in part to the Precambrian granite basement. The basement surface was detected somewhat deeper than expected at sites 2, 3, and 4. However, discussion with *Flawn* [1961] and other geologists familiar with the Ouachita system would indicate that the model depths are not unreasonable and are within tolerances of the expected surface estimates. The granite is not known to exist over the entire traverse; however, the resistive bed is definitely observed, except possibly between sites 3 and 4.

The thin conductive bed overlying the resistive basement region is found to be a necessary part of the MT model. It may represent a severely weathered granite surface or granite wash. Although a conductive granite surface is not seen at sites 1 and 2, it probably exists but to a lesser extent; the resistivity log for borehole W1 penetrated the granite surface some tens of feet and indicated much lower resistivity than would be expected for granite, i.e., approximately 100–200 Ω-m.

Below the resistive bed, penetration depths are thought to extend into the upper mantle. The lower two conductive layers of the model are a source of concern. The parameters associated with these layers are reasonably well defined in the MT inversion and do not appear to be unduly affected by the influence of noise. However, the indicated resistivity grade-off with depth is much more rapid than is found by deep resistivity soundings in most other areas. *Keller et al.* [1966] show a resistivity-depth distribution compiled from the several methods including, in order of increasing depth, borehole electric logs, galvanic resistivity surveys, magnetotelluric sounding, and studies of long-period magnetic field variations by *Lahiri and Price* [1939] and *McDonald* [1957].

Five major zones of electrical conductivity with depth are recognized by *Keller et al.* [1966] and are labeled A to E from the crust to the core of the earth. Zone A is the surface layer of sedimentary, igneous, and metamorphic rocks. Zone B, probably granite and basalt, is highly resistive, 3×10^3 to 10^5 Ω-m and above. Zone C represents the first significant drop in resistivity for $z \geq 35$ km or so to values found as low as approximately 30 Ω-m, determined by MT in the Colorado Plateau area. Zone D is estimated to begin at $z \approx 200$ km, with further decrease in resistivity. Zone E, the mantle resistivity of 1 Ω-m or less, is estimated to exist for $z \geq 600$ km or so, based on magnetic field variation studies. *Swift* [1967] found a similar conductivity profile with deep MT soundings in Arizona and New Mexico. The low-frequency $\rho_{\parallel}$

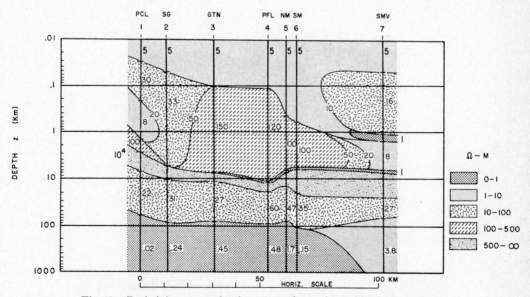

Fig. 13.   Resistivity versus depth cross section estimated by MT soundings.

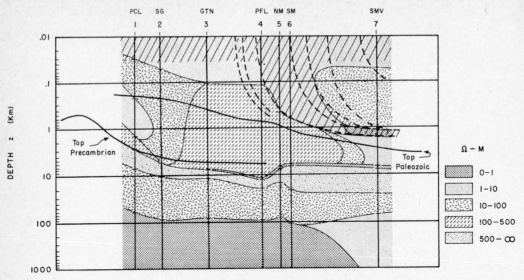

Fig. 14.   Superposed geological section.

results in the Swift sounding have a mild decrease with decreasing frequency and are still 15 to 30 Ω-m or so at 10^{-5} Hz, whereas, for the Texas soundings, $\rho_\parallel \leq 10$ Ω-m at 10^{-4} Hz and generally the $\rho_\parallel$ results have a more rapid decrease with decreasing frequency (see Figures 9 and 10). *Hopkins* [1966] made two MT soundings in the Delaware basin of west Texas at Pecos and Fort Stockton and synthesized layered models to depths in excess of 100 km. The resistivity at Pecos and Fort Stockton was found to decrease to approximately 50 Ω-m at 14.6 and 31 km, respectively, and an ultimate layer of 2.5–10 Ω-m was found at depths of 77 and 106 km, respectively, in reasonable agreement with the results herein.

There is some suggestion of correspondence between the deep (zone *B* and below) resistivity boundaries and seismic boundaries [*Keller et al.*, 1966]; however, there is still insufficient evidence to determine a definite correspondence. According to *Keller* [1966], a decrease in resistivity has been observed at several locations for depth corresponding roughly to the Conrad discontinuity from 10 to 35 km. The resistivity of the basalt or gabbro material below the granite is still placed at over approximately 3000 Ω-m. For the MT section herein, the Conrad discontinuity is possibly within the resistive 500–∞ Ω-m bed. The Mohorovicic discontinuity, estimated by seismic sounding to be at 35–45

km [*Cram*, 1962; *Dietz*, 1964] near the Llano uplift is within the conductive region between 10–100 km (see Figure 13).

The deep-resistivity profiles for sites 4, 5, 6, and 7 of Figure 13 are in particularly good agreement with the Hopkins sounding and possibly with the zone *B* and zone *C* transitions of the *Keller* [1966] model. The conductive substrate appears to be about 1/10 as deep as expected, and there is strong evidence of a rise in both conductive substrata toward the northwest.

Dip axes. The principal axis angle Φ_{z0} was used as an estimate of the pseudo dip-axis bearing, to imply the nature of variations off the cross section. Comparison of Φ_{z0} with the expected subsurface contours in Figures 5, 6, and 7 appears to support the speculation that some of the two- and three-dimensional properties of the sounding were attributable to sloping interfaces between formations.

Figure 5 shows the dip axes for a frequency of 1 Hz on the Paleozoic surface contours. The sounding at 1 Hz is thought to be sensitive to material near to and shallower than the Paleozoic surface. It is affected by an area roughly the size of the circles marking the sites in Figure 5 and is expected to be influenced by very local features perhaps at odds with the regional trend. Correlation with the contours in Figure 5 seems quite good, except for sites 1, 6, and 7. Site 7 seems to be responding to the Wilcox formation, which at site 7 possibly has the apparent dip axis shown by the arrow, if we judge by the shape of the surface outcrop.

Figure 6 shows the dip axes on the Precambrian surface contours for frequencies corresponding to the peak in the mid band excursion (rise or dip) in Φ_{zo} for each site. The mid band excursions are confidently attributable to the resistive bed surface.

Figure 7 shows the dip axis for a frequency of 10^{-4} Hz where the sounding is sensitive to the conductive substrate. The tendency for the dip axis to align with the Llano uplift granite outcrop is interesting, particularly in view of the apparent northwesterly rise in the substrate. This pattern would appear to suggest a rise in the conductive mantle associated with the uplift. Additional soundings would be helpful in a study of the nature of the conductive anomaly. A northwest extension of the sounding traverse is planned for the near future.

CONCLUSIONS

In the present study, the conductivity cross section along the central Texas traverse is obtained by contouring the conductivity profiles obtained at each site. Each conductivity profile is obtained by inverting an impedance-versus-frequency function into an estimate of conductivity versus depth. This inverted conductivity estimate is taken to be representative of the true conductivity profile beneath the recording site. Although a quantitative evaluation of the errors involved in constructing such a cross section is quite intractable, it is possible to make a few statements pertaining to the over-all accuracy.

The sensitivity of the MT inversion process to a given earth model parameter is a complicated function of the conductivity structure and the noise influences. Consequently, resolution of the structure will depend on the situation involved, the amount of data available, and the method of interpretation. For a one-dimensional structure, a tolerance of $\pm 10\%$ in depth, thickness, and conductivity is not uncommon for a layer or region meeting the conditions in $(17a)$.

The one-dimensional inversions are subject to additional error in the presence of multidimensional geology. If the estimates of the impedance tensor give clear evidence of the presence of three-dimensional geometry, little faith can be put in the conductivity estimates

derived by one-dimensional inversion. Fortunately, however, all the MT soundings along the central Texas traverse produce impedance tensors that have a strong tendency to diagonalize over a considerable range of frequencies. This feature signals the presence of a nearly two-dimensional earth. In the general case, however, even two-dimensional inversions are almost completely out of the question. If a two-dimensional model were almost completely known, then one or two parameters of that model may possibly be estimated by iterating the numerical evaluation of the almost specified model, but a full two-dimensional evaluation is far too unwieldy. However, the effects of two-dimensional geometry can, to some extent, be accounted for in the one-dimensional treatment by selecting for inversion only that impedance element that is generated by the electric-field component parallel to the strike of the two dimensionality. This strike direction may, of course, be different for structures at different depths beneath the earth's surface. This can be interpreted from the impedance tensor if the angle for which it diagonalizes is distinctly different at different frequencies. In other words, one could interpret a three-dimensional structure if the three dimensionality in fact consists of stacked two dimensionalities whose two-dimensional properties are manifest in the impedance tensor in different portions of the frequency spectrum. In such cases, however, it is necessary to modify the inversion for the deeper of the two structures, because the shallower feature tends to be interactive with the impedance function features of the deeper one.

We conclude that the MT method in its present state of development can become a valuable tool for geophysical exploration if its capabilities are carefully balanced against its limitations.

Acknowledgments. This work was supported by the Office of Naval Research under contract N00014-67-A-0126-0004 and by the National Science Foundation under grant GA 17457.

REFERENCES

Bailey, R. C., Inversion of the geomagnetic induction problem, *Proc. Roy. Soc. London A, 315,* 185–194, 1970.

Bloomquist, M. G., H. W. Smith, F. X. Bostick, Jr., L. Ames, and A. S. Orange, Propagation characteristics of micropulsations, *Rep. AFCRL-*

67-0540, Elec. Eng. Res. Lab., Univ. of Texas at Austin, Aug. 1967.

Bostick, F. X., Jr., and H. W. Smith, An analysis of the magnetotelluric method for determining subsurface resistivities, *Rep. 120,* Elec. Eng. Res. Lab., Univ. of Texas at Austin, Feb. 1961.

Bostick, F. X., Jr., and H. W. Smith, Investigation of large-scale inhomogeneities in the earth by the magnetotelluric method, *Proc. IRE, 40*(11), 2339–2346, 1962.

Bryunelli, B. E., Magnetotelluric profiling in conditions of horizontally inhomogeneous media, *Akad. Nauk SSSR Geophys.,* 600–604, 1964.

Cagniard, L., Basic theory of the magnetotelluric method of geophysical prospecting, *Geophysics, 18,* 605–635, 1953.

Cantwell, T., and T. R. Madden, Preliminary report on crustal magnetotelluric measurements, *J. Geophys. Res., 65*(12), 4202–4205, 1960.

Cram, I. H., Crustal structure of Texas coastal plain region, *Bull. Amer. Ass. Petrol. Geol., 46* (9), 1721–1727, 1962.

d'Erceville, I., and G. Kunetz, The effect of a fault on the earth's natural electromagnetic field, *Geophysics, 27*(5), 651–655, 1962.

Dietz, R. S., Origin of continental slopes, *Amer. Sci., 52,* 50–69, 1964.

Flawn, P. T., Basement rocks of the Texas gulf coastal plain, *Trans. Gulf Coast Ass. Geol. Soc., 14,* 281–275, 1964.

Flawn, P. T., A. Goldstein, Jr., P. B. King, and C. E. Weaver, The Ouachita system, *Publ. 6120,* Bur. Econ. Geol., Univ. of Texas, Austin, Oct. 1961.

Fournier, H., La spectrographie directionelle magneto-tellurique a Garchy (Nievre), *Ann. Geophys., 19*(2), 1963.

Fowler, P., Faults and folds of south-central Texas, *Trans. Gulf Coast Ass. Geol. Soc., 6,* 37–42, 1956.

Halbouty, M. T., Stratigraphic-trap possibilities in upper jurassic rocks, San Marcos arch, Texas, *Bull. Amer. Ass. Petrol. Geol., 50*(1), 3–24, 1966.

Heirtzler, J., A summary of the observed characteristics of geomagnetic micropulsations, in *Natural Electromagnetic Phenomena Below 30 Kc/ sec,* edited by D. R. Bleil, pp. 351–372, Plenum, New York, 1964.

Hopkins, G. H., Jr., and H. W. Smith, An investigation of the magnetotelluric method for determining subsurface resistivities, *Rep. 140,* Elec. Eng. Res. Lab., Univ. of Texas, Austin, Feb. 1966.

Keller, G. V., L. A. Anderson, and J. I. Pritchard, Geological survey investigations of the electrical properties of the crust and upper mantle, *Geophysics, 31*(6), 1078–1087, 1966.

Kenney, D. M. (Ed.), Basement map of North America, Amer. Ass. Petrol. Geol. and U.S. Geol. Surv., 1967.

Lahiri, B. N., and A. T. Price, Electromagnetic induction in non-uniform conductors, and the determination of the conductivity of the earth from terrestrial magnetic variations, *Phil. Trans. Roy. Soc. London A, 237,* 509–540, 1939.

Mann, J. E., Jr., Magnetotelluric theory of the sinusoidal interface, *J. Geophys. Res., 69*(16), 3517–3524, 1964.

McDonald K. L., Penetration of the geomagnetic secular field through a mantle with variable conductivity, *J. Geophys. Res., 62*(1), 117–141, 1957.

McNish, A., The Aurora, in *Natural Electromagnetic Phenomena Below 20 kc/sec,* edited by D. R. Bleil, pp. 77–93, Plenum, New York, 1964.

Niblett, E. R., and C. Sayn-Wittgenstein, Variation of electrical conductivity with depth—the magneto-telluric method, *Geophysics, 25*(5), 998–1008, 1960.

Orange, A. S., and F. X. Bostick, Jr., Magnetotelluric micropulsations at widely separated stations, *J. Geophys. Res., 70*(6), 1407–1414, 1965.

Parkinson, W. D., The influence of continents and oceans on geomagnetic variations, *Geophys. J., 6,* 441–449, 1962.

Patrick, F. W., Magnetotelluric modeling techniques, Ph.D. thesis, Univ. of Texas at Austin, January 1969.

Pokityanksi, I. I., On the application of the magnetotelluric method to anisotropic and inhomogeneous masses, *Bull. Acad. Sci. U.S.S.R., Geophys. Ser.,* 1050, 1961.

Pospeyev, V. I., Some results of magnetotelluric sounding in Irkutsk amphitheater, *Geofiz. Issled., 5,* 94–96, 1965. (translated from Russian by E. R. Hope, Defense Res. Board, Ottawa, Ontario, Sept. 1966.)

Price, A. T., The theory of magnetotelluric methods when the source field is considered, *J. Geophys. Res., 67*(5), 1907–1918, 1962.

Prince, C. E., F. X. Bostick, Jr., and H. W. Smith, A study of the transmission of plane hydromagnetic waves through the upper atmosphere, *Rep. 134,* Elec. Eng. Res. Lab., Univ. of Texas, Austin, July 1964.

Rankin, D., The magnetotelluric effect on a dike, *Geophysics, 27*(5), 666–676, 1962.

Sims, W. E., Methods of magnetotelluric analysis, Ph.D. thesis, Univ. of Texas, Austin, Jan. 1969.

Spitznogle, R. R., Some characteristics of magnetotelluric fields in the Soviet Arctic, Ph.D. thesis, Univ. of Texas, Austin, 1966.

Srivastava, S. P., Method of interpretation of magnetotelluric data when source field is considered, *J. Geophys. Res., 70*(4), 945–954, 1965.

Srivastava, S. P., J. L. Douglass, and S. H. Ward, The application of the magnetotelluric and telluric methods in Central Alberta, *Geophysics, 28,* 426–446, June 1963.

Swift, C. M., Jr., A magnetotelluric investigation of an electrical conductivity anomaly in the southwestern United States, Ph.D. thesis, Geophys. Dep., M.I.T., Cambridge, Mass., July 1967.

Vozoff, K., H. Hasegawa, and R. M. Ellis, Results

and limitations of magnetotelluric surveys in simple geologic situations, 1, *Geophysics, 28*(5), 778–792, 1963.

Vozoff, K., R. M. Ellis, and M. D. Burke, Telluric currents and their use in petroleum exploration, *Bull. Amer. Ass. Petrol. Geol., 48*(12), 1890–1901, 1964.

Wait, J. R., On the relation between telluric currents and the earth's magnetic field, *Geophysics, 19*(2), 281–289, 1954.

Wait, J. R., Theory of magneto-telluric field, *Radio Sci., 66D*(5), 509–541, 1962.

Weaver, J. T., The electromagnetic field within a discontinuous conductor with reference to geo-

magnetic micropulsations near a coastline, *Can. J. Phys., 41*, 484–495, 1963.

Woods, R. D., The northern structural rim of the Gulf basin, *Trans. Gulf Coast Ass. Geol. Soc., 6*, 3–11, 1956.

Word, D. R., H. W. Smith, and F. X. Bostick, Jr., An investigation of the magnetotelluric tensor impedance method, *Tech Rep. 82*, Elec. Geophys. Res. Lab., Elec. Res. Center, Univ. of Texas at Austin, March 1970.

Wu, F. T., The inverse problem of magnetotelluric sounding, *Geophysics, 33*(6), 972–979, 1958.

Yungul, S. H., Magnetotelluric sounding three-layer interpretation curves, *Geophysics, 26*(4), 465–473, 1961.

DISCUSSION

Madden: You showed a conductive zone in the mantle with resistivity less than 1 ohm-m. Were you working at low enough frequency to measure this? Your longest period was about two hours?

Bostick: All we can say is that we were

heading into a conductive zone. The absolute value of the conductivity is not well defined.

Porath: I am not against the conductive zone in the mantle, but it appears to come too shallow under the Precambrian uplift.

The Thermal Structure of the Continental Crust

DAVID D. BLACKWELL

Department of Geological Sciences
Southern Methodist University, Dallas, Texas   75222

Abstract. In the United States, the surface heat flow can be separated into two main components. The first component is due to radioactive heat sources in the upper crust, and the second component is due to sources in the lower crust and upper mantle. The heat flow from the lower crust and upper mantle is constant over large regions, called heat flow provinces, and the transitions between provinces are very narrow (less than 100 km). High values of mantle heat flow (>1.4 μcal/cm^2 sec) occur in the Basin and Range, Columbia Plateaus, Northern Rocky Mountains, Southern Rocky Mountains physiographic provinces, and in the Franciscan rocks east of the San Andreas fault zone. Normal or near-normal mantle heat flow (0.8) is found in the United States east of the Rocky Mountains, in the Colorado Plateaus, the Southern California batholith, and in the Puget Sound Region. Subnormal mantle heat flow (0.4, the lowest known anywhere) occurs in the Sierra Nevada Mountains. The variations in mantle heat flow are attributed to the thermal effects of sea-floor spreading during the Cenozoic. Measurements on other continents suggest heat flow provinces there with mantle heat flow values similar to those found in the United States. Temperatures are calculated for the three heat flow provinces under a variety of assumptions. The mantle heat flow is the main determinant of crustal temperatures, and thus vertical variations in electrical conductivity due directly or indirectly to temperature effects must vary in depth between the heat flow provinces. The region in the United States with the lowest crustal temperatures in crystalline terrain is the western foothills of the Sierra Nevada; hence this area should be the most favorable for feasibility studies of crustal transmission of electromagnetic waves.

The most important parameters influencing the variation of electrical resistivity with depth in the crust and upper mantle of the earth are fluid content and temperature; composition is relatively unimportant [*Brace et al.*, 1965; *Brace and Orange*, 1968]. *Brace* [1971] discusses in detail the effect of these parameters; he uses some of the temperature information discussed here. The emphasis in this paper will be on review of the distribution of continental heat flow, discussion of the parameters that influence this distribution, and presentation of several different crustal temperature-depth curves for different heat flow provinces. According to preliminary studies of the variation of electrical resistivity with depth in the crust, it appears that conditions are most favorable for the presence of an electromagnetic wave guide when the temperatures are lowest and crystalline rocks are exposed at the surface. Hence, areas of low heat flow in plutonic rocks will be of particular interest, and factors influencing

the possible location of such areas will be discussed. Heat flow in the oceans is rather clearly related to the age of the oceanic lithosphere, and temperatures can be calculated on the basis of the sea floor spreading models [*McKenzie*, 1967; *Sclater and Francheteau*, 1970], at least away from the trench-island arc areas, and will not be discussed here.

Throughout the paper, the units of heat flow used are 10^{-6} cal/cm^2 sec and the units of heat generation used are 10^{-13} cal/cm^3 sec. The units will be referred to as hfu (heat flow units) and hgu (heat generation units), respectively.

DISTRIBUTION OF HEAT FLOW IN THE UNITED STATES

Effect of heat production in the upper crust. Two main factors influence the heat flow measured at the surface in the United States: the heat production from uranium, thorium, and potassium measured in the basement rocks in which (or above which, in the case of such areas as the midcontinent) the heat flow value is measured; and the heat flow from the upper mantle [*Birch et al.*, 1968; *Roy et al.*, 1968a, 1971; *Lachenbruch*, 1968]. The key to the separation of these two components of heat flow for determinations made in plutonic rocks was the discovery that there is a linear relation between heat flow and heat production in such rocks [*Birch et al.*, 1968]. The relationship is

$$Q = Q_0 + A\mathbf{b} \qquad (1)$$

where Q is the surface heat flow, Q_0 (the intercept value) is the heat flow from below some layer whose thickness is related to $\mathbf{b}$, A is the radioactive heat production of the plutonic rocks, and $\mathbf{b}$ (the slope of the line) is a constant. All data relating heat flow and heat production in the eastern United States are shown in Figure 1. The black circles are the original data that *Birch et al.* [1968] used to obtain (1). There are two extreme distributions of heat production versus depth that satisfy this empirical relationship: a constant heat generation from the surface to a constant depth given by $\mathbf{b}$ [*Birch et al.*, 1968]; and a heat generation of the form

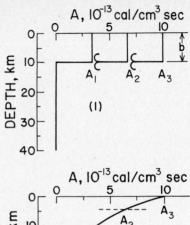

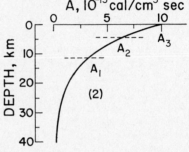

Fig. 2. Two models of heat production versus depth that satisfy the linear relationship of surface heat flow and surface heat production. Model 1 has constant heat generation to the depth $\mathbf{b}$ given by the slope of the line; thus different surface heat generation (A_1, A_2, and A_3, for example) values imply lateral variations in the heat production of the layer. Model 2 has a heat generation decreasing exponentially with depth according to equation 2; different surface heat generation values imply differential erosion and-or geographic variations in the initial surface heat generation A_0.

$$A(x) = A_0 \exp (-x/\mathbf{b}) \qquad (2)$$

where x is depth and A_0 is the measured surface heat production [*Lachenbruch*, 1968; *Roy et al.*, 1968a]. Values of the constant $\mathbf{b}$ range from 7.5 to 10 km. The model with constant heat production versus depth is illustrated in part one of Figure 2. In this model, different surface heat generation values require geographic variation in the heat production of the layer. The model with exponentially decreasing heat production versus depth is illustrated in part two of Figure 2. In the exponential model, different surface heat generation may reflect differential erosion or perhaps a different initial surface heat generation (A_0). If the exponential distribution is applicable, the depth of the layer would apparently have to be 2 or 3 times $\mathbf{b}$ [see *Lachenbruch*, 1970], and so Q_0 would be in effect the

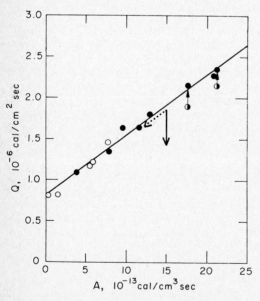

Fig. 1. Heat flow and heat production data for plutons in the New England area (black circles) and the central stable region (white circles) after *Roy et al.*, [1968a]. The line is fitted to both sets of data. The significance of the heavy and dotted arrows is explained in the text.

mantle heat flow. If the constant heat production distribution is applicable, then the heat generation in the crust below the layer must be very low or very uniform for the linear relationship to be observed. Because the lower crust is probably more basic than the upper crust, it seems more reasonable that its heat production is small. *Lachenbruch* [1968] pointed out that the principal argument in favor of the exponential model is that the linear relationship between heat flow and heat production will not be affected by differential erosion.

The two arrows in Figure 1 show the effect of differential erosion at a locality originally having a surface heat production of 15 hgu and a surface heat flow of 1.92 hfu. If erosion affects only that locality, then the point will move relative to the other points, as shown by the dotted arrow if the heat production distribution follows (2) (remaining on the line) or as shown by the black arrow if the heat production is constant with depth. These two models clearly represent the extreme cases as stated by *Roy et al.* [1968a, p. 7], not including a systematic increase in heat production with depth (which seems very unlikely except perhaps in local situations) or a heat production decreasing faster than given by (2) with the proper value for **b**. In several areas, very rapid decrease in heat production with depth is observed or inferred [*Dolgushin and Amshinsky*, 1966; *Jaeger*, 1970; *Roy et al.*, 1971], but the significance of such cases to the linear heat flow-heat production relationship is not yet understood. This situation of very rapid decrease does appear to be rare, however. *Lachenbruch* [1970] added to the constant and exponential models a third heat production-depth relationship. For a given amount of differential erosion, this third model (with a linear decrease of heat production versus depth) diverges from the curve but less rapidly than the constant model. Obviously there are an infinite number of models between the constant and exponential models that diverge more or less rapidly from the curve depending on the rate of decrease of heat production with depth. Thus the introduction of additional models such as the linear one is unnecessary and indeed confusing until we have additional independent data on the behavior of heat production with depth.

Available data on the distribution of uranium, thorium, and potassium with depth in plutonic rocks is inconclusive as to the usual rate of variation with depth (if any). The values of **b** imply decreases in heat production of 7 to 10% per kilometer, and thus vertical sections of several kilometers are necessary to overcome the inherent imprecision of U, Th, and K determinations ($\pm$5–10%) and the effects of natural fluctuations of radioelement content (10–20% from sample to sample [see *Rogers et al.*, 1965]). Surface exposures with vertical variations of 1 to 2 km are rare and occur only over lateral distances of several kilometers and there are few deep boreholes in homogeneous granitic rock. *Lachenbruch* [1968] suggested that the exponential model was valid for the Sierra Nevada, but in the areas where 20 to 30 km of erosion is required (the western foothills) the metamorphic grade is the lowest in the Sierra Nevada [*Clark*, 1960; *Best*, 1963], and in fact nowhere in the Sierra Nevada mountains is there evidence that the granitic rocks were emplaced at a depth of more than 5 to 10 km [*Kerrick*, 1970; *Bateman and Eaton*, 1967]. Finally, it appears that the rocks in the Sierra Nevada foothills are significantly different chemically and petrologically from the rocks along the crest [*Bateman and Dodge*, 1970; *Tilling et al.*, 1970]. Thus, the hypothesis that the rocks in the foothills represent deeply eroded roots of the rocks along the crest [*Lachenbruch*, 1968] appears invalid, and the inference that the exponential decrease in heat production with depth applies to the Sierra Nevada is unsubstantiated.

In a recent study of the distribution of heat production in another Mesozoic batholith of the western cordillera, the Idaho batholith, *Swanberg* [1971] develops strong evidence for a decrease in heat production with depth of emplacement that is consistent with the exponential decrement inferred independently from the heat flow-heat production studies. *Heitanen* [1967, 1968] mapped an extensive area in the northern part of the Idaho batholith and identified several generations of intrusive activity. The oldest rocks were intruded during regional metamorphism at conditions of temperature-pressure near the aluminum silicate triple point (a pressure of about 3.75 kb, according to *Holdaway* [1971]). The fact that the oldest quartz monzonites have primary muscovite indicates a minimum pressure of consolidation of about 4 kb [*Evans*, 1965]. *Swanberg* [1971] considers two groups of data, one using heat production values only from quartz monzonites with K content of between 2.8 and 3.4% and a second using heat production values from an apparently genetically related suite of plutonic rocks ranging in composition from gabbro to granite. In both groups, heat production decreases with increasing depth of emplacement, and the decrease can be closely fitted with a curve of the form of (2) with **b** equal to about 9 km. Thus, there is apparently a strong pressure control in the distribution of U and Th that may not operate for K. Uranium is particularly affected, since Th/U ratios vary from 3 to 5 for the shallow rocks to 6 to 8 for the muscovite quartz monzonite.

The actual distribution is likely to be much more complex and indeed will probably vary from constant to exponential with very low decrements (1 to 2 km). It is probably only in a gross average way that the exponential model holds, much as is shown in a figure of *Roy et al.* [1968a, Figure 4, model 2], particularly since gravity studies suggest that the thickness of individual plutons rarely exceeds 10 km [*Bott and Smithson*, 1967].

Variations in mantle heat flow. The observation that heat flow measurements over large areas are linearly related to the heat production of plutonic rocks implies that the heat flow from below the radioactive layer varies little in areas characterized by a single line e.g., Figure 1. In the exponential model of heat production, Q_0 is in effect the mantle heat flow. The constant model could result in having the linear relationship hold for large areas only if the heat production in the lower crust is very small, and again Q_0 would not be much above the mantle heat flow. Therefore, if we call Q_0 the 'mantle' heat flow, we cannot be far wrong. Thus, the fact that heat flow is linearly related to heat production in the large regions of the United States implies that the mantle heat flow is nearly constant in these areas. The transition zones between regions of different mantle heat flow values are generally very narrow compared with those between the regions of uniform heat flow [*Roy et al.*, 1971].

From a study of the heat flow and heat production in plutonic rocks in the United States, *Roy et al.* [1968a] defined a heat flow province on the basis of its characteristic relationship between heat flow and heat production, computed the results, and identified three provinces: the eastern United States, where $Q_0 = 0.8$ hfu

and **b** = 7.5 km; the Basin and Range province, where $Q_0 = 1.4$ hfu and **b** = 9.4 km; and the Sierra Nevada, where $Q_0 = 0.4$ hfu and **b** = 10 km. From the geophysical point of view, the large variations of the intercept value Q_0, which reflect variations in heat flow from the mantle, are of more interest than the slope **b**, which varies much less from province to province and probably reflects variations in the geochemistry of the upper crust. With the relative importance of Q_0 and **b** in mind, *Roy et al.* [1971] re-defined a heat flow province as a region with the same mantle heat flow (Q_0). Thus, within a single heat flow province there might be sub-areas with heat flow and heat production lines of different slopes but identical intercept values. They presented a map of Q_0's and termed it a 'reduced' heat flow map. That map with additional data in the northwestern United States is shown as Figure 3. A 'reduced' heat flow value is Q_0 calculated from (1) transposed to $Q_0 = Q - bA$. In regions with only a few data points such as the Peninsular Ranges of southern California, the Salinian block, and the Northern Cascades, et cetera, **b** was assumed to be 10 km in calculating values of 'reduced' heat flow.

Figure 3 includes major physiographic provinces as well as heat flow provinces. The locations of reduced values are indicated, except in the Sierra Nevada and Peninsular Ranges, where the data are too closely spaced to be shown. The physiographic provinces make convenient units for discussion, since heat flow and physiographic boundaries often seem to be close to one another.

All available measurements of terrestrial heat flow in the western United States and adjacent portions of the Pacific Ocean are plotted in Figure 4. Although the heat flow contours shown in Figure 4 demonstrate that the western United States is characterized by large areas of differing regional heat flow, the map of reduced heat flow (Figure 3) is more useful and clearly indicates that the regional heat flow patterns are related to significant variations of heat flow from the upper mantle.

As indicated in Figure 3, the Appalachian Highlands, the central stable region, and the southern part of the Canadian Shield in the United States comprise the eastern United States heat flow province. No data are now available for the eastern Great Plains or for the Coastal Plain provinces. The Basin and Range heat flow province has been extended to include the Columbia Plateaus, Northern Rocky Mountains, Southern Rocky Mountains, the southeast part of the Colorado Plateaus, part of the Great Plains, and part of the Cascade Mountains [*Decker,* 1969; *Blackwell,* 1969; *Roy et al.,* 1971]. In California, the Franciscan block east of the San Andreas fault has very high heat flow, whereas the Salinian block west of the San Andreas has a mantle heat flow intermediate between that of the eastern United States and the Basin and Range heat flow provinces. There are no heat flow measurements presently published for the Klamath Mountains, Oregon Coast Ranges, or Olympic Mountains. Apparently normal heat flow is characteristic of most of the Wyoming Basin and Colorado Plateaus, although determinations are sparse.

The most striking feature of the heat flow pattern is the band of normal to low heat flow in the Puget Sound depression, Sierra Nevada Mountains, and Peninsular Ranges, where man-

tle heat flow values (Q_0) are 0.8, 0.4, and 0.7 hfu, respectively, flanked on the east by a broad region of high heat flow in the Basin and Range, Columbia Plateaus and Northern Rocky Mountains provinces and on the west by high heat flow in the Pacific Ocean.

The eastern United States heat flow province has been tectonically stable since early in the Mesozoic in the Appalachians and for somewhat longer in the other physiographic provinces. Thus, the province is considered characteristic of a 'normal' continent [*Roy et al.,* 1968b]. The regions of high heat flow in the west ($Q_0 =$ 1.4 hfu or greater) have all been the sites of extensive Cenozoic volcanism and tectonism. The physiographic provinces in the western United States with mantle heat flow significantly below 1.4 hfu have been less tectonically active in the Cenozoic and in general have had little or no Cenozoic intrusive or extrusive activity, although the Sierra Nevada Mountains and Peninsular Ranges were the sites of voluminous Mesozoic intrusive activity.

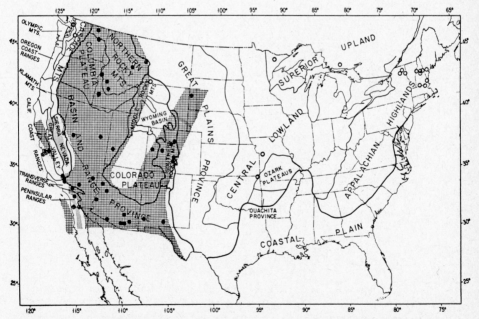

Fig. 3. Physiographic provinces, reduced heat flow values, and heat flow provinces in the United States. Sites indicated by white circles have reduced heat flow values of 0.8 ±0.1; dotted circle sites have values of between 0.9 and 1.3; and black circle sites have values greater than 1.3 hfu. Regions of high reduced heat flow are designated by a square pattern; regions of low reduced heat flow are designated by a dot pattern. The low values in the Sierra Nevada and many determinations in the Pacific Coast provinces could not be plotted because of the small scale of the map (after *Roy et al.,* [1971] with additional data in the northwestern United States).

Roy et al. [1968a, 1971] analyzed available data from other continents where combined heat flow and heat production measurements were available and concluded that at least two of the three heat flow–heat production curves discussed above might have a much broader significance than just relating heat flow and heat production in different parts of the United States. They found data from the shield regions of Canada and Australia that plotted close to

the curve for the eastern United States and a data point from the region of high heat flow in eastern Australia that plotted near the Basin and Range curve. They also pointed out that the similarity of the distribution of heat flow values [*Lee and Uyeda*, 1965] from the stable portions of all the continents (modes of 1.1–1.3 hfu, lowest values of 0.7–0.8 hfu) supports the inference about the applicability of the eastern United States curve to the stable portions of continents.

Jaeger [1970] discussed in detail the combined

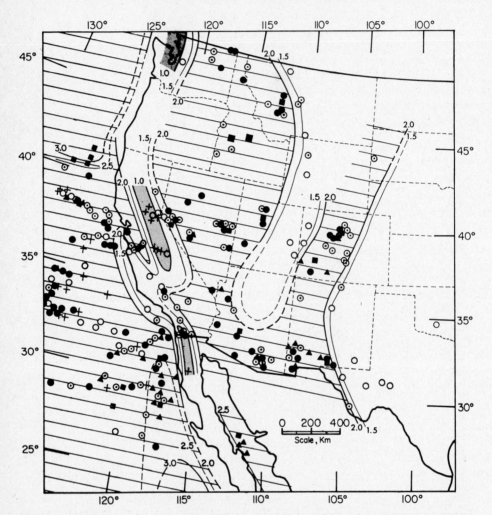

Fig. 4. Contour map of heat flow in the western United States (after *Roy et al.* [1971] with modifications in the northwestern United States). The contours delineate regions of high and low heat flow with average values of flux that would be measured in rocks with heat production within the range of granodiorite. Pluses represent observed heat flow values in the range 0–0.99; white circles, 1.0–1.49; dotted circles, 1.5–1.99; black circles, 2.0–2.49; black triangles, 2.5–2.99; and black rectangles, >3.0. Units are 10^{-6} cal/cm^2 sec.

heat flow and heat production data for Australia and found a linear relationship for the western portion of the Australian shield (using 3 points) with a Q_0 of 0.64 hfu and a **b** of 4.5 km. The rocks have isotopic ages in excess of 2.5 Gy. It will be interesting to see whether the 0.16-hfu difference between the curve for the eastern United States (based primarily on rocks of Grenville age (1.1 Gy) or younger) and for the Australian shield holds up as more data are collected. Most of data from the high heat flow region in eastern Australia fall near the curve for the Basin and Range heat flow province. However, several of the points are in metamorphic or sedimentary rocks and, as *Roy et al.* [1968a] emphasize, such data are not strictly pertinent, because the vertical distribution of heat sources may not be as uniform as in large bodies of plutonic rock.

Without concomitant consideration of the radioactive heat production on the rocks in which the heat flow values are measured, divisions of heat flow based on 'age' of the orogenic province [*Polyak and Smirnov*, 1968; *Hamza and Verma*, 1969; *Verma and Panda*, 1970; *Sclater and Francheteau*, 1970] have no clear significance. In the United States, for example, the late Mesozoic orogenic belt of the Sierra Nevada Mountains has the lowest heat flow. There may be a correlation for Mesozoic and Cenozoic orogenic belts, but the detailed nature of the correlation cannot be determined without much more heat production data. For the older orogenic belts, particularly of Precambrian age, the question of bias must be resolved. As *Jaeger* [1970] points out, a high proportion of mineral deposits in Precambrian rocks is in the greenstone belts, where heat production is low. Therefore, an average of measurements made in holes drilled for mineral exploration (as are most heat flow determinations) will be weighted toward rocks of low heat production.

The average heat production for the 'continental crust' is estimated to be 4.4 hgu by *Heier and Rogers* [1963]. An almost identical value was found by *Phair and Gottfried* [1964] for the area in the western United States underlain by Mesozoic batholiths (nearly 250,000 km²) and by *Shaw* [1967] for a large area of the Precambrian shield area in Canada (4.6 hgu). Thus, there is no geochemical evidence that a systematic decrease in surface heat production

with age exists [*Roy et al.*, 1968a], and a model such as the one presented by *Sclater and Francheteau* [1970, figure 10] that assumes such a decrease is not supported by geochemical data. According to Figure 1, a range in surface heat flow of 1.1 to 1.3 hfu for the eastern United States corresponds to a range in surface heat production of 4.0 to 6.7 hgu. Because 1.1 to 1.3 hfu is the modal value for most continents, *Roy et al.* [1968a] suggested that such values might be the average continental heat flow if anomalous regions such as the Sierra Nevada and Basin and Range provinces were excluded. In turn, the surface heat generation implied to be the most common coincides remarkably with the values found by the geochemical investigations. This internal consistency is another argument for the broad applicability of the eastern United States heat flow-heat production curve. In the following section on temperature calculations, a value of 5.3 hgu is used for the surface heat generation of the models. The assumed heat production corresponds to a surface heat flow for the eastern United States of 1.2 hfu and is not much higher than the average surface heat production values inferred from the geochemical data.

CRUSTAL TEMPERATURES

The heat flow from below the upper crustal layer of heat production and the 'thickness' of this layer (the two quantities obtained from the linear relation between heat flow and heat production) and the limited range of models that can explain the relation provide a much more rational basis for calculation of crustal temperatures than information available in the past. Crustal temperatures are an important parameter bearing on the interpretation or calculation of electrical resistivity profiles, and several different heat production-depth models will be considered in this section for the three heat flow provinces defined by *Roy et al.* [1968a] in order to illustrate the differences in temperature that are compatible with different crustal models of heat production, thermal conductivity, and mantle heat flow. Temperatures were calculated with the assumption of steady state conditions. The validity of this assumption will be discussed below. The temperatures inside a layer of constant heat production and thermal conductivity are given by *Jaeger*

[1965, equation 10]; they are

$$T(x) = T_0 + Qx/K - Ax^2/(2K) \qquad (3)$$

where x is depth, T_0 is surface temperature, Q is surface heat flow, and K and A are, respectively, thermal conductivity and heat production of the layer. The temperatures in a layer with heat production decreasing with depth according to (2) are given by Lachenbruch [1968, equation 4]; they are

$$T(x) = T_0 + Q_0x/K + A_0\mathbf{b}^2(1 - e^{-x/\mathbf{b}})/K \qquad (4)$$

The temperature due to the radioactive layer alone is:

$$A\mathbf{b}^2/K \qquad (5a)$$

for the exponential case and

$$A\mathbf{b}^2/(2K) \qquad (5b)$$

for the constant case. Thus the temperature difference due to the different heat production models at depths of 2 or 3 times $\mathbf{b}$, if we assume constant thermal conductivity and no radioactive heat production in the lower crust, is merely $A\mathbf{b}^2/(2K)$; temperatures are higher for the corresponding exponential model. The temperature from (5) for a surface heat production of 5 hgu in the Sierra Nevada heat flow province is 42°C for the constant model and 84°C for the exponential model ($K = 6.5 \times 10^{-3}$ cal/cm sec °C). The calculated Moho temperatures are about 350°, 500°, and 750°C for the three provinces (Figure 5), and a maximum difference of 42°C (12, 8, and 6% respectively) would be possible, owing to the radioactive surface layer alone (conductivity constant), with the assumed heat generation of 5.3 hgu. On the other hand, a variation of 10% in the mantle heat flow will result in about an 8% change in temperature (25°–30°C) at the Moho for the Sierra Nevada, an 8.5% change (40°–45°C) for the eastern United States, and a 9% change for the Basin and Range (65°–70°C). Similarly, a change in the conductivity values of 10% would result in 10% (35°, 50°, and 75°C) variations in Moho temperature. Therefore, we conclude that the uncertainty of temperature at the crust-mantle boundary is due as much (or more) to different possible values of mantle heat flow or crustal thermal conductivity as it is to different possible heat production-depth models that satisfy (1).

Small variations in crust-mantle temperatures may be caused by geographic variations in surface heat production. Such differences are only 70°C at 7.5 km for the constant heat production model or 140°C at 35 km for the exponential model of heat production with a heat flow variation of 0.8 to 2.0 hfu in the eastern United States. These maximum temperature differences will probably not be reached, because the surface heat production varies laterally as well as vertically and the temperatures at the Moho will reflect some average surface radioactivity rather than point-by-point surface variations.

To illustrate the range of crustal temperatures possible within the broad framework of the heat production-depth models, four different temperature-depth curves have been calculated for each heat flow province. The parameters assumed for each of the four models are listed in Table 1, and the resulting temperatures are listed in Table 2 and plotted in Figure 5. Of the four calculations for a particular province, models 1 and 2 were calculated assuming a layer of constant heat generation of thickness given by the constant $\mathbf{b}$ of (1) and a thermal conductivity of 6.5×10^{-3} cal/cm sec °C. The average thermal conductivity for approximately 100 sites in plutonic rocks in the United States reported by Roy et al. [1968b] is 7.0×10^{-3} cal/cm sec °C. The conductivity of 6.5×10^{-3} cal/cm sec °C allows a small decrease in that average value to take into account the temperature dependence of thermal conductivity. The conductivity of the lower crust was assumed to be 5.0×10^{-3} cal/cm sec °C, about the value for gabbro or granite of temperatures above 200°–300°C [Birch and Clark, 1940]. Surface temperatures different from zero can be included merely by adding the appropriate surface temperature to the temperature at each depth.

A value of 5.3 hgu was assumed for the surface heat generation in all models; this implies surface heat flow values of 0.95, 1.2, and 1.9 hfu for the Sierra Nevada, eastern United States, and Basin and Range provinces, respectively. The two values of heat generation assumed for the lower crust (0 and 1.5 hgu) span the range of likely values. Indeed a lower value (0.5 hgu) must be assumed for the Sierra Nevada model to avoid a mantle heat flow of zero. Model 4 for each province consists of only one layer which

TABLE 1.   Models for Temperature-Depth Calculations
(Units of thermal conductivity are 10^{-3} cal/cm sec °C.)

Model Number	Surf. Heat Flow, hfu	Mantle Heat Flow, hfu	Depth to Layer 2, km	Depth to Mantle, km	Surf. Heat Gen., hgu	Layer 2 Heat Gen., hgu	Thermal Conduct., Layer 1	Thermal Conduct., Layer 2
Sierra Nevada								
1	0.95	0.42	10	40	5.3	0	6.5	5.0
2	0.95	0.25	10	40	5.3	0.5	6.5	5.0
3	0.95	0.37	20	40	5.3	0.69	6.5	5.0
4*	0.95	0.43	···	40	5.3	···	6.5	···
Eastern United States								
1	1.20	0.78	8	35	5.3	0	6.5	5.0
2	1.20	0.37	8	35	5.3	1.5	6.5	5.0
3	1.20	0.73	15	35	5.3	0.69	6.5	5.0
4*	1.20	0.81	···	35	5.3	···	6.5	···
Basin and Range province								
1	1.90	1.37	10	30	5.3	0	6.5	5.0
2	1.90	1.07	10	30	5.3	1.5	6.5	5.0
3	1.90	1.39	19	30	5.3	0.69	6.5	5.0
4*	1.90	1.42	···	30	5.3	···	6.5	···

* There is no second layer for this model.

makes up the entire crust with a conductivity of 6.5×10^{-3} cal/cm sec °C, an A_0 of 5.3 hgu, and an exponential decrement given by **b** calculated from the linear plot for the appropriate province. Model 2 has the same constants as the models discussed by *Roy et al.* [1968a], and the models numbered 4 are in effect the models discussed by *Lachenbruch* [1970], with a conductivity of 6.5 instead of 6.0×10^{-3} cal/cm sec °C. The base of the crust was assumed to be at 40, 35, and 30 km for the Sierra Nevada, eastern United States, and Basin and Range provinces, respectively, although temperatures were extrapolated to 50 km, assuming the same conductivity

TABLE 2.   Temperature* versus Depth for the Models in Table 1

Depth, km	Sierra Nevada				Eastern United States				Basin and Range			
	1	2	3	4	1	2	3	4	1	2	3	4
0	0	0	0	0	0	0	0	0	0	0	0	0
5	63	63	64	64	82	82	84	84	136	136	138	138
10	105	105	116	116	153	152	157	157	252	252	263	263
15	147	146	160	160	230	223	225	225	389	385	381	381
20	189	184	200	200	308	286	309	290	526	511	501	494
25	231	220	247	236	385	342	389	353	663	629	645	605
30	273	253	291	271	463	467	*786*	*715*	800	740	786	715
35	315	384	333	305	*541*	*431*	*542*	*478*	937	847	925	857
40	*357*	*312*	*371*	*339*	618	468	615	558	1074	954	1065	999
45	399	339	408	381	696	505	689	639	1211	1061	1204	1140
50	441	366	445	424	773	543	762	719	1348	1168	1344	1282

* Temperature is given in degrees centigrade. Temperature at the base of the crust in each model is italicized.

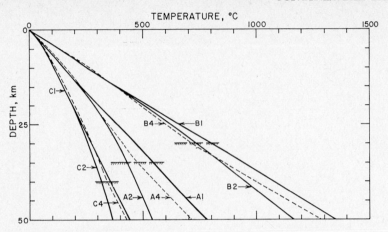

Fig. 5. Temperature-depth curves for the models with parameters listed in Table 1. The designations *A*, *B*, and *C*, refer to the eastern United States, Basin and Range and Sierra Nevada heat flow provinces, respectively. The base of the crust is indicated by hatching. The temperatures from models *A*3, *B*3, and *C*3 were not plotted because they nearly coincide with the temperatures from models *A*1, *B*1, and *C*1, respectively.

in the mantle as in the lower crust and no heat generation. The final model for each province is model 3. This crustal model was used in calculating crust and mantle temperatures by *Roy et al.* [1971] and *Herrin* [1971]. The upper crustal layer is assumed to extend to 2*b*, where a second layer begins with an exponential decrement of 80 km and an '*A*$_0$' of 0.69 hgu (calculated from 5.3e^{-2}). The first layer has an exponential distribution of heat sources according to (2). In the models calculated here, the heat production layer was stopped at the base of the crust. This model suggests a continental crust composed of 15 to 20 km of granitic material and 10 to 20 km of more basic rock with lower heat production and thermal conductivity. This distribution of material is in reasonably close agreement with seismic evidence on the composition of the continental crust [e.g., *Pakiser and Robinson*, 1966]. Quite independently, *Shaw* [1970] suggested a similar model for the heat source distribution in the oceans as used in model 3 [see *Roy et al.*, 1971] for the lower crust and upper mantle. It must be emphasized that all temperatures below the base of the crust in each model are extrapolations assuming the same conductivity as in the lower crust and no heat sources. For the Basin and Range model particularly, these assumptions are unrealistic and can at best hold only until the solidus curve for the mantle is reached.

Differences in Moho temperatures from those shown in Figure 5 due to departure from the assumed condition of steady state may be more important for the Sierra Nevada and Basin and Range provinces than variations in properties. The actual temperature distribution in the Basin and Range may be close to steady state because the source of the anomaly is near the base of the crust and has been operative for several tens of millions of years. In the Sierra Nevada, temperatures in the deep crust are undoubtedly underestimated, since the low mantle heat flow is interpreted to be a transient effect from a sink that ceased to operate 10 to 30 m.y. ago [*Roy et al.*, 1971]; see the following section.

The first and most obvious result of the calculations is that the temperature differences between the three provinces are much larger than the uncertainties in calculated temperatures introduced by the several possible models used, and a second result is that the differences within a province in temperature at the crust-mantle boundary are of the same order of magnitude or less than the possible differences due to variations in lower crustal heat production or assumed thermal conductivity values. The temperatures at the crust-mantle boundary can be calculated to as close as ± 50°C regardless of the model of radioactivity distribution assumed, if the heat production at the surface is known. *Brace* [1971] used the temperatures from model 2 for each province in his discussion of electrical resistivity in the crust. His values will be an upper limit for depths in excess of 25 km for the eastern United States and Sierra Nevada provinces because model 2 predicts the lowest temperatures of the four models. Also the depth at which mineral conduction becomes important may be slightly shallower than his calculations show.

DISCUSSION

The local variability of heat flow in crystalline terrains (on the scale of a few kilometers) is due primarily to lateral variations in upper crustal heat production. On the other hand, regional variations are due to differences in mantle heat flow. In the United States, variations in mantle heat flow are found in the west in provinces of late Mesozoic and Cenozoic tectonic activity. Electrical resistivity values in the shallow part of the crust of 10^5 to 10^6 ohm-m, which seem to be required for long distance transmission of electromagnetic radiation [*Levin*, 1968] might be reached only in the areas with lowest temperature and fluid content.

Thus, regions with high surface radioactivity and normal mantle heat flow or of high mantle heat flow, such as the northern Appalachians, and the Basin and Range heat flow province (see Figures 3 and 4) are certainly least favorable for feasibility studies of electromagnetic wave transmission in the crust. Where the mantle heat flow is normal, the most favorable locations for a high resistivity crustal layer are in areas of low surface heat production and crystalline rock exposures, such as the anorthosite terrain in New York. An area like the mid-continent gravity high, where the crust is apparently composed predominantly of basalt or gabbroic material, might also be a possibility. The region in the United States with the lowest temperatures in the crust is the western foothills of the Sierra Nevada, where both mantle heat flow and surface heat production are low and crystalline rocks are exposed at the surface. Other areas of low to normal mantle heat flow and low surface radioactivity are in the Peninsular Ranges and in the Puget Sound region.

Important questions to be answered are the extent to which the distribution of heat flow found in the United States is typical of continental heat flow and the extent to which similar areas on other continents will have similar heat flow. To answer this question we must know the origin of the heat flow pattern in the United States. The most striking feature of the heat flow distribution is the couple of low heat flow in the Puget Sound-Sierra Nevada-Southern California areas and the high heat flow in the immediately adjacent Cordilleran Thermal Anomaly zone [*Blackwell*, 1969]. This couple is attributed to the thermal effects of sea floor spreading during the Cenozoic [*Roy et al.*, 1971]. During most of the Cenozoic (and probably much of the Mesozoic), the North American continent drifted toward the East Pacific rise, and a subduction zone existed along the coast [*Bullard et al.*, 1965; *Atwater*, 1970]. Thus the western margin of the continent was a continental island-arc system. The characteristic heat flow pattern associated with an island-arc system is shown on the left side of Figure 6. A band of low heat flow (100–300 km wide) is found oceanward of a much broader band of high heat flow [*McKenzie and Sclater*, 1968; *Yasui et al.*, 1970; *Hasebe et al.*, 1970].

The zone of low heat flow has usually been ignored in discussions of the thermal effects of subduction [*McKenzie*, 1969; *Oxburgh and Turcotte*, 1970] but does appear to be present. For example, the low heat flow in Puget Sound is above a subduction zone that operated at least as recently as a few thousand years ago [*Dickinson*, 1970]. Such a zone of low heat flow eliminates the possibility that the subduction zone acts as a heat source along its entire length. *Minear and Toksoz* [1970] calculate several models that illustrate the maximum extent of low heat flow that might be associated with a subduction zone. *Hasebe et al.* [1970] present calculations that fit the observed data best. They assume that no heat is generated along the fault above a vertical depth of about 100 km. Thus, the low heat flow band is a conduction anomaly. However, at some depth, the oceanic crust along the upper part of the sinking lithospheric block begins to melt (with or without help from heat generated at some point along the subduction zone) and penetratively convects into the upper mantle and

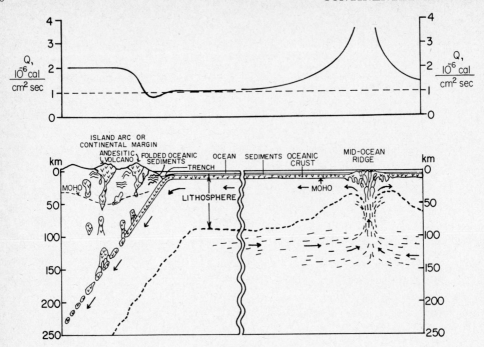

Fig. 6. Heat flow pattern and block movements associated with sea floor spreading. Not to scale.

crust to form calc-alkaline intrusives and andesitic volcanoes [see *Dickinson*, 1970]. The convection results in the sharp boundary between the low and high heat flow regions. *Hasebe et al.* [1970] generated the convective portion of the model by assuming a very high effective thermal conductivity.

Thus, during the late Mesozoic, a trench existed at the site of the Franciscan terrain in western California [*Ernst*, 1965, 1970; *Hamilton*, 1969]. Inland from the trench, melting along a subduction zone fed the batholiths forming in the Peninsular Ranges, Salinian block, Sierra Nevada, and Klamath Mountains, et cetera. Near the beginning of the Cenozoic, the direction, dip, or rate of underthrusting changed so that the region of high heat flow shifted inland and the crust beneath the Sierra Nevada and Peninsular Ranges began to be cooled by conduction of heat into the cold sinking block of lithosphere [*Roy et al.*, 1971]. The high heat flow in the Basin and Range was probably established with something near its present boundaries by early Oligocene.

Within the past 10 to 30 m.y., the continent

has interfered with the spreading from the rise and the pattern has become more complex; the San Andreas began to function as a transform fault, subduction ceased between the north end of the Acapulco trench and the Mendocino Fracture zone, and part of the continent was split by a new branch of the East Pacific rise (in the Gulf of California). This interaction has been summarized in detail by *Atwater* [1970]. However, because of the thermal time constant of the crust, the pattern that was established during the early and middle Cenozoic can still be recognized in the Sierra Nevada and Peninsular Ranges.

A cross section illustrating variations in heat flow in the far western United States at about 38°–39°N is shown in Figure 7; it illustrates the occurrence of low mantle heat flow in the Sierra Nevada next to the broad zone of high heat flow in the Basin and Range province. The heat flow in the Pacific Ocean is variable but usually high. If we refer to Figure 6, it is clear that as a trench and rise approach each other the temperatures in the sinking block will be progressively higher when it first starts

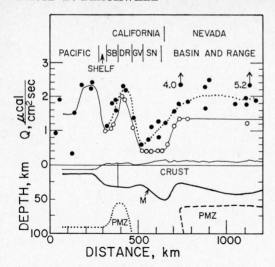

Fig. 7. Cross section of part of the western United States at 38°–39° N (adapted from *Roy et al.* [1971, Figure 18]). Black circles represent observed heat flow at the surface and white circles represent reduced heat flow. PMZ is the inferred partial melt zone. M represents the Mohorovicic discontinuity. The abbreviations for the Pacific Coast provinces are: SB is Salinian block; DB is Diablo Range; GV is Great Valley; SN is Sierra Nevada.

to sink. Thus the high heat flow in the Pacific Ocean off the West Coast reflects the youth of the sea floor generated along the East Pacific Rise before spreading stopped. The heat flow is appropriate for the age of the oceanic crust [*Sclater and Francheteau*, 1970]. The high temperatures in the oceanic block (due to its youth) that are now being destroyed between the still active Juan de Fuca rise and the North American plate may explain the lack of deep earthquakes along the proposed zone of subduction there. The high and very high heat flow in the Salinian block and Diablo range (Franciscan block) are probably due to recent changes in the thermal pattern associated with formation of the San Andreas fault and the northward translation of the Salinian block. The source in the Franciscan block must be within the crust because of the sharp boundaries of the anomaly. The mantle heat flow in the Basin and Range heat flow province is probably near the maximum for a broad region of a continent [*Roy et al.*, 1968a], since the partial melt zone in the mantle (based on the long range P wave profiles of *Archambeau et al.*

[1969] rises very close to the crust. If it rose higher, melting of the crust would begin and buffer the additional heat input. At the present time, 10% of the measured surface heat flow in the Basin and Range heat flow province is attributed to penetrative convection of material from the mantle to a shallow level in the crust [*Blackwell*, 1970].

Thus the three heat flow provinces defined for the United States may be general types. The curve found for the eastern United States may be typical of the stable portions of continents. The Basin and Range heat flow province may be typical of both high heat flow regions behind subduction zones and above continental extensions of rise systems, because it has the presumed uppermost limit of mantle heat flow possible for continental regions, and, finally, the Sierra Nevada heat flow province may be an example of a region of low mantle heat flow due to subduction of the oceanic lithosphere. Hence, tectonic and geochemical provinces similar to those in the United States which are favorable to shallow zones of high electrical resistivity in the crust should be favorable in other continents as well. Unknown at present, however, is whether such zones may be present in the oceans and how continental and oceanic zones might be connected across the continental shelves where there are no heat flow data.

Acknowledgments. This paper is based on ideas developed in cooperative work with R. F. Roy and E. R. Decker. E. Herrin, M. J. Holdaway, E. C. Robertson, and R. F. Roy read the manuscript and made useful suggestions.

The work was supported in part by National Science Foundation grant GA11351.

REFERENCES

Archambeau, C. B., E. A. Flinn, and D. G. Lambert, Fine structure of the upper mantle, *J. Geophys. Res., 74,* 5825–5865, 1969.

Atwater, T., Implications of plate tectonics for the Cenozoic tectonic evolution of western North America, *Bull. Geol. Soc. Amer., 81,* 3513–3536, 1970.

Bateman, P. C., and F. W. Dodge, Variations of major chemical constituents across the central Sierra Nevada batholith, *Bull. Geol. Soc. Amer., 81,* 409–420, 1970.

Bateman, P. C., and J. P. Eaton, Sierra Nevada batholith, *Science, 158,* 1407–1417, 1967.

Best, M. G., Petrology and structural analysis of metamorphic rocks in the southwestern Sierra

Nevada Foothills, California, *Univ. Calif. Publ. Geol. Sci., 42,* 111–158, 1963.

Birch, F., and H. Clark, The thermal conductivity of rocks and its dependence upon temperature and composition, 1, *Amer. J. Sci., 238,* 529–558, 1940a.

Birch, F., and H. Clark, The thermal conductivity of rocks and its dependence upon temperature and composition, 2, *Amer. J. Sci., 238,* 613–635, 1940b.

Birch, F., R. F. Roy, and E. R. Decker, Heat flow and thermal history in New England and New York, in *Studies of Appalachian Geology: Northern and Maritime,* edited by E-an Zen, W. S. White, J. B. Hadley, and J. B. Thompson, Jr., pp. 437–451, Interscience, New York, 1968.

Blackwell, D. D., Heat flow in the northwestern United States, *J. Geophys. Res., 74,* 992–1007, 1969.

Blackwell, D. D., Heat flow near Marysville, Montana (abstract), *Eos Trans. AGU, 51,* 824, 1970.

Bott, M. P., and S. B. Smithson, Gravity investigations of subsurface shape and mass distributions of granite batholiths, *Bull. Geol. Soc. Amer., 78,* 859–878, 1967.

Brace, W. F., Resistivity of saturated crustal rocks to 40 km based on laboratory measurements, in *The Structure and Physical Properties of the Earth's Crust, Geophys. Monogr. Ser.,* vol. 14, edited by J. G. Heacock, AGU, Washington, D.C., this volume, 1971.

Brace, W. F., and A. S. Orange, Further studies of the effect of pressure on electrical resistivity of rocks, *J. Geophys. Res., 73,* 5407–5420, 1968.

Brace, W. F., A. S. Orange, and T. M. Madden, The effect of pressure on the electrical resistivity of water saturated crystalline rocks, *J. Geophys. Res., 70,* 5669–5678, 1965.

Bullard, E. C., J. E. Everett, and A. G. Smith, The fit of the continents around the Atlantic, *Phil. Trans. Roy. Soc. London, 258,* 41–51, 1965.

Clark, L. D., Foothills fault system, western Sierra Nevada, California, *Bull. Geol. Soc. Amer., 71,* 483–496, 1960.

Decker, E. R., Heat flow in Colorado and New Mexico, *J. Geophys. Res., 74,* 550–559, 1969.

Dickinson, W. R., Relations of andesites, granites and derivative sandstones to arc-trench tectonics, *Rev. Geophys. Space Phys., 8,* 813–860, 1970.

Dolgushin, S. S., and N. N. Amshinsky, Uranium distribution in certain Altay granitoid intrusives, *Geokhimiya, 9,* 1081–1086, 1966.

Ernst, W. G., Mineral parageneses in Franciscan metamorphic rocks, Panoche Pass, California, *Bull. Geol. Soc. Amer., 76,* 879–914, 1965.

Ernst, W. G., Tectonic contact between the Franciscan mélange and the Great Valley sequence, crustal expression of a late Mesozoic Benioff zone, *J. Geophys. Res., 75,* 886–902, 1970.

Evans, B. W., Application of a reaction-rate method to the breakdown equilibria of muscovite

plus quartz, *Amer. J. Sci., 263,* 647–667, 1965.

Hamilton, W., Mesozoic California and the underflow of the Pacific mantle, *Bull. Geol. Soc. Amer., 80,* 2409–2430, 1969.

Hamza, V. M., and R. K. Verma, The relationship of heat flow with age of basement rocks, *Bull. Volcanol., 33,* 123–152, 1969.

Hasebe, K., N. Fujii, and S. Uyeda, Thermal processes under island arcs, *Tectonophysics, 10,* 335–355, 1970.

Heier, K. S., and J. J. W. Rogers, Radiometric determinations of thorium, uranium, and potassium in basalts and in two magmatic differentiation series, *Geochim. Cosmochim. Acta., 27,* 137–154, 1963.

Heitanen, A., Scapolite in the Belt Series in the St. Joe-Clearwater Region, Idaho, *Geol. Soc. Amer. Spec. Pap. 86,* 56 pp., 1967.

Heitanen, A., Belt series in the region around Snow Peak and Mallard Peak, Idaho, *U.S. Geol. Surv. Prof. Pap. 344-E,* 34 pp., 1968.

Herrin, E., A comparative study of upper mantle models: Canadian Shield and Basin and Range Provinces, in *The Nature of the Solid Earth,* edited by E. C. Robertson, McGraw-Hill, New York, in press, 1971.

Holdaway, M. J., Stability of andalusite and the $Al_2 SiO_5$ phase diagram, *Amer. J. Sci., 271,* 97–131, 1971.

Jaeger, J. C., Application of the theory of heat conduction to geothermal measurements, in *Terrestrial Heat Flow, Geophys. Monogr. Ser.,* vol. 8, edited by W. H. K. Lee, pp. 7–23, AGU, Washington, D.C., 1965.

Jaeger, J. C., Heat flow and radioactivity in Australia, *Earth Planet. Sci. Lett., 81,* 285–292, 1970.

Kerrick, D. M., Contact metamorphism in same areas of the Sierra Nevada, California, *Bull. Geol. Soc. Amer., 8,* 2913–2933, 1970.

Lachenbruch, A. H., Preliminary geothermal model for the Sierra Nevada, *J. Geophys. Res. 73,* 6977–6989, 1968.

Lachenbruch, A. H., Crustal temperature and heat production: Implications of the linear heat flow relation, *J. Geophys. Res., 75,* 3291–3300, 1970.

Lee, W. H. K., and S. Uyeda, Review of heat flow data, in *Terrestrial Heat Flow, Geophys. Monogr. Ser.,* vol. 8, edited by W. H. K. Lee, pp. 87–190, AGU, Washington, D.C., 1965.

Levin, S. B., Model for electromagnetic propagation in the lithosphere, *IEEE Trans., 56,* 799–804, 1968.

McKenzie, D. P., Some remarks on heat flow and gravity anomalies, *J. Geophys. Res., 72,* 6261–6273, 1967.

McKenzie, D. P., Speculations on the consequences and cause of plate motions, *Geophys. J. Roy. Astron. Soc., 18,* 1–32, 1969.

McKenzie, D. P., and J. G. Sclater, Heat flow inside the island arcs of the northwestern Pacific, *J. Geophys. Res., 73,* 3173–3179, 1968.

Minear, J. W., and M. W. Toksoz, Thermal regime of a downgoing slab and new global tectonics, *J. Geophys. Res., 75,* 1397–1420, 1970.

Oxburgh, E. R. and D. L. Turcotte, Thermal structure of island arcs, *Bull. Geol. Soc. Amer., 81,* 1665–1688, 1970.

Pakiser, L. C. and R. Robinson, Composition of the continental crust as estimated from seismic observations, in *The Earth Beneath the Continents, Geophys. Monogr. Ser.,* vol. 10, edited by J. S. Steinhart and T. J. Smith, pp. 620–626, AGU, Washington, D.C., 1966.

Phair, G., and D. Gottfried, The Colorado Front Range, Colorado, U. S. A., as a uranium and thorium province, in *The Natural Radiation Environment,* edited by J. A. S. Adams and W. M. Lowder, pp. 7–38, Rice University Press, Houston, 1964.

Polyak, B. G., and Ya. B. Smirnov, Relation between terrestrial heat flow and the tectonics of continents, *Geotectonics,* 205–213, 1968.

Rogers, J. J. W., J. A. S. Adams, and B. Gatlin, Distribution of thorium, uranium and potassium concentrations in three cores from the Conway granite, New Hampshire, U. S. A., *Amer. J. Sci., 263,* 817–822, 1965.

Roy, R. F., D. D. Blackwell, and F. Birch, Heat generation of plutonic rocks and continental heat flow provinces, *Earth Planet. Sci. Lett., 5,* 1–12, 1968*a*.

Roy, R. F., E. R. Decker, D. D. Blackwell, and F. Birch, Heat flow in the United States, *J. Geophys. Res., 73,* 5207–5222, 1968*b*.

Roy, R. F., D. D. Blackwell, and E. R. Decker, Continental heat flow, in *Nature of the Solid Earth,* edited by E. C. Robertson, McGraw-Hill, New York, in press, 1971.

Sclater, J. G., and J. Francheteau, The implications of terrestrial heat flow observations on current tectonic and geochemical models of the crust and upper mantle of the Earth, *Geophys. J. Roy. Astron. Soc., 20,* 493–509, 1970.

Shaw, D. N., U, Th, and K in the Canadian Precambrian shield and possible mantle compositions, *Geochim. Cosmochim. Acta., 31,* 1111–1113, 1967.

Shaw, H. R., Earth tides, global heat flow, and tectonics, *Science, 168,* 1084–1087, 1970.

Swanberg, C. A., The measurement, areal distributon, and geophysical significance of heat generation in the Idaho batholith and adjacent areas in eastern Oregon and western Montana, *Ph.D. thesis, Southern Methodist Univ.,* Dallas, Texas, 1971.

Tilling, R. I., D. Gottfried, and F. C. W. Dodge, Radiogenic heat production of contrasting magma series: Bearing on interpretation of heat flow, *Bull. Geol. Soc. Amer., 581,* 1447–1462, 1970.

Verma, R. K., and P. K. Panda, Further study of the correlation of heat flow with age of basement rocks, *Tectonophysics, 10,* 301–321, 1970.

Yasui, M., D. Epp, K. Nagasaka, and T. Kishii, Terrestrial heat flow in the seas around the Nansie Shoto (Ryukyu Islands), *Tectonophysics, 10,* 225–234, 1970.

DISCUSSION

Madden: The complete absence of deep focus earthquakes in the Sierra Nevada makes a Benioff zone look incongruous. Not only is it not active, it seems to me it is not there.

Blackwell: The whole pattern was disrupted—there is no spreading anymore. I do not know how long you want to follow the last bit of rock sinking under the Basin and Range province, but that is all gone now and the mantle under the Sierras has readjusted.

Evernden: When was this zone active?

Blackwell: Perhaps as recently as 5 to 10 million years ago.

Evernden: How do you explain the continuity of Sierra rocktype halfway out into the San Joaquin Valley?

Blackwell: The Basin and Range and the Sierra anomalies are part of one and the same problem and we are trying to explain both together. We do not particularly like our explanation, and you have brought up a serious

difficulty, but we have not been able to come up with anything better.

Brace: I am interested in possible correlation of seismicity with the boundaries of heat flow provinces. It seems you can make a case for this along the San Andreas, on each side of the Sierras, and along the Utah boundary. The earthquakes are shallow, of course, and the heat flow differences go much deeper.

Blackwell: The heat sources have to be relatively shallow because of the rapid transition between provinces—of the order of 20 to 50 km—but of course the earthquakes are shallower than that. Perhaps the temperatures are so high that you get plastic deformation at depth.

Kennedy: In a single province you measure heat flow and radioactivity. Then you correct your heat flow on the assumption that the surface radioactivity extends uniformly down to 10 km and find a uniform heat flow from below. That is very surprising, as it proves that all the rocks at the surface extend to 10 km depth. It is incredible considering the variability in radioactivity.

Blackwell: On the exponential model you get the same thing with the heat production distributed through a thicker layer and decreasing with depth. We take our measurements in large granitic bodies, where we have some hope that we are looking at a valid sample of what is happening at depth. We avoid metamorphic rocks because there is likely to be a very complicated distribution of radioactivity.

Higgins: How does the thermal time constant influence the heat flow in the Sierras if you take some kind of erosion rate into account?

Blackwell: Any normal history of erosion tends to increase the heat flow. The important thing about the Sierra Nevada is that the mantle heat flow is low there now.

Higgins: Is that why you fix the time of cessation of the underthrusting as recently as you do? It would not take very long to establish an anomalously high heat flow with the kinds of erosion rates that people talk about.

Blackwell: That is right. The cooling effect of a block of lithosphere 75 km thick in the mantle beneath the Sierra would only persist for 20 to 30 million years.

An in Situ Method of Determining the Pressure Dependence of Phase-Transition Temperatures in the Crust

ANTHONY F. GANGI AND NEAL E. LAMPING

Department of Geophysics, Texas A&M University, College Station, Texas   77843

Abstract. A multidisciplinary geophysical and geological approach has been undertaken to determine the temperatures and pressures at various velocity discontinuities (in particular, the Mohorovicic discontinuity, or Moho) in the earth's crust. One purpose of the investigation is to determine whether the Moho is the interface between two isochemical solid phases, such as basalt and eclogite. If this condition holds, the temperatures and pressures at the Moho would be linearly related as expressed by the Clausius-Clapeyron equation. Three different radioactive heat production models are used to calculate the temperatures at the Moho. An empirical relationship between surface heat flow and radioactive heat production provides a constraint on the calculated temperature values. The pressure is determined by using the thicknesses and densities determined from seismic data. An empirical velocity-density relationship is used to obtain the density from the measured seismic velocities. These crustal layer densities and thicknesses are constrained by the requirement that they be consistent with the measured gravity values. A modified Monte Carlo method is used to determine the variances in the determined temperatures and pressures caused by the variances in the values of the parameters used in the calculations. The area chosen as the test site for determining the Moho temperatures and pressures is the central United States because it is expected that a one-dimensional, steady-state thermal and kinetic model will be valid in this tectonically stable area. A calculation for an ideal (or representative) model of the central United States yields temperature and pressure values that are consistent with expected Moho values.

A multidisciplinary geophysical and geological approach has been undertaken to determine the pressures and temperatures at various velocity discontinuities in the earth's crust to test the hypothesis that these velocity discontinuities correspond to first-order, solid-solid, isochemical phase transitions. If these velocity discontinuities are first-order phase transitions, then a relationship should exist between the temperatures and the pressures that is consistent with the Clausius-Clapeyron equation. The program was designed primarily to test the Moho as a possible solid-solid phase transition; however, the method can readily be applied to other velocity discontinuities that appear to be as widespread as the Moho (e.g., the 'Conrad' discontinuity and the 'Reil' discontinuity).

Shortly after *Mohorovicic* [1910] described the velocity discontinuity that now bears his name, *Fermor* [1913, 1914] recognized that eclogite (with garnet and pyroxene) is a high pressure form of basalt (with plagioclase and pyroxene), and he proposed that the Moho is the interface between a basaltic lower crust and an eclogitic upper mantle. His proposal was supported by several researchers, including *Goldschmidt* [1922] and *Holmes* [1926, 1927], who recognized the significant tectonic implications of uplift and subsidence of the earth's crust by means of the basalt to eclogite transformation. The solid-solid phase transformation hypothesis received added impetus from the research by *Birch* [1952] on the geochemical and elastic properties of peridotite and eclogite and from the experimental synthesis of several eclogite mineral components by Coe [in *Roy and Tuttle,* 1956]. Later experimental work by *Robertson et al.* [1957] and *Kennedy* [1956, 1959] produced preliminary results that were highly favorable to the hypothesis that the Moho is an isochemical phase transformation from a basalt in the lower crust to an eclogite in the upper mantle.

In recent years, many researchers have taken

positions for and against the phase-change hypothesis. These contributors include *Lovering* [1958], *Kennedy* [1959], *MacDonald and Ness* [1960], and *Wetherill* [1961], who favored the hypothesis, and *Bullard and Griggs* [1961], *Ringwood* [1962a, b, 1966], and *Ringwood and Green* [1964, 1966], who opposed it. Calculations of the amounts of uplift and depression of the crust due to the effects of a phase change have been made by *MacDonald and Ness* [1960], *Wetherill* [1961] and, more recently, by *van de Lindt* [1967], *Joyner* [1967], and *O'Connell and Wasserburg* [1967].

Compromise hypotheses [*Wyllie*, 1963; *Pakiser*, 1965] have been advanced that incorporate the favorable aspects of a phase transformation (especially the tectonic consequences) and exclude or modify it where apparent contradictions occur. These hypotheses include the possibilities that: first, a different phase transformation occurs beneath the continents than beneath the oceans; second, the oceanic Moho is not a phase change, whereas the continental Moho is; or, third, a phase change occurs throughout the earth but it is not necessarily at the level of the accepted Moho everywhere.

The uncertainties exist primarily because: it was necessary to estimate the thermal properties of the crust, since accurate local values have not been known; the crust's structure has been idealized by, at most, two- or three-layer models; and it was necessary to use extrapolations of the pressure-temperature relations for the phase transitions of various minerals or mineral assemblages.

OBJECTIVE AND OUTLINE OF METHOD

The method used here follows the work of *Pakiser* [1965] closely. However, we have not yet decided that a compromise hypothesis is necessary, we have not assumed, a priori, that the Moho is a basalt-eclogite transition, and we have limited our study to a particular kind of geological environment, namely, ancient sedimentary basins in the central (stable) United States (i.e., the Great Plains province). In contrast, Pakiser compared the pressures (or depths) and temperatures of the Moho between the Basin and Range province and the Great Plains province of the United States. Also, because of the lack of data, Pakiser had to assume values for the thermal properties in these pro-

vinces. However, the values he chose were reasonable.

We have limited our study to ancient sedimentary basins in the central United States because: they represent a single type of geologic setting; accurate values of heat flux, thermal conductivities, velocities, densities, and thicknesses of the sedimentary layers are available there from well logs; good seismic data exist from which thicknesses and velocities can be determined for the basement layers as well as for the sedimentary layers; these are the most likely regions in which flat layers represent the geology with reasonable accuracy (so that one-dimensional solutions can be attempted with some hope of success); and these are the regions that are most likely to be close to isostatic and thermal equilibrium (so that static solutions can be expected to be reasonably accurate).

The most probable temperatures and pressures (and their variances) at various Moho depths are determined by using seismic data, heat-flow data, well-log data, and an empirical relationship between the P-wave velocity and the density (the velocity-density relationship is used primarily for the lower crustal, or basement, layers). A computer program, as it is now written, can incorporate 15 layers for the crust on top of an infinite mantle (see Figure 1). The layers are assumed to be flat, and we rely on well data and seismic reflection and refraction data to delineate the velocities and thicknesses of the crustal layers.

The pressure at the Moho, $P_M = P(z_M)$, is determined by

$$P_M = P(z_M)$$
$$= \int_{z_s}^{z_M} \rho(z) g(z) \, dz \cong \sum_{i=1}^{N} \rho_i g_i h_i \qquad (1)$$

where $\rho(z)$ and ρ_i are the densities at depth z and in the ith layer, respectively, $g(z)$ and g_i are the gravity values at depth z and in the ith layer, respectively, z_M is the depth to the Moho, z_s is the surface elevation, and h_i is the thickness of the ith layer. The summation is a valid approximation if the densities are constant within the layers. To compute pressure, we set all $g_i = g_0$, the value of gravity at sea level, without loss of precision in equation 1, considering the precision with which we determine ρ_i and h_i.

The densities and thicknesses used in equation 1 must be consistent with the measured gravity value at the surface, $g(z_s)$, defined as follows:

$$g(z_s) = g_s(z_R) + 2\pi\gamma \int_{z_s}^{z_R} \rho(z)\, dz$$
$$\cong g_s(z_R) + 2\pi\gamma \sum_i \rho_i h_i \qquad (2)$$

where $g_s(z_R)$ is the gravitational attraction at the surface due to the mass below some reference level (say, $z_R = 80$ km below the reference geoid) and γ is the gravitational constant ($\gamma = 6.67 \times 10^{-8}$ in cgs units).

The densities to be used in the above calculations come from either well data or an empirical velocity-density relationship such as that shown in Figure 2. The data shown in Figure 2 are a compilation of data compiled by *Woollard*

[1962], Nafe and Drake [in *Grant and West*, 1965, p. 200], and *Bateman and Eaton* [1967].

The temperature at the velocity discontinuity of interest is obtained from three different models of radioactive heat production in the basement layers of the crust. The first postulates two basement layers above the Moho with constant but different radioactive heat productions, velocities, densities, and thermal conductivities; the second has a single basement layer in which the radioactivity decreases linearly with depth, and the third assumes an exponential variation of the radioactive heat production with depth in the basement and mantle. This third model, which we prefer, is based on the work of *Lachenbruch* [1968] and *Roy et al.* [1968]. It can be shown that radioactivity in the sedimentary layers is negligible if large amounts of shale are

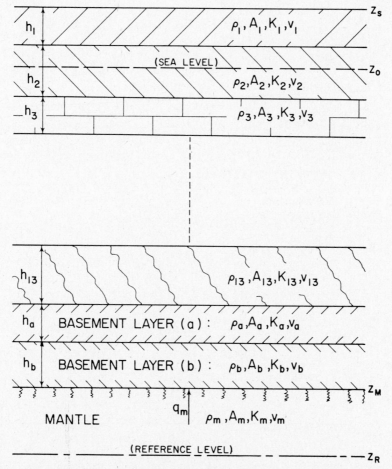

Fig. 1. Model crust used in computations of temperature and pressure.

not present; however, the computer program used to make the temperature and pressure calculations can incorporate the effect of sediment radioactivity.

For the third model, we obtain for the temperature at the Moho (assuming steady-state thermal and kinetic conditions):

$$T_M = T_s + q_s \sum_i h_i/K_i + (D^2 A\rho/K)$$

$$\cdot [1 - \exp(-h/D)] + q_m h/K \qquad (3)$$

where T_s is the temperature at the surface (in °C), q_s and q_m are the heat fluxes (in cal/cm²-sec) at the surface and from deep in the mantle, respectively, h_i and h are the thicknesses of the ith sedimentary layer and basement, respectively, K_i and K are the thermal conductivities (in cal/cm-sec°C) in the ith sedimentary layer and basement, respectively, D is

the empirical exponential decrement ($D = 8.3$ km, from *Roy et al.* [1968]) of the heat production depth distribution, A is the heat production (in cal/g-sec) at the top of the basement layer, and ρ is the density of the basement layer.

Equation 3 is constrained by the empirical relationship [see *Roy et al.*, 1968] between the surface heat flux and the 'surface' radioactivity

$$q_s = DA\rho + q_m \qquad (4)$$

Each variable in equations 1 and 3 for P_M and T_M is assigned bounds, that is, maximum and minimum values bracketing the measured (or derived) 'most probable values.' The most probable values of the Moho temperature and pressure are calculated by using expressions similar to the above, and their standard deviations are determined by using both Monte Carlo and variation of parameter type methods.

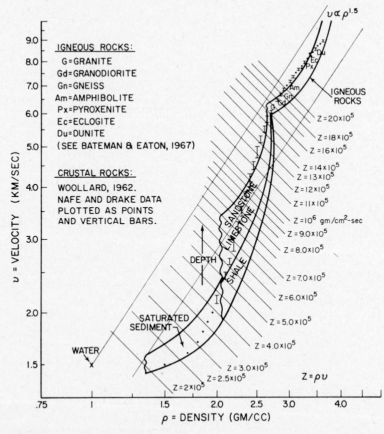

Fig. 2. Velocity and density distribution of crustal and upper mantle materials (elastic impedance $Z = \rho v$; a parameter).

The most probable values of the Moho temperatures and pressures obtained for different geographic regions are plotted on a P-T diagram. This diagram includes all phase-transition curves proposed in the past. The most probable values of Moho pressures and temperatures, along with error bounds, permit the testing of the hypothesis that the Moho is a solid-solid phase transition and the determination of whether the transition is one of those proposed earlier.

The ancient sedimentary basins chosen have different depths to the Moho and, consequently, they have different Moho pressures and temperatures. If the Moho is a first-order phase transition, then the temperatures and pressures should be related by the Clausius-Clapeyron equation [Swalin, 1967]; note that Swalin calls this the Clapeyron equation:

$$dT_M/dP_M = \Delta V/\Delta S = T \Delta V/\Delta H = \alpha \quad (5)$$

where dT_M/dP_M is the slope of the phase-transition curve, T_M and P_M are the phase-transition temperatures and pressures, respectively, ΔV is the change in volume, ΔS is the change in entropy, and ΔH is the change in enthalpy or the latent heat between the two phases.

Experimental results for a solid-solid phase transition over limited ranges of high temperatures and pressures should be closely represented by the integrated Clausius-Clapeyron equation:

$$T_M = \alpha P_M + T_0 \quad (6)$$

The values of α and T_0 obtained from this data will give a measure of the change in density and-or entropy of the phases and some bounds on the geothermal gradient.

SAMPLE CALCULATION, AVERAGE CENTRAL U. S. MODEL

An average or ideal central United States model of the crust has been constructed. The details of the methods used to obtain the model are given by Lamping [1970]. This crustal model is representative of the central portion of the Great Plains of the United States, including Oklahoma, Kansas, Nebraska, Iowa, and Missouri. Thus, it is in the central stable heat flow province of Roy et al. [1968] with heat flow parameters $D = 8.3 \pm 0.5$ km and $q_m = 0.76 \pm 0.04$ μcal/cm²-sec. The average surface heat flux for the area is 1.50 ± 0.03 μcal/cm²-sec, the average surface elevation is 0.6 km, the

SURFACE temp: 20.5 ±1.0°C elev: 0.6km q_s: 1.50 ±0.03		LAYER VARIABLES					
		h	V	ρ	g	K	A
QUATERNARY		0.32 ±0.01		2.15 ±0.04	979.9 ±0.1	3.87 ±0.08	
PLEISTOCENE		0.88 ±0.01		1.93 ±0.04	979.9 ±0.1	4.42 ±0.09	
CRETACEOUS		1.07 ±0.02		2.25 ±0.05	980.1 ±0.1	4.48 ±0.09	
LOWER MESOZOIC		0.65 ±0.01		2.52 ±0.05	980.2 ±0.1	6.75 ±0.13	
PALEOZOIC		1.63 ±0.08		2.75 ±0.14	980.4 ±0.1	7.20 ±0.36	
CAMBRIAN		0.58 ±0.03	5.65	2.69 ±0.13	980.5 ±0.1	4.75 ±0.24	
PRECAMBRIAN crystalline basement D: 8.3 ±0.5		11.26 ±0.56	6.25	2.72 ±0.14	981.1 ±0.3	5.60 ±0.30	3.28 ±0.16
LOWER CRUST		22.75 ±1.14	7.10	2.91 ±0.15	982.9 ±0.3	5.00 ±0.25	
MANTLE Moho depth:39.1 km q_m: 0.76 ±0.04			>8.00	3.32			

Fig. 3. Crustal model for an ideal central U. S. crust. Lithologic column is schematic and not to scale. Thickness h and logarithmic decrement D in km, P-wave velocity V in km sec⁻¹; density ρ in g cm⁻³; gravity g in cm sec⁻²; conductivity K in 10^{-3} cal cm⁻¹ sec⁻¹ °C⁻¹; heat production A in 10^{-13} cal g⁻¹ sec⁻¹; surface and mantle heat flux (q_s and q_m) in 10^{-6} cal cm⁻² sec⁻¹.

average surface temperature is 20.5 ± 1.0°C, the average depth to the Moho is 39.1 ± 1.8 km, and the average gravity value is 979.932 ± 0.010 gal.

The crustal model is composed of six sedimentary layers, with thicknesses, velocities, densities, and thermal conductivities as given in Figure 3. The surface gravity and the Moho temperature and pressure were calculated by using these values. The calculated gravity value is within two milligals of the initially assumed value (979,932 mgal) and the Moho temperature and pressure are (assuming an exponential radioactive heat production dependence with depth) $T_M = 736 \pm 37$°C and $P_M = 10.74 \pm 0.62$ kb, respectively. These values are of the order we would expect for the Moho, and their standard deviations are fairly small, i.e., of the order of 5 per cent.

The above temperature and pressure values have been compared with Ringwood and Green's [1966] data for the basalt-eclogite transformation. It is found that this data point is on the extension to lower temperatures and pressures of their transition curve between garnet granulite and eclogite (where garnet granulite is an intermediate phase between the basalt, or gabbro, and eclogite phases).

Acknowledgments. We would like to thank Professor James E. Case for his help and encouragement at the inception of this investigation, Professor Davis A. Fahlquist for his help and interest, and Mr. D. E. Powley of the Pan American Petroleum Corporation for providing us with data and his thermal conductivity results.

REFERENCES

Bateman, P. C., and G. P. Eaton, Sierra Nevada, *Science, 158*, 1407–1417, 1967.

Birch, F., Elasticity and constitution of the earth's interior. *J. Geophys. Res., 57*, 227–286, 1952.

Bullard, E. C., and D. T. Griggs, The nature of the Mohorovicic discontinuity, *Geophys. J. 6*, 118–123, 1961.

Fermor, L. L., Preliminary note on garnet as a geological barometer and on an infra-plutonic zone in the earth's crust, *Rec. Geol. Surv. India, 43*, 1913.

Fermor, L. L., The relationship of isostasy, earthquakes and vulcanicity to the earth's infraplutonic shell, *Geol. Mag., 51*, 65–67, 1914.

Grant, F. S. and F. G. West, *Interpretation Theory in Applied Geophysics*, p. 200, McGraw-Hill, New York, 1965.

Goldschmidt, V. M., Uber die Massenverteilung im Erdineren, . . . , *Natur wissenschaften, 10*, 918–920, 1922.

Holmes, A., Structure of the continents, *Nature, 118*, 586–587, 1926.

Holmes, A., Some problems of physical geology and the earth's thermal history, *Geol. Mag., 64*, 263–287, 1927.

Joyner, W. B., Basalt-eclogite transition as a cause for subsidence and uplift, *J. Geophys. Res., 72*, 4977–4998, 1967.

Kennedy, G. C., Polymorphism in the felspars at high temperatures and pressures, *Bull. Geol. Soc. Amer., 67*, 1711–1712, 1956.

Kennedy, G. C., The origin of continents, mountain ranges, and ocean basins, *Amer. Sci., 47*, 491–504, 1959.

Lachenbruch, A. H., Preliminary geothermal model of the Sierra Nevada, *J. Geophys. Res., 73*, 6977–6989, 1968.

Lamping, N. E., The Mohorovicic discontinuity as a phase transition, Ph.D. thesis, Dept. Geophysics, Texas A&M University, College Station, Texas, 1970.

Lovering, J. F., The nature of the Mohorovicic dis-

continuity, *Eos Trans. AGU, 39*, 947–955, 1958.

MacDonald, G. J. F., and N. F. Ness, Stability of phase transitions in the earth, *J. Geophys. Res., 65*, 2173–2190, 1960.

Mohorovicic, A., Das Beben vom 8. X. 1919, *Jahrb. Meteorol. Observ. Zagreb fur das Jahr 1909, 9* (pt. 4, sec. 1), 1–63, 1910.

O'Connell, R. J., and G. J. Wasserburg, Dynamics of the motion of a phase change boundary to changes in pressure, *Rev. Geophys. Space Phys., 5*, 329–410, 1967.

Pakiser, L. C., The basalt-eclogite transformation and crustal structure in the western United States, *U. S. Geol. Surv., Prof. Pap., 525-B*, 1965

Ringwood, A. E., A model for the upper mantle, 1, *J. Geophys. Res., 67*, 857–867, 1962a.

Ringwood, A. E., A model for the upper mantle, 2, *J. Geophys. Res., 67*, 4473–4477, 1962b.

Ringwood, A. E., The chemical composition and origin of the earth, in *Advances in Earth Sciences*, edited by P. M. Hurley, p. 287, M.I.T. Press, Cambridge, Mass., 1966.

Ringwood, A. E., and D. H. Green, Experimental investigations bearing on the nature of the Mohorovicic discontinuity, *Nature, 201*, 566–567, 1964.

Ringwood, A. E., and D. H. Green, An experimental investigation of the gabbro-eclogite transformation and some geophysical implications, *Tectonophysics, 3*(5), 383–427, 1966.

Robertson, E. C., F. Birch, and G. J. F. MacDonald, Experimental determinations of jadeite stability relations to 25,000 bar, *Amer. J. Sci., 255*, 115–137, 1957.

Roy, R., D. Blackwell, and F. Birch, Heat generation of plutonic rocks and continental heat flow provinces, *Earth Planet. Sci. Lett., 5*, 1–12, 1968.

Roy, R., and O. F. Tuttle, Investigations under hydrothermal conditions, *Physics and Chemistry of the Earth*, vol. 1, edited by L. H. Ahrens, K. Rankama, and S. K. Runcorn, pp. 138–180, Pergamon, London, 1956.

Swalin, R. A., *Thermodynamics of Solids*, p. 75, John Wiley, New York, 1967.

van de Lindt, W. J., Movement of the Mohorovicic discontinuity under isostatic conditions, *J. Geophys. Res., 72*, 1289, 1967.

Wetherill, G. W., Steady-state calculations bearing on geological implications of a phase-transition Mohorovicic discontinuity, *J. Geophys. Res., 66*, 2983–2993, 1961.

Woollard, G. P., *Rep. 62-9, AF contract AF23 (601)-3455*, Univ. of Wisconsin, for USAF Aeronautical Chart and Information Center, St. Louis, Missouri, Dec. 1962.

Wyllie, P. J., The nature of the Mohorovicic discontinuity: A compromise, *J. Geophys. Res., 68*, 4611–4619, 1963.

DISCUSSION

Healy: Most reflections or possible reflections from the Moho suggest it is a multi-boundary—that there are a number of layers more or less like a sedimentary basin. The Soviets showed some beautifully coherent reflections over a considerable distance in one particular area at a meeting in Leningrad two years ago. There were two boundaries, one where they expected the Moho and one somewhat deeper—10 or 20 km deeper, I think, but this was dependent on velocities which were poorly determined because of the near vertical incidence.

Kennedy: A sodic basalt will give you two sharp transitions in density and velocity; the first where garnet comes in with a velocity jump to 7.5 km/sec and the second where the jadeitic feldspar comes in with a jump to 8.1 or 8.2 km/sec. If the basalt is not sodic the second step may be masked. The second step may be either sharp or diffuse depending on the soda content of the basalt. The first transition will occur at the order of 20 to 25 km depth; the second, at 40 to 50, may be as shallow as 35 km depending on the temperature.

Hill: How sharp is the transition?

Kennedy: The first is always sharp—within a kilobar. The second may be either sharp or diffuse.

Hales: When one gets clear phases from the intermediate layer the amplitudes are always large. It is difficult to explain this other than in terms of a transition in velocity over a fair range of distance, say 5 to 6 km, or in terms of multiple reflections from a whole host of discontinuities that are quite close together. Must the upper discontinuity take place within 5 km, say, or would 10 km be allowable?

Kennedy: It occurs within the precision of the laboratory experiment, that is about 1 kb or 3 km. The density jump from 3 to 3.2 or 3.25 takes place over an interval that is at least as narrow as 3 km and maybe much narrower.

Hales: If you are well away from the melting point, is there any possibility that the time constant of the reaction could result in its being spread?

Kennedy: Of course. Spatially and temporally.

Sutton: The biggest difference in conditions for the transition we now call the Moho is between continents and oceans. If you compare continents and oceans you do not need a very high precision of some of these numbers, assuming of course that this is the same transition.

Gangi: That is the problem—there may be a different transition for the 'Moho' under the oceans. However, I would not expect much change in the chemical composition under the oceans themselves and, consequently, the temperatures and pressures obtained for the oceanic Moho ought to represent the equilibrium curve for a first-order phase change also.

3. PHYSICAL PROPERTIES OF THE OCEANIC CRUST

Widespread Occurrence of a High-Velocity Basal Layer in the
Pacific Crust Found with Repetitive Sources and Sonobuoys[1]

G. H. SUTTON, G. L. MAYNARD, AND D. M. HUSSONG

Hawaii Institute of Geophysics, University of Hawaii, Honolulu, Hawaii 96822

Abstract. Recent innovations in seismic field techniques make it possible to identify a high-velocity basal crustal layer (about 7.1 to 7.7 km/sec, averaging 7.4 km/sec) at several locations in the Pacific Ocean basin. This layer, which usually is masked and is primarily identified from second arrivals, has been only infrequently reported from seismic measurements. Use of a repetitive airgun source, however, shows that it occurs widely in the Pacific Ocean: in the Central basin southwest of the Hawaiian Islands; off the California coast; near the Murray fracture zone at about 155°W; and on the Fiji plateau. Twenty-one crustal columns analyzed to date have sufficient penetration to define deep crustal parameters. All these columns exhibit normal deep oceanic velocity structure above the high-velocity basal layer. Furthermore, all the stations except those off California and in the Fiji area have normal oceanic abyssal depths. The thickness of the high-velocity basal layer averages about 3.1 km and results in an increase in the total thickness of the crust compared to that obtained from interpretations that do not include this layer. In the past, a high-velocity basal layer had occasionally been reported from widely scattered locations in the oceans, generally on or adjacent to geologic structural features, and the layer was identified variously as either anomalous crust, upper mantle, or the result of misleading data. However, similar high-velocity layering under continents is well substantiated. Indications are that the layer may well exist under most or all ocean basins. The widespread occurrence of such a layer would have major significance, for both interpretation of other geophysical data (gravity and surface waves) and speculations on the geologic processes that resulted in the formation of the earth's crust.

Largely as the result of newly improved seismic refraction techniques, we have accumulated strong evidence for the existence of a widespread layer of high seismic velocity (7.4 ± 0.3 km/sec) at the base of the crust under the Pacific Ocean at the stations shown in Figure 1 and listed in Table 1. The evidence is being obtained with a single ship using sonobuoys, repetitive seismic energy sources, and simple visual phase correlation of refraction signals on a continuous precision echo-sounding

recorder. For convenience, we shall call this method Asper for Airgun and Sonobuoy Precision Echo-Recorder method.

In 1967, the Hawaii Institute of Geophysics began a program of single-ship wide-angle reflection and refraction measurements aimed primarily at defining sediments and the upper crustal layers; a 7000-joule sparker was used for a signal source. In 1968, an airgun source was added to the system.

In January 1969 we ran a series of Asper refraction stations in water depths in excess of 5 km that were located well away from any

---

[1] Hawaii Institute of Geophysics Contribution 381.

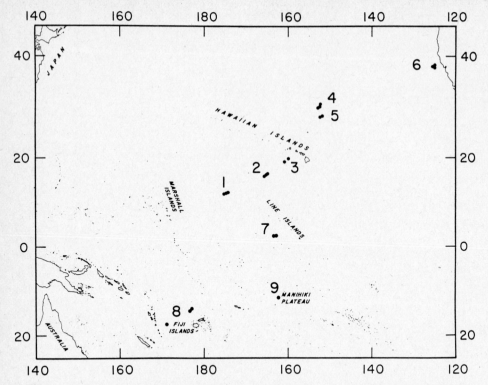

Fig. 1. Location of Asper stations conducted in 1969–1970 by the authors and discussed in the text. A high-velocity basal crustal layer was observed at all stations except 9, which had insufficient penetration. The stations are divided into nine groups to aid identification in the text and in other figures (also, see Table 1).

obvious structural transition or topographic high; we used sonobuoys as seismic receivers and a 20-cubic-inch airgun (having the equivalent energy per impulse of 0.02 pounds of 60% dynamite at a depth of 30 feet [*Kramer et al.,* 1968]) as a repetitive signal source. For the first time, we were able to obtain penetration through the entire crust to the upper mantle. These records clearly show a layer of about 7.4-km/sec velocity lying between the normal oceanic crustal layer (about 6.6 km/sec) and the top of the mantle (about 8.3 km/sec) [*Sutton et al.,* 1969; *Maynard et al.,* 1969; *Maynard,* 1970]. The stations were spaced along a line extending almost a thousand kilometers over a deep ocean basin from near the Marshall Islands to near the Hawaiian Islands. Since January 1969, the layer has been found at all locations where the Asper technique was used and where penetration was sufficiently deep; we have 21 observations to date. Using a similar technique, *Ewing and Houtz* [1969] reported

the observation of wide-angle reflections from the Moho at two locations in the Pacific.

Raitt [1963], on the basis of a large number of seismic refraction measurements that were made and analyzed by many different workers, established a nomenclature and a generalized crustal section for ocean basins that are still considered typical and standard. At normal ocean depths, beneath layer 1 (the unconsolidated sediments), this crustal section [after *Raitt,* 1963] is as follows:

Layer	Veloc., km/sec	Thickn., km
2	5.07 ± 0.63	1.71 ± 0.75
3 (oceanic)	6.69 ± 0.26	4.86 ± 1.42
4 (Moho)	8.13 ± 0.24	...

The sedimentary layer 1 is the most variable layer; it ranges in thickness from zero to over a kilometer and in velocity from less than the velocity of water to about 3 km/sec. Many refractors of widely varying velocities, as well as both positive and negative and linear and nonlinear

TABLE 1. Locations and Seismic Sections from Asper Profiles

Profile	Receiver Coord.	Azimuth,[a] deg.	Water Depth, km	Velocity[b] (km sec^{-1}) and Thickness[c] (km)							
									Oceanic	Basal	Moho
Group 1											
M	12°05′N, 175°05′W	069	5.4	1.5^d / 0.1	3.6 / 0.6	4.9 / 0.7			6.8 / 3.6	7.6 / 3.3	8.5
N	12°13′N, 174°44′W	069	5.4	1.5^d / 0.1	3.6 / 0.7	4.5 / 0.7			6.7 / 4.1	7.6 / 0.5	8.5
O	12°20′N, 174°25′W	069	5.4	1.5^d / 0.2		4.4 / 0.5		6.1 / 1.1	6.6 / 3.8	7.8 / 3.6	≥8.3
Group 2											
P	16°10′N, 165°42′W	059	5.3	1.5^d / 0.1		4.2 / 1.1			6.9 / 1.6	7.3 / 5.1	7.9
Q	16°26′N, 165°14′W	059	5.3	1.5^d / 0.3		4.5 / 0.4			6.2 / 2.1	7.1 / 5.1	≥8.5
R	16°29′N, 165°04′W	059	5.3		2.7 / 0.6		5.7 / 1.0		6.6 / 1.6	7.2 / 5.3	8.4
Group 3											
S	19°13′N, 160°49′W	055	4.9	1.5^d / 0.2	3.7 / 0.1	4.4 / 0.1	5.8 / 1.2		6.6 / 1.0	7.1 / 4.8	8.1
T	19°50′N, 159°53′W	050	4.5		3.5 / 1.1		5.9 / 1.1		6.9 / 3.6	8.0^e	
Group 7											
7	02°14′N, 163°07′W	065	5.2	1.6^d / 0.2		4.0 / 1.2			6.5 / 2.6	7.4 / 2.2	8.5
8	02°23′N, 162°50′W	245	5.2	1.6^d / 0.1	3.6 / 1.4				6.5 / 2.7	7.1 / 2.3	8.4
Group 4											
42	31°03′N, 152°13′W	164	5.3	1.6^d / 0.1		4.4 / 1.9			6.9 / 2.4	7.4 / 3.2	8.3

TABLE 1. (Continued)

Profile	Receiver Coord.	Azimuth,[a] deg.	Water Depth, km	Velocity[b] ($km\ sec^{-1}$) and Thickness[c] (km)					Oceanic	Basal	Moho
44[f]	30°30'N, 152°27'W	255	5.1	1.6 / 0.1		2.9 / 0.4	4.4 / 0.9		6.8 / 3.7	7.4 / 2.1	8.2
45[f]	30°24'N, 152°54'W	080	5.3	1.6 / 0.2		2.9 / 0.0	4.4 / 1.4		6.8 / 2.9	7.4 / 3.4	8.2
Group 5											
46	28°41'N, 151°49'W	260	5.4	1.6[d] / 0.1			4.4 / 1.0	6.5 / 1.4	6.9 / 2.8	7.5 / 2.6	8.2
48	28°29'N, 152°18'W	166	5.5	1.6[d] / 0.1			4.5 / 1.3		6.8 / 2.8	7.3 / 2.6	8.0
Group 6											
OBS[g]	38°10'N, 124°54'W	085	4.0		2.1 / 0.4	2.7 / 1.0	4.2 / 0.9		6.7 / 3.4	7.8 / 2.9	8.3
OBS[f]	38°10'N, 124°54'W	252	4.0		2.1 / 0.3	2.7 / 0.8	4.5 / 1.2		6.8 / 4.0	7.6 / 1.4	8.3
OBS27[f]	38°06'N, 125°27'W	082	4.1		2.1 / 0.2	2.7 / 0.9	4.5 / 1.2		6.8 / 1.9	7.6 / 2.9	8.3
OBS[f]	38°10'N, 124°54'W	185	4.0		2.1 / 0.3	2.7 / 1.0	4.7 / 1.0		6.9 / 3.3	7.5 / 2.4	8.3
OBS28[f]	37°52'N, 124°54'W	005	3.8		2.1 / 0.3	2.7 / 0.8	4.7 / 1.1		6.9 / 2.7	7.5 / 2.7	8.3

TABLE 1. (Continued)

Profile	Receiver Coord.	Azimuth,[a] deg.	Water Depth, km	Velocity[b] (km sec⁻¹) and Thickness[c] (km)							Oceanic	Basal	Moho
OBS[f]	38°10′N, 124°54′W	337	4.0	1.6[d]/*0.2*	2.1/*0.3*	2.7/*1.0*	4.7/*1.0*				6.8/*3.3*	7.6/*2.4*	8.2
OBS30[f]	38°24′N, 125°04′W	157	3.8	1.6[d]/*0.3*	2.1/*0.3*	2.7/*1.1*	4.7/*1.1*				6.8/*3.8*	7.6/*2.1*	8.2
Group 8													
20	14°29′S, 176°39′E	130	2.7	1.6[d]/*0.2*		2.6/*0.1*		5.3/*2.9*			6.9/*2.5*	≤7.8	
22	14°15′S, 176°53′E	210	3.0	1.6[d]/*0.3*					5.6/*2.4*		6.6/*1.3*	7.4	
30	17°26′S, 171°10′E	180	3.1	1.6[d]/*0.2*			3.6/*0.3*	4.9/*0.8*	5.6/*1.8*		6.9/*2.3*	7.5	
Group 9													
10	11°46′S, 162°28′W	163	2.6	1.5[d]/*0.1*	1.9/*0.3*	2.3/*0.4*	4.9/*0.3*	5.4/*4.4*	5.6/*1.9*	6.1/*3.5*	6.9		

[a] Direction of refraction line from receiving station.
[b] The velocity is the value in the upper half of the vertical pairs below.
[c] The thickness is the value in the lower half of the vertical pairs below. It is printed in italics.
[d] Velocity assumed from other measurements.
[e] Interpreted as basal crustal layer because of depth compared to nearby data (see text).
[f] One end of reversed profile. Pairs are joined by brackets.
[g] Columbia Ocean Bottom Seismometer.

velocity gradients, are found within this layer. Layer 2 is almost always identified and exhibits a fairly wide range of velocities. Layer 3, the oceanic layer, is nearly universally observed, is usually quite precisely measured, and has heretofore been assumed to be the bottommost layer of the crust. In general, the most persistent, strong, first seismic refraction arrivals are observed from the 6.7-km/sec layer 3.

In many respects, the difficulty in identifying the basal crustal layer is analogous to the difficulty that was encountered in identifying the (approximately) 5.0-km/sec layer 2 in the early days of seismic refraction work at sea. The layer was regularly found by workers in the Pacific and in the eastern Atlantic. But, in the western Atlantic, where it was relatively thin and overlain by a heavy overburden of sediments, layer 2 arrivals were masked and were not usually identified. Eventually the layer was found when investigators began to use denser shot spacing at crucial ranges and specifically searched for it in their records. Since then, the layer has been regularly found and included in seismic sections. We feel that a somewhat similar situation exists today and that our Asper method is making apparent a major crustal layer that was previously obscured.

Occasionally, a layer of roughly 7.1- to 7.6-km/sec seismic velocity has been reported in marine seismic profiles, but usually it was assumed to be either layer 3 with anomalously high velocity or a low-velocity mantle. This assumption was logical, because in the majority of cases either one or both of the 6.7-km/sec and the 8.1-km/sec layers were missed or marginally determined. In those cases in which the 6.7-km/sec oceanic layer and the mantle were both well determined and in which there was also a layer of intermediate velocity between them, the profile was almost invariably associated with some prominent topographical feature, such as a trench, an island arc, a continental margin, or a mid-ocean ridge. Figure 2 depicts the various locations where the layer has been observed. (The appendix lists references used to prepare Figure 2.) In constructing this map, we first looked for seismic sections containing three layers corresponding to the 6.7-km/sec layer 3, our basal crustal layer, and layer 4, the mantle. Although the sequence of layers was the main guide in selecting the

sections, a somewhat arbitrary range of velocity values was also used as a criterion for putting each layer in one of the three categories. Thus, the velocity of layer 3 must be at least 6.0 km/sec, that of the basal crustal layer must be 7.0 km/sec or more, and that of the Moho must be not less than 7.7 km/sec. We feel that a layer of at least 6.0 km/sec must be identified above the basal crustal layer to exclude the possibility that the velocity of the latter is from an anomalously high-velocity layer 3. Also, the basal crustal layer should ideally overlie a layer with a velocity of 7.7 km/sec or more, to exclude the possibility of confusing the basal layer with an anomalous mantle velocity. Stations meeting these requirements are marked by circles on the map. In some locations, a $6.0(+)$ -km/sec layer and a $7.0(+)$ -km/sec layer were observed, but no Moho velocity was found, although Moho often could be inferred from nearby crustal sections in apparently the same geological setting; these locations are marked by triangles. Finally, the cases of a layer 3 with abnormally high velocity (7.0 km/sec or greater) between apparently normal layers 2 and 4 are marked by squares. It is our opinion that at least some of these cases marked by squares could be examples of arrivals from the basal crustal layer that mask arrivals from layer 3, rather than the usual reverse situation in which the basal crustal arrivals are masked. Velocities near 7.5 km/sec are commonly found at the base of the seismic section under mid-ocean ridges (i.e., no higher velocities are observed at greater depth). The mid-ocean ridge locations are omitted from the map.

Of the above observations, the ones that most resemble ours are those of *Furumoto et al.* [1968] on the southwestern flank of the Hawaiian Island Ridge, and of *Den et al.* [1969] on the Shatsky rise. It can be argued, however, that Furumoto's observations are within the influence of the Hawaiian ridge, whereas Den's structural section is over a rise.

We now have enough observations of the basal crustal layer in different places in deep parts of the Pacific Ocean to warrant new study of the distribution of the layer and of the earlier explanations for its occasional identification. In addition to the first series (groups 1, 2, and 3 in Figure 1) in the Central basin between the Hawaiian and Marshall islands, we have data from the same basin west of the Line Islands (group 7, Figure 1) and from the deep ocean north of Hawaii (groups 4 and 5, Figure 1). Using the Asper technique, we have also observed the layer on the continental rise off California (group 6, Figure 1) and on the Fiji plateau (group 8, Figure 1). The last two localities are geologically similar to many of the sites where other investigators have identified a 7.1- to 7.7-km/sec layer with conventional seismic techniques.

It is obvious that the layer appears in several rather widely spaced locations. We believe it could exist quite generally under the oceans and that it was missed in earlier work because of limitations in the data.

EXPERIMENTAL AND ANALYTICAL PROCEDURES

The repetitive signal source employed in the Asper method is an airgun, often used in conjunction with a 7000-joule sparker. The airgun can be deployed with a 20-, 40-, 80-, or 120-cubic-inch chamber. The sources are synchronized so that the beginnings of their bottom reflections coincide. The repetition rate for the sources is usually 10 seconds, which at a speed of 6 knots is the equivalent of having a shot spacing of a little over 30 meters. Ship speed is often reduced to about 6 knots to increase the data density and to get the towed airgun deeper into the water and thereby to increase its low-frequency seismic energy output.

Seismic signals are received by a free floating,

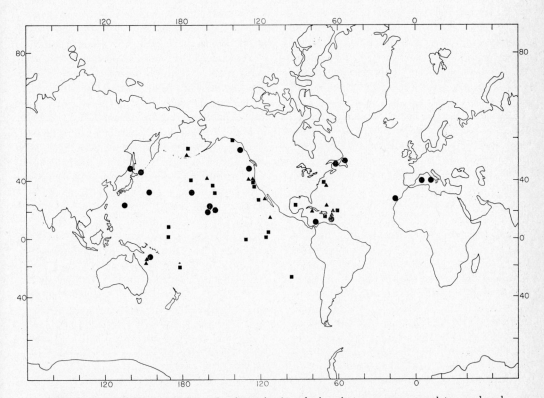

Fig. 2. Locations where a layer having seismic velocity that may correspond to our basal crustal layer has been reported by other workers (for references, see appendix). Circles denote stations that have a crustal layer of at least 6.0 km/sec overlying the basal crustal layer with velocity of about 7.4 km/sec, overlying a probable Moho with velocity of at least 7.7 km/sec. Triangles denote stations where two crustal layers of at least 6.0 km/sec and at least 7.0 km/sec were observed and where Moho velocity was either missing or inferred from other data. Squares denote stations where only one velocity that could be called layer 3 was identified but is abnormally high (7.0 km/sec or greater) and could be our basal crustal layer; again, the layer beneath has velocity of at least 7.7 km/sec.

expendable sonobuoy supplied by the United States Navy (model SSQ41A). This sonobuoy uses a hydrophone system which can be set to suspend the sensing element at a depth of about 60 feet. The sonobuoy relays the signals to the ship with an FM transmitter. The shipboard receiving system is comprised of directional antennas with preamplifiers and United States Navy sonobuoy receivers. From the receiver, the signal is filtered into three passbands: approximately 10–35; 50–120; and 500–1500 Hz. Each of these channels is independently amplified to appropriate levels, and the three channels are recombined. The high-frequency band is adjusted to detect the direct water-wave used to accurately determine range, the intermediate band is most useful for detection and resolution of reflections and refractions from upper sediment layers, and the low-frequency band is tuned for optimum reception of the refraction arrivals from the deeper crust and upper mantle.

The mixed signal is clipped, half-wave rectified, and amplified so that the ambient noise level is nearly the same as the clipping level. The signal is then recorded on the same precision echo sounding recorder used for vertical reflection work, generally on a 10-second sweep. This procedure produces a record having relatively little contrast in over-all darkness; the signals of interest are strongly correlated in phase.

Since the Asper method produces a time-versus-distance plot of seismic arrivals, it is possible to make rough but meaningful plane-interface refraction interpretations directly from the record by using standard formulas of refraction seismology. The accuracy of direct analysis of the record is limited because of paper warp (the result of using wet-paper type recorders) and changes in ship's speed relative to the buoy. For final analysis, the records are picked at intervals of about 0.5 second of water-wave time, and the resultant travel-time plot is analyzed using least-square solutions similar to analysis of conventional refraction runs. During this process, corrections for bottom and sub-bottom topography can also be applied.

Figure 3 is an Asper record obtained at station M (group 1 in Figure 1) in the Central basin. It can be thought of as a travel-time plot folded back and superimposed on itself at 10-second intervals. The usual initial reaction to

this type of record is confusion because of the many arrivals. This complication is at once both the principal value and the persistent difficulty with these records. Considerable care must be taken during interpretation to properly identify and disassociate reflections, refractions, and side echoes. Interference patterns from nearly simultaneous arrivals also complicate the record. Nevertheless, recorded this way, seismic signals that are completely indistinguishable when displayed on a wiggly-line record are readily traced on the Asper record. When reflections and refractions are properly identified, critical angles can be located and some properties of the interfaces can be studied. For instance, in cases of gradational interfaces, the reflection and refraction may come from different levels and have different time delays near the critical angle. In addition to showing a number of interfaces within the crust, the record often shows irregularities along individual curves, which are probably a result of geologic structure. The ability to follow individual phases, even through intersections with other higher amplitude signals, is perhaps the most useful characteristic of the method.

For example, on the record from line M (Figure 3), the 6.8-km/sec refraction line is a conspicuous first arrival, but prominent phases can readily be seen arriving after it. Most prominent of these is the arrival from the 7.6-km/sec basal crustal layer near its calculated critical range. The mantle arrival is also quite strong. These arrivals are seen to converge near the extreme right side of the record.

Ranges of error for both velocity and thickness are, for the crustal layers, generally smaller from Asper records than are those encountered in conventional refraction recordings, although presently the Asper method is too limited by range and source energy to yield consistently good Moho arrivals. The crustal velocities, when represented by first arrivals even for very short segments, are usually good to ± 0.1 km/sec. When picked entirely as a second arrival, where interference patterns sometimes confuse the record, the velocity can occasionally be in error by as much as ± 0.4 km/sec. Fortunately, most of the exclusively second-arrival refractions are from the sediments and sometimes from layer 2, and the interpretation of these shallow layers can be constrained and improved by analysis of the wide-angle and vertical reflection arrivals. The Moho velocity, when obtained, is usually calculated from a clear second arrival and is thought to be accurate to within ±0.2 km/sec.

By the nature of the Asper records, a precise velocity can be obtained from emergent refraction arrivals by correlating a strong phase other than the beginning of the first cycle of the arrival. In determination of thicknesses, if an error of one cycle is made in picking the beginning of the arrival, the resultant depth error for a typical oceanic crust ranges from about ±0.2 km for the 5.0- and 6.7-km/sec velocity layers to ±0.4 km for the 7.4-km/sec velocity layer and the Moho.

If errors in both the slope (velocity) and the first break are such that they add at the zero intercept, the resultant error in depth determi-

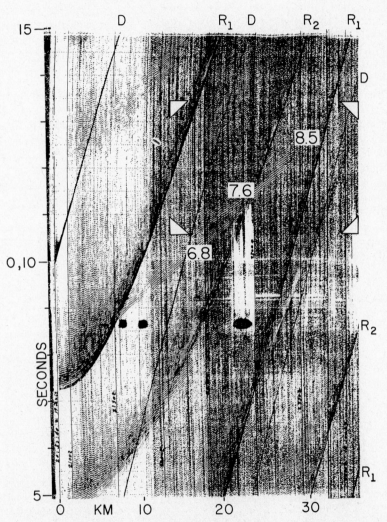

Fig. 3. Precision echo sounder record of Asper station M. Location shown in Figure 1, Group 1; coordinates in Table 1. Presentation is similar to a standard travel-time versus distance diagram except arrivals are folded back at 10-second intervals. (Middle of record represents 0, 10, 20, ··· sec travel time. Prominent refractions arrive in the interval 5–15 sec.) D, R_1, and R_2 are direct water wave, first bottom reflection and second bottom reflection, respectively. The numbers 6.8, 7.6, and 8.5 indicate location of prominent refraction arrivals having those apparent velocities (km/sec). The white triangles denote the section of the record enlarged in Figure 4.

nation can be considerably greater. In such
cases, again using station M as an example, the
error in depth to the 7.6-km/sec layer could be
±0.9 km, and to the Moho it could be ±1.1 km.
Of course, if a layer is undetected, errors in the
thickness of the adjacent shallower layer and
depths to deeper interfaces can be even greater.

Figure 4 is an enlargement of a portion of
the record from station M. In this portion, the
most prominent signal is the refraction from
near the top of the 7.6-km/sec layer. Within
this distance range, the 7.6-km/sec refraction
is a later arrival following the refraction from
the 6.8-km/sec layer, which is seen most clearly
in the lower left-hand portion of the figure. The
8.5-km/sec refraction starts near its calculated
critical range and with careful inspection can be
followed (intermittently) across the record.

Energy from the 7.6–8.5-km/sec interface appears to increase near the critical range, and
energy from the 6.8–7.6-km/sec interface increases at somewhat less than critical range. The
apparent velocity from the 6.8–7.6-km/sec interface near the critical range is considerably higher
than expected. Although this could be the result
of local topography on the interface, similar
discrepancies are found on other records, and
we are still searching for an adequate explanation. There are other details in these records,
also not yet fully understood, that may help
in studies of the nature of the interfaces.

Figure 5 is a theoretical travel-time plot showing reflection and refraction curves calculated
for the constant-velocity, flat-layer model interpreted from station M. It can be seen that
most of the arrivals from the 7.6-km/sec layer
are masked as secondary arrivals. Where the
refraction line does emerge as a first arrival, at
a range where explosives techniques often call
for large but widely spaced charges, the reflection and refraction curves from the 6.8-, 7.6-,
and 8.5-km/sec layers are very tightly grouped.
The 7.6-km/sec layer is represented by first arrivals for less than 10 km, and they precede the
6.8- and 8.5-km/sec arrivals by a maximum of
about 0.06 second.

OBSERVATIONAL RESULTS

Twenty-two crustal refraction measurements,
all made by using the Asper method, are reported here (Figure 1). Locations and results are
summarized in Table 1. The basal crustal layer
was observed in twenty-one of these measurements; the exception is the station on the
Manihiki plateau (group 9, Figure 1), which,
although it had insufficient penetration to reach
a possible basal crustal layer, is included for
peripheral interest and will be discussed later.

Central basin. Seismic sections were obtained
for the Central basin from 10 stations in normal
oceanic depths: 8 stations between the Marshall
and Hawaiian Islands (groups 1, 2, and 3,

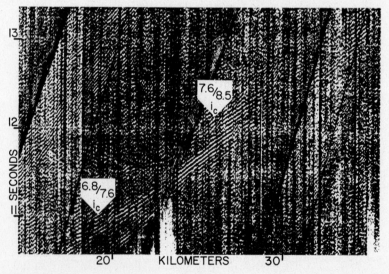

Fig. 4. Enlargement of portion of record from Asper station M, shown in Figure 3. Points at
bottom of inserts indicate calculated critical range for the interfaces of the basal crustal layer.

Figure 6); and 2 stations just southwest of the Line Islands (group 7, Figure 10).

The Asper stations in Figure 6 were made in January 1969, using a 20-cubic-inch airgun that was augmented by explosives on station S. The seismic sections are all the result of single-ended interpretations. The extension of the axis of the Line Islands intersects the ship's track between the stations in group 1 and group 2 with no apparent manifestation on the ocean bottom.

The uppermost sediment velocity was not observed but is inferred to have roughly the same velocity as sea water from velocimeter measurements on cores taken by the Hawaii Institute of Geophysics and by the *Glomar Challenger* on the Deep-Sea Drilling Project. Using normal-incidence and wide-angle measurements, which tend to corroborate the velocimeter findings, we estimate that the thickness of these very soft sediments averages a little more than 100 meters over this track.

Beneath the surficial layer of soft sediment is a system of one, two, and sometimes three layers which correspond roughly to layer 2 and which average about 1.3 km in total thickness. Velocities of about 3.6, 4.5, and 5.8 km/sec show up repeatedly but with no consistent discernible pattern. It is possible that all three layers are present in each of the sections but are not always resolved. Layer 3 has an average velocity of between 6.6 and 6.7 km/sec with an average thickness of 2.7 km if the 6.9-km/sec layer in the seismic section of station T is included and an average thickness of 2.5 km if it is excluded. There is some question regarding the interpretation of station T. Approximately 40 km to the north and parallel to this station, *Furumoto et al.* [1968] shot a conventional reversed profile, using explosives, which defined layer 3 at a depth much the same as that found at station T but having a somewhat lower velocity with the layer dipping to the southwest. They also found a 7.5-km/sec layer dipping to the southwest corresponding in depth to the unreversed 8.0-km/sec layer at station T. Therefore, it is likely that the section from station T defines only the two crustal layers and that the mantle was not reached. It should also be noted that this profile may be within the influence of the Hawaiian Islands.

The basal crustal layer has significantly

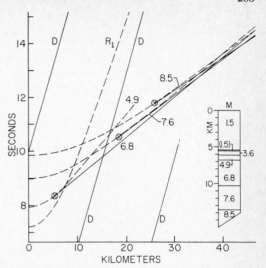

Fig. 5. Theoretical travel-time versus distance plot obtained from a ray path solution for the velocity model from station M. The diagram is plotted on roughly the same scale as the Asper record from station M to aid comparisons. For clarity, only the direct water-wave (D), the first bottom reflection (R_1), and the reflection curves and refraction lines for the 6.8, 7.6, and 8.5 km/sec interfaces are shown. The critical ranges are also marked.

higher velocity and greater depth in group 1 than in group 2. When all 21 sections are considered, there is some indication of the same velocity-depth relationship. (Error in measurement of slope could be partially responsible for such a trend.) The Moho arrival was weak at stations N, O, and Q and, as mentioned above, was probably not observed at station T. At station N, the Moho velocity is well determined, but the beginning of the refraction arrival is uncertain, so the intercept time, and therefore the depth to the mantle, is not well established.

Group 7 stations (Figure 10) were obtained in 1970 in 5.2 km of water about 400 km west of Christmas Island in the Line Islands. These seismic sections have a basal crustal layer that has low velocities in relation to the layer depth. The layer is not, however, as well defined here as elsewhere. Whereas the velocities of the basal crustal layer at our other stations are generally determined to ±0.1 km/sec, the data near the Line Islands, obtained there solely from second arrivals, yield velocities that are good to only about ±0.3 km/sec. The thickness error is therefore about ±1.0 km, as compared to about ±0.4 km at most sites, where a small segment of the refraction comes in as a first arrival or where a clear set of second arrivals exists. These sections are, however, quite similar to those from other deep-water stations reported in this paper.

For the deeper stations in the Central basin (groups 1, 2, and 7), the velocity of Moho appears to be significantly higher than average for the oceans: 8.38 ± 0.07 km/sec, or, eliminating the one lowest value, 8.44 ±0.03 km/sec.

Murray fracture zone. The stations in groups 4 and 5 (Figures 7 and 8) are located about 100 km north and south, respectively, of the axis of the Murray fracture zone, in about 5.2 km of water on either side. The Asper measurements were augmented with explosives and standard wiggly-line recording. Figure 7 shows the locations of the four profiles; one is parallel and one is perpendicular to the fracture zone on each side. The basal crustal layer appears in all four sections. One of the profiles was reversed, and the dips obtained on this profile give some indication of the errors in depth that may be encountered at unreversed stations with fairly simple geologic structure. In this case, the apparent velocities of the unreversed lines are essentially the same as the reversed velocity under stations 44 and 45.

The structures of both sides of the fracture zone are very similar at this longitude. If we study the sections in detail, we find, beneath the thin veneer of soft sediments, layer 2 (4.4 km/sec), overlying the oceanic layer (6.8 or

6.9 km/sec) except at station 46. The oceanic layer averages about 2.8 km in thickness. At Station 46, an additional layer is found overlying the 6.9-km/sec material and having a velocity of 6.5 km/sec and a thickness of 1.4 km. This layer might also exist (unresolved) under the other stations. Evidence for this layer at station 46 is a very short segment near the intersection of the crustal refraction with the ocean bottom reflection. This refraction arrival does not carry on as a second arrival after the 6.9-km/sec refraction begins as a first arrival and may represent a zone of increasing velocity at the top of the 6.9-km/sec layer. The basal crustal layer in this region has an average velocity of 7.4 km/sec, is about 2.8 km thick, and overlies Moho, which has an average velocity of 8.2 km/sec.

With the arrangement of profiles parallel and perpendicular to the ENE trending Murray fracture zone, this study is not optimum to detect the mantle anisotropy which *Morris et al.* [1969] and *Meyer et al.* [1969] found south of the fracture zone near the Hawaiian Islands. From their results, the maximum effect would be expected along N-S and E-W lines, with a velocity maximum in the E-W direction, and we would expect to find about 0.2 km/sec of anisotropy in mantle velocity with our orientation. In our observations, the lack of reversals for three of the four profiles and lateral velocity variations could disguise a difference in velocity resulting from anisotropy. The lines south of the fracture zone show approximately the expected velocity variation in the proper sense.

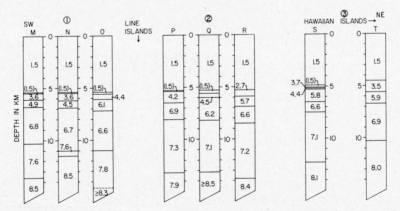

Fig. 6. Seismic velocity sections from stations made in January, 1969, with a 20-cubic-inch airgun. Velocities are in km/sec. Locations are shown in Figure 1.

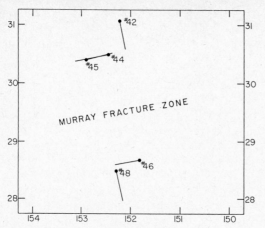

Fig. 7. Location of Asper stations on either side of the Murray fracture zone.

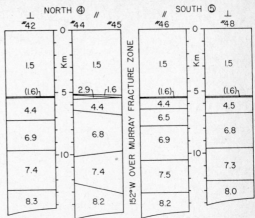

Fig. 8. Seismic velocity sections from stations, located in Figure 7, oriented perpendicular (indicated by ⊥) and parallel (indicated by ||) to the Murray fracture zone. These stations were made in July, 1969, with a 20-cubic-inch airgun and explosives. Velocities are in km/sec here and in Figures 9 and 10.

North of the fracture zone, however, the rather small difference, if it results from anisotropy, is of the opposite sense.

California continental rise. The profiles comprising group 6 were made with the Asper method and were reversed by the Columbia University ocean bottom seismometer, using a 40-cubic-inch airgun and explosives [*Hussong et al.*, 1969]. The seismometer is located in about 4 km of water on the Delgada fan, a portion of the continental rise off Point Arena, near San Francisco. When the shallower water and the geologic structure are taken into account, it is not surprising, from past work, that we found the basal crustal layer there.

Figure 9 shows two perpendicular seismic sections centered over the ocean bottom seismometer site. Except for about 1 km of overlying sediments having rather high velocity (probably turbidites), the section is much the same as those obtained in the deep ocean. The north, south, and west legs radiating from the ocean bottom seismometer were reversed; the eastern leg was not. Information from the horizontal components of the ocean bottom seismometer aided greatly in identifying the shear arrivals. The shear velocities in the sediment layers were approximated by using delay times from *P-S* conversions at the base of the sediments under the ocean bottom seismometer. The deeper crustal shear velocities were determined from observed refractions. A complete analysis of these records is being prepared for another paper.

Fiji plateau. The seismic sections of group 8 (Figure 10) were obtained on the Fiji plateau in water depths of about 3 km. A high-velocity basal crustal layer could have been anticipated here on the basis of the earlier work (Figure 2). From our sections, depths to the interface which we identify as the top of the basal crustal layer average 8 km below sea level. Although our measurements did not reach Moho, if we assume that the area is approximately in isostatic equilibrium, its depth can be estimated.

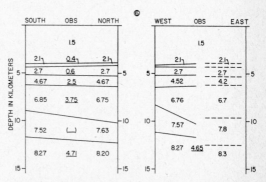

Fig. 9. Seismic velocity sections made in four directions over the Columbia University ocean bottom seismometer (located at 38°09'N, 124° 54'W; see Figure 1) and reversed on all sides except the east side by the Asper method (using 40-cubic-inch airgun). Underlined velocities are those of shear waves through the respective layers.

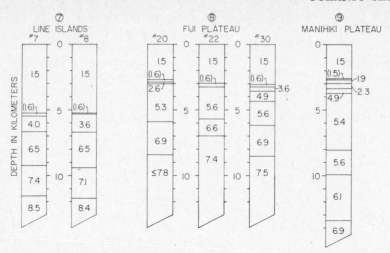

Fig. 10. Preliminary interpretation of Asper records just obtained from a 1970 cruise, using an 80- or 120-cubic-inch airgun. Locations shown in Figure 1.

Through use of water depth and normal density contrasts and use of the empirical relation established by *Woollard* [1959], we find that the Moho (if it exists as a sharp discontinuity) should be at a depth of about 18 km below sea level. If we use the relation established by *Demenitskaya* [1959], the Moho would be about 12.5 km below sea level. Velocities and thicknesses of the sediments and layers 2 and 3 are also defined; those for layer 2 are somewhat greater than the average for normal ocean depth.

It should be noted that in the majority of cases in deep water, layer 3 produces a first arrival for most of the record, with the signal from the basal crustal layer arriving behind it. However, in the case of station 22, on the Fiji plateau, the situation is reversed. In this case, the basal crustal layer produces a prominent first arrival throughout most of the record, and the signal from the oceanic layer above it is usually a later arrival. With the Moho too deep to interfere and with the arrival from the oceanic layer masked for most or all ranges, it becomes clear why earlier studies [e.g., *Raitt,* 1956] found only a layer of velocity approximately 5.5 km/sec over a layer of velocity 7.1 km/sec (or slightly higher), with the deeper layer being identified as layer 3 or a 'transitional' layer. Also, in certain geological settings, if the basal crustal layer has a fairly high velocity it may be erroneously identified as Moho.

More recent interpretations of Fiji plateau data from the *Nova* expedition by investigators from Scripps Institution of Oceanography [*Shor et al.,* 1971] also placed the Mohorovicic discontinuity at a depth of 8 to 9 km below sea level with only 2 km of water. This seismic model is not consistent with the gravity in the area, but, as pointed out in their paper, Shor et al. feel that the inconsistency could be resolved by including a masked 7.5 km/sec layer in the section.

Manihiki plateau. This station (group 9, Figure 10) is included because of its spectacular multiplicity of crustal layers. It is a good example of the amount of detail that can be obtained from the Asper method. On the original record, all the layers are clearly seen as distinct refracted arrivals. With conventional shooting techniques there would have been little chance of obtaining this resolution. At this station, penetration was insufficient to detect the basal crustal layer.

DISCUSSION AND CONCLUSIONS

Because of its high data density and its capability for phase correlation of low-level signals, the Asper method has yielded more detail in crustal velocity interpretations, through the use of one ship and expendable sonobuoys as seismic receivers, than have the more expensive marine refraction methods, which used multiple ships and explosives. The measurements are made while the ship is underway, with the result that crustal refraction data can be continuously collected simultaneously with gravity and magnetic data and with reflection seismic profiles. If repeated stations are made along a track, the resultant almost continuous seismic refraction section can eliminate most of the uncertainties that hamper interpretation of single-ended profiles. The method is presently limited in distance range because of the line-of-sight FM telemetry system used and because of the limited useful energy of the present airgun source. The largest airgun array (an 80-cubic-inch and a 120-cubic-inch airgun fired simultaneously) used to date yields energy equivalent to only 0.2 pounds of 60% dynamite [Kramer et al., 1968]. The short range sometimes makes an accurate determination of Moho difficult, and can (in areas with thick crustal sections, such as on shallow oceanic 'plateaus') prevent determination of deep crustal layers. When a complete section is determined, however, it is accomplished in a much shorter distance than is possible by older methods and is therefore likely to represent a more reliable interpretation in areas of complex structure.

Although the Asper method, compared to other refraction techniques, is able to get far greater effective penetration of seismic energy using signal phase correlation, the resultant records lack the amplitude information that is so important to analysis of wiggly-line records. Spot checks and extensions of profiles with explosives have verified the Asper arrivals, but the use of explosives overloads and disrupts the continuity of the Asper records and is generally kept to a minimum. With a repetitive source of greater energy, simultaneous wiggly-line and variable density precision echo sounding recording (e.g., Figures 3 and 4) would be feasible, and subsequent analysis would be easier and more positive.

Another development that would aid in analysis of these records is a truly impulsive seismic energy source (or an equivalent time-compressed source) in place of one that generates a bubble-pulse train of several cycles. To date, the only successful methods of eliminating the bubble-pulse train have resulted in unacceptable loss of seismic energy. The complicated records resulting from the present prolonged source

have been especially troublesome in our search for low-velocity zones and cusps resulting from gradients and have made it more difficult to separate nearly coincident signals.

In every one of the 21 seismic refraction stations where sufficient range and energy transmission were obtained using the Asper method, a basal crustal layer averaging about 7.4 km/sec in seismic velocity has been identified. (The twenty-second station reported here, on the Manihiki plateau, did not penetrate to the basal crustal layer but exhibited an exceptionally great number of refractors above layer 3.) Many of our deep refraction profiles showing the basal crustal layer have been located in Pacific Ocean basins. Contrary to the results and interpretations of earlier workers (Figure 2), this indicates that the layer is not confined to the trenches, island arcs, oceanic rises, continental margins, and other areas of transition between differing geologic provinces. The inescapable conclusion obtained by combining our results with those of others is that the layer exists widely under the Pacific Ocean and possibly exists worldwide as a part of normal oceanic crustal structure. It is our opinion that as the Asper method becomes more widely used the layer will be reported frequently.

It would be useful to search some of the older data for evidence of second arrivals or short segments of first arrivals from the 7.4-km/sec basal crustal layer both from different regions of the Pacific Ocean and from other oceans. If nothing else, reinterpretation could determine whether a basal crustal layer of significant thickness could be included in the seismic models without violating the available data. This would be similar to what was done 15 years ago after layer 2 was identified. It is also important that the Asper method be applied in other areas of the Pacific and in other oceans. If, for example, the layer is not found in the Atlantic, then it would be important to determine why it is missing there but is widely distributed in the central Pacific. If, as we believe, the layer is widespread, it should greatly affect theories of crust-mantle formation and interaction.

Finally, three points warrant emphasis. First, we know that a high-velocity basal crustal layer with important geophysical implications exists under the deep ocean in several places, but we need to explore many more locations to discover whether it exists everywhere, to determine possible regional variations in thickness and velocity, and to study any possible correlations with other geological parameters, e.g., geomagnetic age, distance from sea-floor spreading centers, heat flow, and gravity. Second, we are now finding layers and interfaces, not only in the deep crust but also within the sediments and layer 2, that usually have not been resolved in explosion refraction work. Third, we badly need more energy and less complicated source signals to make possible more detailed analysis.

APPENDIX

The seismic measurements used in Figure 2 were obtained from the following works: *Berry and Knopoff* [1967], *Bosshard and MacFarlane* [1970], *Dainty et al.* [1966], *Den et al.* [1969], *Edgar et al.* [1971], *J. Ewing and Houtz* [1969], *J. Ewing et al.* [1960], *M. Ewing et al.* [1954, 1970], *Furumoto et al.* [1968], *Gainanov et al.* [1968], *Houtz and Ewing* [1963], *Murauchi et al.* [1968], *Officer et al.* [1959], *Payo* [1969], and *Raitt* [1956].

Additional unpublished results of oceanic seismic refraction work performed by the Scripps Institution of Oceanography were provided to the authors by Dr. Russell W. Raitt.

Acknowledgments. We are particularly grateful to Loren W. Kroenke for his continued interest and assistance in developing the techniques that made the collection of these data possible. The work of William N. Ichinose on the sonobuoy receiving system and that of Robert C. Mitiguy on the airgun energy source are gratefully acknowledged. We appreciate the help of Mark E. Odegard, who provided a ray tracing computer program and valuable assistance in its use. We would also like to thank the officers and crew of the R. V. *Mahi* and all the scientific staff, too numerous to identify individually, who helped collect the data. George P. Woollard, Director of the Hawaii Institute of Geophysics, provided continuing guidance and encouragement throughout the work and valuable suggestions concerning the manuscript.

This research was supported under Office of Naval Research contracts Nonr 3748(05)NR 083 603 (1967–1968, 1968–1969) and N00014-70-A-0016-0001 (1969–1970) and under National Science Foundation contracts NSF GA 1639 and GA 15792.

REFERENCES

Berry, M. J., and L. Knopoff, Structure of the upper mantle under the western Mediterranean basin, *J. Geophys. Res.*, *72*(14), 3613–3626, 1967.

Bosshard, E., and D. J. MacFarlane, Crustal structure of the western Canary Islands from seismic refraction and gravity data, *J. Geophys. Res.*, *75*(26), 4901–4918, 1970.

Dainty, A. M., C. E. Keen, M. J. Keen, and J. E. Blanchard, Review of geophysical evidence on crust and upper mantle structure on the eastern seaboard of Canada, in *The Earth Beneath the Continents, Geophys. Monogr. Ser.*, vol. 10, edited by J. S. Steinhart and T. J. Smith, pp. 349–369, AGU, Washington, D. C., 1966.

Demenitskaya, R. M., The method of research of the geological structure of the crystalline mantle of the earth. *Sovi. Geol. 1*, 92–112, 1959.

Den, N., W. J. Ludwig, S. Murauchi, J. I. Ewing, H. Hotta, N. T. Edgar, T. Yoshii, T. Asanuma, K. Hagiwara, T. Sato, and S. Ando, Seismic-refraction measurements in the Northwest Pacific Basin, *J. Geophys. Res.*, *74*(6), 1421–1434, 1969.

Edgar, N. T., J. I. Ewing, and J. Hennion, Seismic refraction and reflection in the Caribbean Sea, *Bull. Amer. Ass. Petrol. Geol.*, *55*(6), 1971.

Ewing, J., and R. Houtz, Mantle reflections in airgun-sonobuoy profiles, *J. Geophys. Res.*, *74* (27), 6706–6709, 1969.

Ewing, J., J. Antoine, and M. Ewing, Geophysical measurements in the western Caribbean Sea and in the Gulf of Mexico, *J. Geophys. Res.*, *65*(12), 4087–4126, 1960.

Ewing, M., G. H. Sutton, and C. B. Officer, Jr., Seismic refraction measurements in the Atlantic Ocean, 6, Typical deep stations, North American Basin, *Bull. Seismol. Soc. Amer.*, *44*(1), 21–38, 1954.

Ewing, M., L. V. Hawkins, and W. J. Ludwig, Crustal structure of the Coral Sea, *J. Geophys. Res.*, *75*(11), 1953–1962, 1970.

Furumoto, A. S., G. P. Woollard, J. F. Campbell, and D. M. Hussong, Variation in the thickness of the crust in the Hawaiian Archipelago, in *The Crust and Upper Mantle of the Pacific Area, Geophys. Monogr. Ser.*, vol. 12, edited by L. Knopoff, C. L. Drake, and P. J. Hart, pp. 94–111, AGU, Washington, D. C., 1968.

Gainanov, A. G., S. M. Zverev, I. P. Kosminskaya, Y. V. Tulina, M. Kh. Livshitz, P. M. Sichev, I. K. Tuyezov, E. E. Fotiadi, A. P. Milashin, O. N. Soloviev, and P. A. Stroev, The crust and the upper mantle in the transition zone from the Pacific Ocean to the Asiatic continent, in *The Crust and Upper Mantle of the Pacific Area, Geophys. Monogr. Ser.*, vol. 12, edited by L. Knopoff, C. L. Drake, and P. J. Hart, pp. 367–378, AGU, Washington, D. C., 1968.

Houtz, R. E., and J. I. Ewing, Detailed sedimentary velocities from seismic refraction profiles in the western North Atlantic, *J. Geophys. Res.,* *68*(18), 5233–5258, 1963.

Hussong, D. M., A. A. Nowroozi, M. E. Odegard, and G. H. Sutton, Crustal structure under an ocean bottom seismometer using explosive sources (abstract), *Eos Trans. AGU, 50*(11), 644, 1969.

Kramer, F. S., R. A. Peterson, and W. C. Walter (Eds.), *Seismic Energy Sources 1968 Handbook,* 57 pp., Bendix-United Geophysical Corp., 1968.

Maynard, G. L., Crustal layer of seismic velocity 6.9 to 7.6 kilometers per second under the deep oceans, *Science, 168,* 120–121, 1970.

Maynard, G. L., G. H. Sutton, and D. M. Hussong, Seismic observations in the Solomon Islands and Darwin Rise regions using repetitive sources (abstract), *Eos Trans. AGU, 50*(4), 206, 1969.

Meyer, R. P., L. M. Dorman, and L. Ocola, The search for anisotropy in the upper mantle—Experiments off Hawaii in 1966 (abstract), *Eos Trans. AGU, 50*(4), 246, 1969.

Morris, G. B., R. W. Raitt, and G. G. Shor, Jr., Velocity anisotropy and delay-time maps of the mantle near Hawaii, *J. Geophys. Res., 74*(17), 4300–4316, 1969.

Murauchi, S., N. Den, S. Asano, H. Hotta, T. Yoshii, T. Asanuma, K. Hagiwara, K. Ichikawa, T. Sato, W. J. Ludwig, J. I. Ewing, N. T. Edgar, and R. E. Houtz, Crustal structure of the Philippine Sea, *J. Geophys. Res., 73*(10), 3143–3171, 1968.

Officer, C. B., J. I. Ewing, J. F. Hennion, D. G. Harkrider, and D. E. Miller, Geophysical investigations in the Eastern Caribbean: Summary of 1955 and 1956 cruises, in *Physics and Chemistry of the Earth,* vol. 3, edited by L. H. Ahrens, F. Press, K. Rankama, and S. K. Runcorn, pp. 17–109, Pergamon, New York, 1959.

Payo, G., Crustal structure of the Mediterranean Sea, 2, Phase velocity and travel times, *Bull. Seismol. Soc. Amer., 59*(1), 23–42, 1969.

Raitt, R. W., Seismic refraction studies of the Pacific Ocean Basin, *Bull. Geol. Soc. Amer., 67,* 1623–1640, 1956.

Raitt, R. W., The crustal rocks, in *The Sea,* vol. 3, edited by M. N. Hill, 85–102, Interscience, New York, 1963.

Shor, G. G., Jr., H. H. Kirk, and H. W. Menard, Crustal structure of the Melanesian area, *J. Geophys. Res., 76*(11), 2562–2586, 1971.

Sutton, G. H., G. L. Maynard, and D. M. Hussong, Marine crustal seismic refraction studies using repetitive sources and sonobuoys (abstract), paper 15 in *Program of Annual Meeting of the Seismological Society of America,* St. Louis, 1969.

Woollard, G. P., Crustal structure from gravity and seismic measurements, *J. Geophys. Res., 64*(10), 1521–1544, 1959.

DISCUSSION

Madden: You reported in your Figure 9 some very low *S* wave to *P* wave velocity ratios. With a 2 km/sec velocity for the *P* wave and a 0.4 km/sec velocity for the *S*—that is a ratio of 5:1.

Sutton: That must be correct in the gross, as we have independent confirmation from the *P* to *S* converted wave from earthquakes on the ocean bottom seismograph.

Oliver: And from surface waves; they also demand very low shear velocities.

Healy: That is true for continental shales also. The Pierre shale has a Poisson's ratio of 0.4. It is a curious property of sediments. What is the equivalent explosive size of your source?

Meyer: We estimated that our 2000-cubic inch airgun running at 2000 psi was equivalent to about five pounds of explosive, but the frequency content is lower, and other parameters are changed. You can scale from that.

Hales: I am glad you had the courage to report some rather unusual mantle velocities.

Sutton: There are good sub-Moho velocity determinations up to 8.8 km/sec.

Reflections of P'P' Seismic Waves from 0 to 150 km Depth under the Ninety East Ridge, Indian Ocean, and the Atlantic-Indian Rise[1]

JAMES H. WHITCOMB

Seismological Laboratory, California Institute of Technology, Pasadena California 91109

Abstract. P'dP' phases, that is, reflections of P'P' seismic waves at depth d, are investigated for $0 < d < 150$ km near the Ninety East ridge, Indian Ocean, and the Atlantic-Indian rise, south of the Cape of Good Hope. The P'P' epicentral range is 55° to 80°. Conversion of P'dP' travel times to depth values strongly depends on the times and relative amplitudes of the main P'P' branches, which until now have been uncertain. This conversion is made by comparing the observed travel time with the appropriate P'P' branch and computing the depth of reflection d from the time difference, using a reasonable velocity distribution. The P'P' data studied here best fit the times of Bolt's AB and DF branches and the time of Adams and Randall's GH branch. The largest-amplitude branch is found to be GH between 55° and 62.5° epicentral distance, AB between 62.5° and 72°, and DF between 72° and 80°. The largest amplitude of the P'P' phase reflecting in an ocean area can be a reflection from the ocean surface or ocean bottom; reflections from the ocean bottom are the more common. The time separation between these two reflections can be up to 8 seconds for deep oceans. The beginning of the ocean-bottom reflection may be picked earlier if a slightly deeper reflector, such as the Moho, is present. The data indicate that some of the earliest largest-amplitude P'P' arrivals delineate a discontinuity (possibly the Moho discontinuity) under part of the Ninety East ridge area. A depth for this discontinuity of 23 km is found under the ridge and beneath a shoal feature just to the west of the ridge. Errors in these depth estimates depend mostly on the velocity model used to reduce the data. Reflectors 2° north of the Atlantic-Indian Rise area are at 21 km, possibly shallowing toward the rise, and at 9 km. A deeper reflecting zone is seen in the Ninety East ridge area at about 15°S latitude. Its lower bound is at 102 km 6.5° west of the ridge, deepens to 137 km 3.5° west of the ridge, and shallows to 87 km under the ridge both at 15°S and 7°S latitude. This feature is believed to be related to the tectonics that formed the ridge itself. Magnetic and bathymetric evidence precludes the possibility that the Ninety East ridge is a crustal-spreading feature. Several possibilities could explain the depth variation of the reflecting zone. If the reflection zone is a partial-melting zone, the depth variation could be caused by migration of water or partial melt upward, leaving behind more solid rock and effectively shallowing the zone under the ridge, or thicker sediments at the sides of the ridge may act as a thermal blanket and raise temperatures underneath, thus lowering the melting zone to the west. If the base of the zone is the base of the lithosphere, the depth variation could be caused by compressive buckling of the lithosphere or by sinking of a lithospheric slab dipping west.

Interpretations of observed precursors to P'P' (PKPPKP) as seismic-wave reflections in the crust and upper mantle have been made by *Gutenberg* [1960], *Adams* [1968], *Engdahl and Flinn* [1969], and *Whitcomb and Anderson* [1970]. This leads to the possibility of using a new reflection technique to map crustal and sub-crustal structures. The P'P' precursor reflections arrive before the main P'P' seismic phase with the same apparent slowness, or $dt/d\Delta$, of 1.2 to 4.0 sec/degree. The subsurface reflection of P'P' will be designated P'dP' as in *Whitcomb and Anderson* [1970], where d is the depth of reflection. The ray paths traveled by P'P' and P'dP' are schematically illustrated in Figure 1.

As noted by *Whitcomb and Anderson* [1970], the interpretation of P'dP' for shallow depths is

[1] California Institute of Technology Contribution 1999.

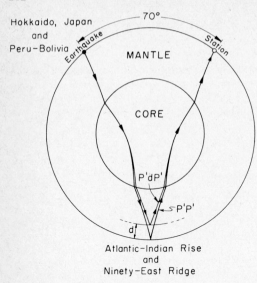

Hokkaido, Japan
and
Peru–Bolivia

Fig. 1.   Ray-path diagrams for the $P'P'$ and $P'dP'$
seismic phases.

strongly influenced by large lateral velocity variations in the crust and upper mantle, including changes of ocean depth and sediment depth. They also noted that there was no general agreement on the travel times, number of branches, or amplitudes of the various branches of P' (PKP) which must be used to reduce the $P'dP'$ reflection data. In this paper, the travel times and largest-amplitude branch for $P'P'$ at a given distance are inferred from earlier studies of P' [*Adams and Randall*, 1964; *Bolt*, 1968] and from $P'P'$ data. The $P'P'$ data are correlated with bathymetry and expected velocity distributions in the crust and upper mantle in order to improve the definition of the lateral variation of reflectors from 0 to 150 km in depth under the oceans. The oceanic areas studied are near the Ninety East ridge, Indian Ocean, and the Atlantic-Indian rise, south of the Cape of Good Hope, shown in Figures 2 and 3, respectively.

DATA

Earthquakes from the northern part of the Japanese arc and South America were used; their parameters are given in Table 1. The events are grouped according to their geographical reflection points, which are shown in Figures 2 and 3. Seismic recordings were read from several stations in California and from the Tonto

Forest array in Arizona (in one case). In addition, in order to better determine the $dt/d\Delta$ or apparent wave slowness of the phases, the large-aperture seismic array (Lasa) installation in Montana was used as a beam-forming array. With the use of Lasa, $dt/d\Delta$ was read directly from contour plots of beam signal power as a function of time and $dt/d\Delta$; these plots are called 'vespagrams' (from 'velocity spectrum analysis') and are described by *Kelly et al.* [1968].

The majority of the data were read as signal onsets on individual seismograms. In order to avoid bias, no cross-referencing of seismograms with each other or with travel-time curves was made during the reading. Because of this, it is expected that, if two signals arrive within one or two periods of each other (for $P'P'$ this is within about three seconds), then they are read as one arrival with the earlier arrival time. The second arrival may be missed even if it is larger, because both phases will appear as a single emerging wave train. This effect is important but is now unavoidable when the reflection depth of the largest-amplitude $P'P'$ phase is calculated. If there is a strong reflector in the crust as much as 20 km deeper than the largest-amplitude reflector, the two phases may merge and the deeper interface will be erroneously designated as the strongest reflector.

The data were read to the nearest second and were graded according to the following scale:

4   is for the beginning of the largest-amplitude $P'P'$ phase on the record.

3   is for a sharp pulse or a wave train with a sharp beginning.

2   is for an emergent beginning but a definite signal, usually a wave train.

1   is for a small wave train with an emergent beginning; character (difference in period or amplitude) distinguishes it from the prevailing noise level.

Samples of individual records and vespagrams are shown in Figure 4 for events 2, 6, 8, and 5. The data were reduced to surface-focus events on a spherically symmetrical earth in the same manner as in *Whitcomb and Anderson* [1970]. However, some alteration was made in the choice of the largest-amplitude $P'P'$ branch used to determine the reductions.

TIMES AND LARGEST AMPLITUDES OF P'P'

Knowledge of the largest-amplitude branch at a given station is useful because: first, it is unambiguous; second, it is independent of amplitude variations caused by local structure from station to station; and third, it indicates which is the most likely branch to reflect at depth and give a $P'dP'$ phase.

The third property is most helpful when direct measurement of $dt/d\Delta$ is not possible, such as at a single station. When a specific branch of P' is mentioned, the notation suggested by *Adams and Randall* [1964] is used. Adams and Randall imply that the largest-amplitude $P'P'$ branches are $P'_{DF} P'_{DF}$ at distances greater than 70°, $P'_{GH} P'_{GH}$ between 70° and 54°, and $P'_{AB} P'_{AB}$ at distances

less than 54°. $P'P'$ largest-amplitude data are shown as black symbols in the travel-time plot of Figure 5. In all the figures with points as a function of time, the black symbols represent grade 4 data according to the scale above, and the white symbols represent grades 3 (largest symbols) through 1 (smallest symbols). The type of symbol represents the event; the key is given in Table 1. Several stations were used for each event, so that the same symbol is seen at different distances. Figure 5 indicates that the amplitude statements of Adams and Randall are correct except in the range of 62.5° to 72°. In this range, it is apparent that the largest amplitudes follow the AB branch from 62.5° to the end of the B caustic at about 72°. Measure-

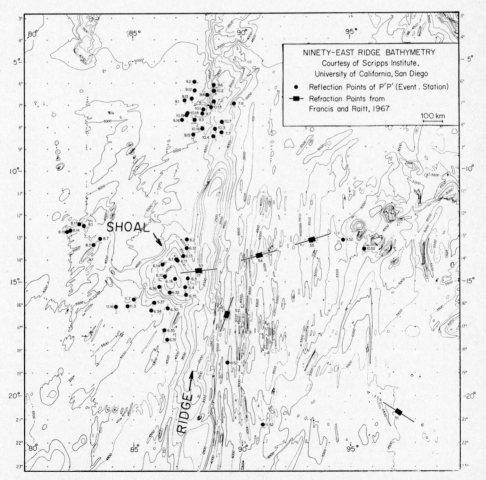

Fig. 2.   Map of the reflection points at the Ninety East ridge, events 6, 8, and 11 (southern group) and events 7, 9, and 10 (northern group). Bathymetry was provided by the University of California, San Diego. Contour interval is 500 meters.

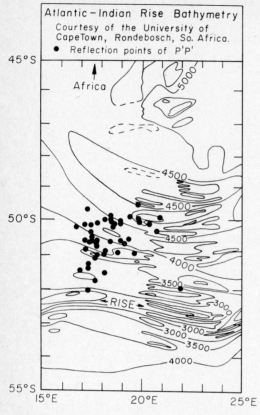

Fig. 3. Map of the reflection points at the Atlantic-Indian rise, events 1, 2, 3, 4, 5, and 12. Bathymetry is from unpublished data of E. S.W. Simpson and Erica Forder, University of Cape Town, Rondebosch, South Africa. Contour interval is 500 meters.

ments of $dt/d\Delta$ from Lasa are indicated by sloping lines through the data. The $dt/d\Delta$ measurements show that the largest-amplitude branches are GH at 58° and 62°, AB at 69°, and DF at 76°, as predicted.

The predicted times in this paper of $P'P'$ arrivals at continental stations are based on P' studies by *Adams and Randall* [1964] and *Bolt* [1968]. The $P'P'$ data of Figure 5 fit Adams and Randall's GH branch best and Bolt's AB and DF branches best. However, the largest amplitudes sometimes miss the predicted times for that branch by several seconds, such as the Lasa data at 58°, 69°, and 76° (symbols pierced by short lines indicating $dt/d\Delta$). The most likely source of scatter of the largest amplitudes appears to be reflections from the ocean surface, ocean bottom, and Moho and their lateral variations at the reflection point. For investigation of the effects of these factors, the data have been divided into three groups based on their reflection location: Ninety East ridge, events 6, 8, and 11 (southern group in Figure 2); Ninety East ridge, events 7, 9, and 10 (northern group in Figure 2); and the Atlantic-Indian rise (Figure 3). The time-distance plots of $P'P'$ and precursors for each of these groups are shown in Figure 6. In order to compare all $P'P'$ data with each other, irrespective of what branch they belong to, all the times are reduced by subtracting the predicted time of the largest-amplitude $P'P'$ branch at that distance (GH between 55° and 62.5°, AB between 62.5° and 72°, and DF

TABLE 1.   Event Data[1]

No.	Date	Origin Time	Location, deg	Depth, km	Mag.	Symbol in Figs.
1	May 31, 1964	00h40m36.4s	43.5N, 146.8E	48	6.3	square
2	June 23, 1964	01h26m37.0s	43.3N, 146.1E	77	6.2	circle
3	June 11, 1965	03h33m44.9s	44.7N, 148.7E	47	6.0	triangle
4	Oct. 25, 1965	22h34m24.3s	44.2N, 145.3E	180	6.2	diamond
5	Jan. 29, 1968	10h19m05.6s	43.6N, 146.7E	40	7.0	crossed square
12	May 16, 1968	10h39m01.5s	41.5N, 142.7E	33	7.0	hourglass
6	Nov. 3, 1965	01h39m02.5s	9.1S, 71.4W	583	6.2	square
7	May 11, 1967	15h05m16.8s	20.3S, 68.5W	67	6.1	circle
8	Sept. 3, 1967	21h07m30.8s	10.6S, 79.8W	38	6.5	triangle
9	Dec. 21, 1967	02h25m21.6s	21.8S, 70.0W	33	6.3	diamond
10	Dec. 27, 1967	09h17m55.7s	21.2S, 68.3W	135	6.4	crossed square
11	June 19, 1968	08h13m35.0s	5.6S, 77.2W	28	6.4–6.9	hourglass

[1] Events 1, 2, 3, 4, 5, and 12 are reflections from the Atlantic-Indian rise; epicenters are in the Japanese arc. Events 6 through 11 are reflections from the Ninety East ridge; epicenters are in South America.

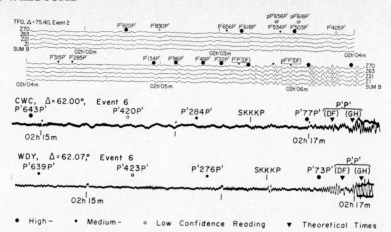

Fig. 4a. Short-period vertical record examples of $P'P'$ and $P'dP'$ (where d is the depth of reflection) from events 2 (TFO) and 6 (CWC, WDY). Station corrections are TFO (Tonto forest, Arizona) = 0 sec, CWC (Cottonwood, California) = 12.7 sec, WDY (Woody, California) = 29.4 sec.

between 72° and 80°). It is to be remembered that these predicted times are based on crust and upper-mantle velocity models corresponding to continents, since that is where most of the P' data are gathered.

There is a wavelength dependent limit in the ability to resolve dimensions of objects using propagating waves. In order to estimate the resolvability of the $P'dP'$ reflection at shallow depths, the Fresnel diffraction pattern [*Born*

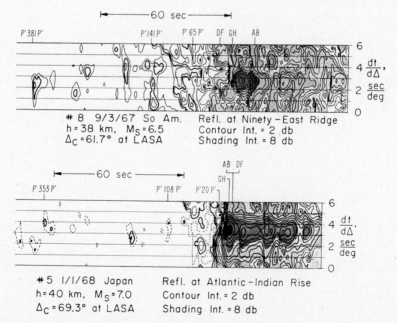

Fig. 4b. Vespagrams (velocity spectra) or contour plots of beam signal power in decibels as a function of time and $dt/d\Delta$ for events 8 and 5. Depths of reflection and theoretical main branch times are shown; main branch times are from *Adams and Randall* [1964, *GH*] and *Bolt* [1968, *AB* and *DF*]. Note the double arrival at the $dt/d\Delta$ for the *GH* branch in event 8. The first is interpreted as the ocean-bottom reflection and the second is the ocean-surface reflection.

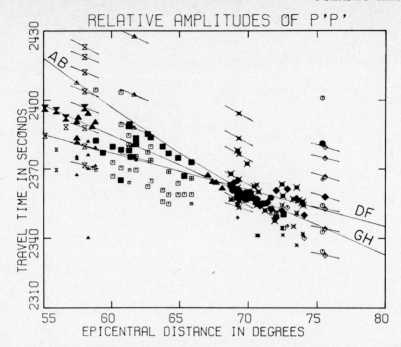

Fig. 5. Main $P'P'$ phase travel times and largest amplitudes (shown as black symbols) as a function of epicentral distance. Sloping lines through symbols indicate the $dt/d\Delta$ measured at Lasa. Largest-amplitude branches are GH between 55° and 62.5°, AB between 62.5° and 72°, and DF between 72° and 80°. Points close to a branch that is not the largest-amplitude branch are not used as $P'dP'$ data. In this and all subsequent figures with seismogram readings, the following is the symbol key. All stations recording the same event have the same type of symbol, which is described in the last column of Table 1; the large black symbols indicate the beginning of the largest-amplitude $P'P'$ phase on the record, and the remaining symbols are white with three size grades (described in the text) indicating the quality of the reading.

and Wolf, 1965, p. 433] of a wave reflecting from a semi-infinite plane with a straight edge was calculated to estimate the ability to resolve the edge. With the criterion that an amplitude drop of one-half (intensity drop of one-fourth) from the peak value is the point at which readings are missed by the observer and the edge is 'seen,' it is estimated that the peak- to one-half amplitude drop takes place within 170 km at the reflector. A wavelength of 10 km and an observer distance of 12,000 km was used. Thus, one cannot expect to see changes in a reflector over lateral distances much less than 170 km or about 1.5°.

In order to compute depths corresponding to a particular tectonic province, three models of the crust and upper mantle are used as shown in Figure 7. The continental model is CIT 208, from Johnson [1969]. For oceanic models, two types of crust were chosen based on seismic re-

fraction work done by Francis and Raitt [1967] in the deep ocean (5.3-km water depth) and on the Ninety East ridge (2.0-km water depth), here called Oc and 90°E, respectively. All three models were made essentially the same below 56 km.

Bathymetry charts for the reflection areas vary in coverage (Figures 2 and 3). The best data are from the Ninety East ridge, events 6, 8, and 11. Figure 8 shows a plot of reduced times as a function of water depth for events 6, 8, and 11 (zero time corresponds to a surface reflection, $P'0P'$, under a continent). The broken lines in Figure 8 indicate the times calculated for a reflection from the ocean surface and ocean bottom for model Oc, with a 5.3-km water depth, and model 90°E, with a 2-km water depth (the models are shown in Figure 7). A linear interpolation of times is assumed between the two. Figure 8 indicates that a fairly good fit of largest-amplitude arrivals can be made to surface- and bottom-reflection times if the velocity of the Oc and 90°E models in the crust and upper mantle is lowered to increase the times of arrival by 2.5 seconds (or 1.25 seconds one-way time). This fit, shown as the continuous lines in Figure 8, is made mainly on the basis of ocean-surface reflections, which should have the least scatter.

The implications of this time shift can be better evaluated when a similar analysis can be made on $P'P'$ reflections under a continent for comparison.

With the adjustment of the velocity models to fit the water-surface reflection times, it is seen that most of the largest-amplitude data in Figure 8 are earlier than the ocean-bottom time;

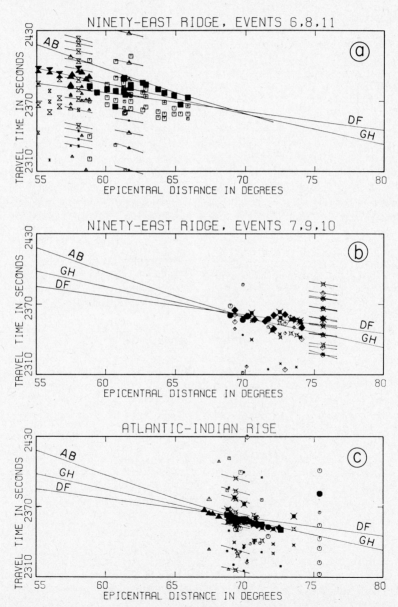

Fig. 6.   Travel times as a function of epicentral distance for each reflection area. Points with a cross in the center and points with the wrong $dt/d\Delta$ compared to the largest-amplitude branch for that range are not used. For the symbol explanation, see the legend of Figure 5.

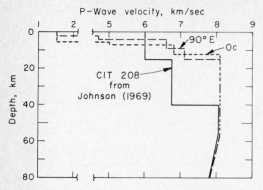

Fig. 7. *P*-wave velocity models CIT208 [from *Johnson,* 1969], 90° E, and Oc.

some are close to the ocean-surface time. Within the expected limits of accuracy (±1 second), there is a gap between the ocean-bottom and ocean-surface times, as would be expected; that is, no reflection should occur within the water. The three late points near 2-km ocean depth are probably due to using the wrong branch to reduce their times. They are near the right time for *AB* but are at a distance less than 62.5°

away, so that the *GH* branch was used for reduction. The largest amplitudes arriving earlier than the ocean-bottom reflection are believed to be a merging of the ocean-bottom reflection with a deeper reflection which is probably not the strongest reflector, as discussed in an earlier section.

Events 7, 9, and 10 on the Ninety East ridge are too closely spaced (Figures 2 and 6) and the bathymetry data in that area are too sparse to provide a useful comparison of reduced times with water depth.

The Atlantic-Indian rise plot of reduced time as a function of water depth is shown in Figure 9. The broken lines correspond to the Oc and 90° E velocity models (Figure 7), and the continuous lines correspond to the adjusted models, just as in Figure 8. The largest-amplitude data also show a prominent gap between the ocean-surface and ocean-bottom reflections, but the times do not fit the slope of the calculated ocean-surface reflection as well. The bathymetry data are based on very few ship tracks, and the water depths probably tend to be too deep, since any smoothing caused by lack of data will wipe out

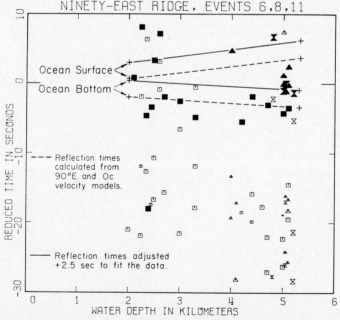

Fig. 8. Reduced times as a function of water depth for Ninety East ridge, events 6, 8, and 11 data. The broken lines represent times for ocean-surface and ocean-bottom reflections which are extrapolated between velocity models 90° E (2.0-km water depth) and Oc (5.3-km water depth). The continuous lines represent the same times adjusted to fit the data. For the symbol explanation, see the legend of Figure 5.

the high bottom features and displace the points toward too great a water depth, as they appear in Figure 9. Nonetheless, the trend of a widening gap with deeper water is easily seen in the data. As in Figure 8, most of Figure 9 largest amplitudes are read earlier than the ocean-bottom reflection time, but the ocean bottom is probably the strongest reflector.

SUBBOTTOM REFLECTIONS

Because the major crustal structures in all reflection areas are ridges, the reduced times from the preceding section are plotted as a function of perpendicular distance from the Ninety East ridge and the Atlantic-Indian rise in Figures 10, 11, and 12. Velocity models Oc and 90°E (Figure 7), modified to fit the surface reflection times as described above, represent a reasonable range of velocity distribution to be expected in the reflection areas, and they are used to calculate depths shown in the figures corresponding to the reduced times. Differences between depths calculated from the two models are less than a

kilometer below depths of 6 km and are considered insignificant.

The Ninety East ridge events 6, 8, and 11 data in Figure 10 show two main features. The first is a shallow reflector which is delineated by the largest-amplitude data (black symbols). The reflector is 6 km deep at 4° to 6° west of the ridge and deepens to 23 km at about 2° west of the ridge. The area at 4° to 6° west of the ridge is deep ocean, so that the reflector is probably the ocean bottom. The area at 2° west of the ridge is shown on bathymetric charts as a large shoal which is probably volcanic in origin [*Francis and Raitt*, 1967]. The two data sets (both from Lasa readings) at 1.2 and 2.9 km east of the ridge in Figure 10 are in deep ocean again and do not show the 23-km reflector. The second feature of Figure 10 is the sharp onset (going from earlier to later times) of data points at −20 to −35 seconds after a long interval of no readings earlier than this range. The depth of this onset varies from 102 km at 6° west of the ridge to 137 km at 3° west of the ridge to

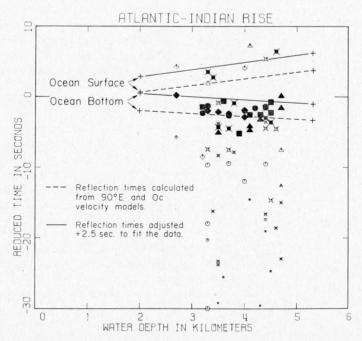

Fig. 9. Reduced times as a function of water depth for Atlantic-Indian rise data. The broken lines represent times for ocean-surface and ocean-bottom reflections which are extrapolated between velocity models 90°E (2.0-km water depth) and Oc (5.3-km water depth). The continuous lines represent the same times adjusted to fit the data. For the symbol explanation, see the legend of Figure 5.

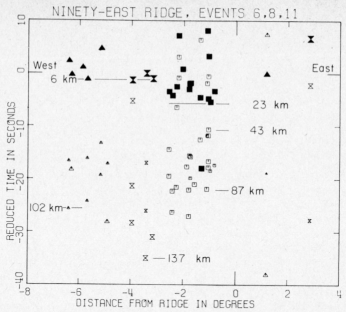

Fig. 10. Reduced times as a function of perpendicular distance from the Ninety East ridge for events 6, 8, and 11. Zero reduced time corresponds to a $P'P'$ reflection at the surface of a continent. For the symbol explanation, see the legend of Figure 5. Calculated depths are shown.

87 km almost at the ridge. The upper bound of this zone is similar in shape and is at 43 km near the ridge. No readings were encountered above this depth until the shallow reflector readings discussed in the preceding paragraph.

Figure 11 shows the reduced time as a function of distance from the Ninety East ridge for events 7, 9, and 10. Even though the reflection points for this group of events are several hundred kilometers north of those in Figure 10, there are remarkable similarities between the two data sets. The onset of the largest-amplitude points is at the same level, 23 km. The lower bound of the deep zone is shallower, 75 km, but follows the trend of Figure 10 because it is closer to the ridge. The upper bound of the zone is at the same 43-km depth as that in Figure 10 nearest the ridge. The shallow largest-amplitude points between 6.0 and −1.0 seconds in Figure 11 are completely bracketed by the times of ocean-surface and ocean-bottom reflection for the adjusted Oc model with 5.3 km of water. Thus, these readings are consistent with reflections from the ocean bottom and ocean surface with varying depths of water. However, as stated earlier, bathymetric control is not good enough

in this area to correlate the data with water depth as was done in Figures 8 and 9.

The reduced time for the Atlantic-Indian Rise as a function of distance from the rise is shown in Figure 12. The onset of largest-amplitude data is at 21 km. However, the largest-amplitude data at this depth are limited to 1° west or farther from the ridge axis. Closer points appear to shallow considerably, although the ability to resolve depth variations over the lateral range of 3° shown in Figure 12 may be questionable because the range is close to the 1.5° minimum limit calculated in an earlier section. The only deeper onset of significance is at 39 km. The depth of this reflector does not appear to vary with distance from the ridge. Below this level, the data are scattered and appear to have no significant onset shallower than the lowest point of Figure 12, which is 156 km deep. What these scattered data represent is difficult to say in the absence of lateral correlation in plots like Figure 12 because of the semi-statistical nature of the interpretation. Because each point represents only a pick of a signal on a seismic trace (except for Lasa data) the scattered points could represent a real $P'dP'$ phase which ap-

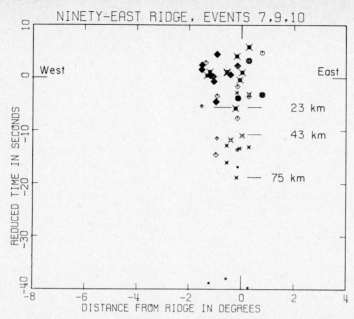

Fig. 11. Reduced times as a function of perpendicular distance from the Ninety East ridge for events 7, 9, and 10. For the symbol explanation, see the legend of Figure 5. Calculated depths are shown.

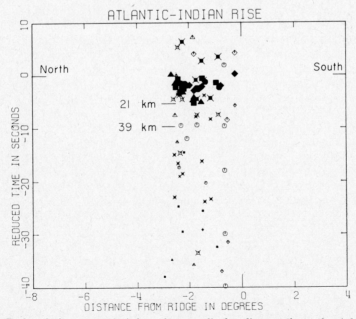

Fig. 12. Reduced times as a function of perpendicular distance from the Atlantic-Indian rise. For the symbol explanation, see the legend of Figure 5. Calculated depths are shown.

pears scattered because it is weak, another seismic body phase, or random noise. It is interesting to note in Figure 12 that the gap between the ocean-surface reflection and the bottom-reflection points widens to the north in a systematic manner. This relation indicates a water deepening to the north and agrees in general with the bathymetry in Figure 3.

The analysis of dipping reflection surfaces has not been done because the $P'dP'$ energy from a dipping reflector is assumed to be small. This is assumed for two reasons. First, energy from a sloping reflector would be reduced because the surface area of a reflector with a given angle of dip should be small for an earth that is nearly spherically symmetrical. Second, if the first P' leg of $P'dP'$ goes through a focusing part of the core, chances of the second P' leg traveling through a focusing part of the core are small because the seismic parameter is changed at the dipping reflector.

DISCUSSION

It is not surprising to note that, in Figures 8 and 9, the largest amplitude can be associated with either the ocean-bottom or ocean-surface reflection. Calculations show that if there is a fairly hard bottom with P-wave velocity greater than about 3.5 km/sec (density about 1.8 gm/cc), the ocean-bottom reflection will be larger than the ocean-surface reflection. Conversely, the amplitude of the bottom reflection is easily reduced below that from the surface reflection by an appreciable sediment layer; sediment velocities are usually much less than 3.5 km/sec. Interference effects can also be effective in reducing the ocean-bottom reflection. For example, a sediment layer that is one-fourth wavelength thick (0.2 to 0.5 km) overlying a crystalline-rock stratum will cause destructive interference of the ocean-bottom reflection. The detailed sediment data needed to test this hypothesis are not available. However, studies of *Ewing et al.* [1969] show that sediment thicknesses in the Ninety East ridge's northern group area can be up to 0.5 km thick adjacent to the ridge and in the southern group up to 0.3 km thick adjacent to the ridge. Sediment thicknesses in the vicinity of the Atlantic-Indian rise can be up to 0.5 km. An additional and perhaps very effective mode of reducing the amplitude of ocean-bottom reflections is scattering due to large bathymetric

relief. A non-horizontal reflecting surface changes the seismic parameter or $dt/d\Delta$ of an incoming ray and prevents the ray from continuing on its second P' leg through the same level of the earth's core. This effect tends to scatter the reflecting energy, as mentioned above. Thus, it is reasoned that ocean-bottom type has a large effect on surface and near-surface reflection amplitudes. In summary, one expects the ocean-surface reflection amplitude to be largest in regions that have an ocean bottom with thick sediments, sediments of one-quarter wavelength (about 0.2 to 0.5 km thick), or strong relief. For regions with bottoms that are flat with thin sediment cover, one expects the ocean-bottom reflection to be the largest-amplitude $P'P'$ phase. Its beginning may be picked earlier, if a slightly deeper reflector is present such as the oceanic Moho discontinuity. Most of the largest-amplitude data here reflect from the ocean bottom.

The shallow reflector for the Ninety East ridge groups at 23 km under the ridge and beneath a shoal feature (probably volcanic) is at a depth that may be reasonable for the Moho under shallow oceans. Under an oceanic rise, a Moho discontinuity as it is usually defined either does not exist or is too deep to be found by normal seismic refraction techniques [*Talwani et al.*, 1965]. Therefore, the Atlantic-Indian rise reflection at 21 km 2° north of the rise and the reflector at 39 km must be related to the special dynamic situation that exists at oceanic rises. It is interesting to note the similarity of these depths with those of the rock-type boundaries of the Case II model for the mid-Atlantic rise structure in *Talwani et al.* [1965, p. 348], who used seismic and gravity data to construct the structure.

In order to interpret the deeper reflections at the Ninety East ridge, an examination of other geophysical parameters was made in order to deduce the general tectonics of the region. Figure 13 combines both north and south Ninety East ridge sets of reduced times as a function of distance from the ridge. The depth exaggeration is about 2 times. Bathymetry, shown above the reduced times, illustrates the relation of bottom topography along 15°S latitude (see Figure 2) to the reflection structures at depth. The shoal between 1° and 2° west of the ridge is nearly equidimensional on the bathymetry map of Figure 2, so that a section north or

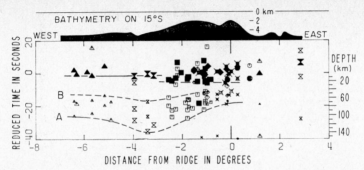

Fig. 13. All reduced times from the Ninety East ridge areas as a function of perpendicular distance from the ridge. Vertical exaggeration of depths is about 2 times. Bathymetry along 15°S is shown above. For the symbol explanation, see the legend of Figure 5.

south of 15°S would not show a shoal in that position.

The bathymetry of Figure 2 shows a general asymmetry of the Ninety East ridge; the west slope is more gradual than the east slope, and a linear north-south low lies at the base of the east slope. Further, the average ocean depth east of the ridge is slightly deeper than depths west of the ridge. This asymmetry is also seen in *Le Pichon and Heirtzler* [1968, profile *CD*, Figure 2]. A free-air gravity profile across the ridge along 18°S just south of the shoal area is given in *Le Pichon and Talwani* [1969]. A portion of this profile is shown in Figure 14 with accompanying bathymetry along 18°S. The free-air anomaly generally decreases from +30 mgal ten degrees west of the Ninety East ridge to about −30 mgal twenty degrees east of the ridge, which mirrors the average ocean depth in each region. The variations in the anomaly, mainly a peak of +50 mgals over the ridge, can be ex-

plained by variations of bathymetry and do not show any obvious indication of active tectonic movement and associated disturbance of the isostatic equilibrium.

Heat flow data along the ridge, although sparse in coverage, show values above 2 μcal/cm² sec just west of the Ninety East ridge at 3°N, just east of the ridge at 12°S, and just west of the ridge at 27°S [*Langseth and Taylor*, 1967]. Thus, the ridge might be associated with a zone of higher than average heat flow.

It appears possible to eliminate the hypothesis of the Ninety East ridge as being a crustal-spreading center. Along the ridge, the bathymetry shows no transform faults, wide symmetric bulge of the ocean bottom, or symmetry of the ridge itself (Figure 2). Also, no parallel, symmetric magnetic lineations are seen on either side of the ridge [*Le Pichon and Heirtzler*, 1968]. All the foregoing properties are usually found in crustal-spreading centers. The ridge is

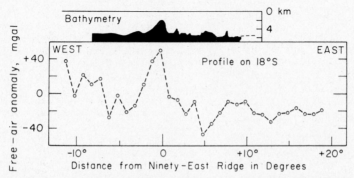

Fig. 14. The free-air gravity anomaly and bathymetry along 18°S. The gravity values have been averaged over 1° squares and are from *Le Pichon and Talwani* [1969]. For the symbol explanation, see the legend of Figure 5.

generally believed to be a zone of compression [*Le Pichon and Heirtzler*, 1968; *Dietz and Holden*, 1970].

The deeper reflecting zone in Figure 13 is striking because of its abrupt onset (the lower edge is the onset because reflections are from the bottom) and its agreement in position relative to the ridge at both 7°S and 15°S, a separation of over 800 km. The boundaries of the zone have been labeled *A* and *B*. Boundary *A* is the deeper and thus is responsible for an earlier precursor to *P'P'*. Because of this pattern and the lack of data immediately earlier, confidence in the *A* horizon is stronger than that in the *B* horizon, which may be either a real reflector or merely the cessation of the wave train from the *A* reflection. There are several possibilities that might cause the depth changes of the *A–B* zone.

First, consider the *A–B* zone to be a range of depth where partial melting exists, implying that the zone is the low-velocity zone [*Anderson and Sammis*, 1970]. The shallowing and thinning of the zone under the ridge could be a result of the migration of water or partial melt upward, leaving behind more solid rock and effectively shallowing the low-velocity zone under the ridge. The upward migration of partial melt would explain the assumed volcanism along the ridge and the high heat flow. An alternate explanation consistent with the partial-melt zone hypothesis is the effect of varying sediment distribution. Thicker sediments along the sides of the ridge [*Ewing et al.*, 1969] may act as a thermal blanket and increase the deeper temperatures 3.5 km west of the ridge. This in turn would lower the depth of partial melting just as the horizon *A* is lowered 3.5 km west of the ridge in Figure 13.

Secondly, consider the horizon *A* to be the base of the lithosphere, that is, the top of the low-velocity zone. The observed changes in the depth of *A* in an area of supposed crustal compression can mean that the lithosphere is buckling down 3.5° west of the Ninety East ridge, or the down-dipping surface *A* can indicate the downturn of a lithospheric slab that is projecting into the mantle. Because the ridge is not outlined by active seismicity [*Sykes*, 1970] and gravity does not indicate active tectonism, it would then be a dormant island-arc structure and may be close to thermal equilibrium with its surroundings. If this solution is chosen, asym-

metry of bathymetry across the ridge indicates that a lithospheric slab dipping to the west is preferable. Future investigations of seismicity, earthquake focal mechanisms, gravity, and magnetics of the Ninety East ridge area may enable a choice to be made among the four proposed models.

Although it does not appear possible to observe details within the oceanic crust with the accuracy of the data here, an experiment designed to observe *P'dP'* reflecting under a continental structure may delineate features within the thicker continental crust. Attempts are currently being made to attain this goal.

Acknowledgments. I wish to thank Don Anderson for many helpful discussions and for reading the manuscript. J. Sclater and R. Fisher, at the University of California, San Diego, kindly provided unpublished bathymetric data of T. Hilde. I also thank E.S.W. Simpson and E. Forder, of the University of Cape Town, Rondebosch, South Africa, for allowing me to use their bathymetry data. The Lincoln Laboratory, Cambridge, made their facilities available for the use of Lasa data.

This research was supported by the Advanced Research Projects Agency of the Department of Defense and was monitored by the Air Force Office of Scientific Research under contract F44620-69-C-0067.

REFERENCES

Adams, R. D., Early reflections of *P'P'* as an indication of upper mantle structure, *Bull. Seismol. Soc. Amer.*, *58*, 1933, 1968.

Adams, R. D., and M. J. Randall, The fine structure of the earth's core, *Bull. Seismol. Soc. Amer.*, *54*, 1299, 1964.

Anderson, D. L., and C. Sammis, Partial melting in the upper mantle, *Phys. Earth Planet. Interiors*, *3*, 41, 1970.

Bolt, B. A., Estimation of *PKP* travel times, *Bull. Seismol. Soc. Amer.*, *58*, 1305, 1968.

Born, M., and E. Wolf, *Principles of Optics*, 3rd ed., 808 pp., Pergamon, New York, 1965.

Dietz, R. S., and J. C. Holden, Reconstruction of Pangaea: Breakup and dispersion of continents, Permian to present, *J. Geophys. Res.*, *75*, 4939, 1970.

Engdahl, E. R., and E. A. Flinn, Seismic waves reflected from discontinuities within the upper mantle, *Science, 163*, 177, 1969.

Ewing, M., S. Eittreim, M. Truchan, and J. I. Ewing, Sediment distribution in the Indian Ocean, *Deep Sea Res., 16*, 231, 1969.

Francis, T. J. G., and R. W. Raitt, Seismic refraction measurements in the southern Indian Ocean, *J. Geophys. Res., 72*, 3015, 1967.

Gutenberg, B., Waves reflected at the 'surface' of

the earth: *P'P'P'P'*, *Bull. Seismol. Soc. Amer.*, *50*, 71, 1960.

Johnson, L. R., Array measurements of *P* velocities in the lower mantle, *Bull. Seismol. Soc. Amer.*, *59*, 973, 1969.

Kelly, E. J., L. T. Fleck, and P. E. Green, Special methods for detailed analysis of individual events, in *Seismic Discrimination, Semi-annual Tech. Sum., June 30, 1968*, Lincoln Lab., MIT, Lexington, Mass., 1968.

Langseth, M. G., Jr., and P. T. Taylor, Recent heat flow measurements in the Indian Ocean, *J. Geophys. Res., 72*, 6249, 1967.

Le Pichon, X., and J. R. Heirtzler, Magnetic anomalies in the Indian Ocean and sea-floor spreading, *J. Geophys. Res., 73*, 2101, 1968.

Le Pichon, X., M. Talwani, Regional gravity anomalies in the Indian Ocean, *Deep Sea Res., 16*, 263, 1969.

Sykes, L. R., Seismicity of the Indian Ocean and a possible nascent island arc between Ceylon and Australia, *J. Geophys. Res., 75*, 5041, 1970.

Talwani, M., X. Le Pichon, and M. Ewing, Crustal structure of the mid-ocean ridges, 2, Computed model from gravity and seismic data, *J. Geophys. Res., 70*, 341, 1965.

Whitcomb, J. H., and D. L. Anderson, Reflection of *P'P'* seismic waves from discontinuities in the mantle, *J. Geophys. Res., 75*, 5713, 1970.

DISCUSSION

Sutton: In two of your figures you have interpreted the largest-amplitude reflections as coming from the Moho. How does that come about when the impedance contrast is not as large as at the other discontinuities in the section?

Whitcomb: I interpret the largest-amplitude reflections which begin at Moho depths to be a combination of Moho reflection and ocean-bottom reflection, where the latter has the true largest amplitude as expected. The explanation for my reading the largest-amplitude data at Moho depth times lies in the short separation, about two seconds or one cycle, between the ocean-bottom and Moho-reflection arrivals for an oceanic crust. The separation is not enough to allow the trace to die down after the Moho reflection. Thus, the two appear as one phase arriving at the earlier or Moho reflection time.

Heacock: Do you have an explanation for the broad vertical scatter of points as a function of distance away from the ridge?

Whitcomb: There are several possible causes for the vertical scattering within zones in the reduced-time plots. One possibility is that the scatter represents the real scatter distribution of reflectors in a zone. Another is that I have not been able to consistently read the beginning of a phase but have read a later time when the amplitude is larger. One of the stronger possibil-

ities is that the scatter is due to side reflections from slightly dipping interfaces at similar depths. Because the *P'dP'* phase is a minimum time path for a horizontal reflector, side reflections from a dipping interface would arrive later. I feel that the problem of scatter will be reduced by looking at more array beaming results. For example, the 630-km reflection has consistently appeared in the six vespagrams that I have made so far.

Heacock: And you interpret this to mean that you have a continuous discontinuity?

Whitcomb: Yes. I have looked at two widely separated oceanic locations, each with spreads of about 10°, and the 630-km discontinuity appears at the same depth, plus or minus about 4 km, on the vespagrams. However, there is still some evidence for a depth variation of about 40 km of this discontinuity based on the individual record data.

Heacock: What resolution do you get in the crust?

Whitcomb: I do not believe I can see any structure within an oceanic crust which is typically 6 km thick, representing a two-way travel time of only about two seconds. Under continents, however, thicknesses of the order of 30 km may permit the resolution of gross features within continental crusts using this method.

The Electrical Conductivity of the Oceanic Lithosphere

CHARLES S. COX

Scripps Institution of Oceanography, University of California, San Diego
California   92037

Abstract. Methods for learning the distribution of electrical conductivity under the ocean are based on: 1, estimates of temperature and composition of the lithosphere; 2, measurements of natural electromagnetic fields; and 3, direct studies from drill holes or of drilled samples. The first method suffers from uncertainties in the temperature distribution associated with poor knowledge of the thermal conductivity, the influence of heat transported by flow of magma, and of heat produced by plastic flow in the asthenosphere. It is conceivable that the plastic flow in the asthenosphere can produce as much as 25% of the average heat flow and can be concentrated in spots at the base of the lithosphere. The existence of pore fluids greatly increases electrical conductivity. The second method is sensitive only to those parts of the lithosphere whose conductivity approaches that of sea water. There are indications off California and Peru that the subocean lithosphere reaches a conductivity one-tenth that of sea water at a depth of 30 km. The third method is becoming practical through recent development of deep drilling techniques.

There are three methods by which the electrical conductivity of the subocean lithosphere has been inferred. The most uncertain method derives from hypothesis as to the temperature and composition of the lithosphere. Uncertain knowledge of both quantities results in very large uncertainties in electrical conductivity. The temperature coefficient of conductivity is large for dry rocks in the lithospheric temperature range, and the existence of small amounts of water increases the conductivity of rocks by many powers of ten in the lower part of that range by conduction in pore fluids [*Brace and Orange,* 1968]; it increases the conductivity to a lesser, but still large, extent at higher temperatures (Figure 1). Hence, even if the composition is known to be basic or ultrabasic, the temperature-conductivity relation is still uncertain.

1.   TEMPERATURE AND COMPOSITION OF THE OCEANIC LITHOSPHERE

The temperature of the subocean lithosphere is inferred from the observed geothermal heat flow at the sea floor, together with estimates of the thermal conductivity, thermal history, radioactive content, and frictional heating of the lithosphere.

The average heat flow through the sea floor has been shown to decrease with increasing age of the sea floor province since formation at a spreading ridge [*Sclater and Francheteau,* 1970]. The decrease has been associated with the gradual cooling of the initially molten or partially molten material that forms the lithosphere at the ridge. If conduction of heat from below and radioactivity decay are the only other operative processes, then the temperature profile in the lithosphere is calculable. On the whole, the distribution of sea-bottom heat flow in this model is in agreement with smoothed values of observed heat flow, but some difficult questions remain. What is the cause of the large scatter of heat flow measurements within a single province? Is it due to slumping of sediments and local refraction of heat flux by irregularities at the contact between poorly conducting sediments and highly conducting igneous rocks, or is it partly caused by magmatic transport of heat, by irregularities of thermal conductivity within the deeper lithosphere, or even by irregularities in the heat flow at the base of the lithosphere?

Some indications of the possible influence of frictional heating are contained in the suggestion by *Elsasser* [1967] that the sea floor plates are driven by differential stresses induced by sinking of dense lithospheric rocks into less dense asthenosphere. The suggestion is strengthened by the model fitting of *Press* [1969, 1970] which implies the existence of a lithosphere more dense by several tenths of a gram/cm³ than the underlying asthenosphere. If such a high density contrast is maintained within sinking slabs of lithosphere, then the rate of conversion of potential energy per unit volume of sinking material, $\Delta\rho\, g\, w$ (where w is the vertical component of velocity), is very large. Some of the conver-

sion will cause heating at the edges of the sinking slab, but some may be available to drag the lithosphere over the resisting asthenosphere and will correspondingly cause heating at the base of the oceanic lithosphere. Suppose the length and thickness, respectively, of a sinking slab are L_s, measured along the downgoing slab in a vertical plane normal to the trench, and H, measured normal to the Benioff plane. Let L be the total length of the plate measured in the direction of motion. Then the average energy flux from potential energy of the sinking slab converted into frictional heat along the entire

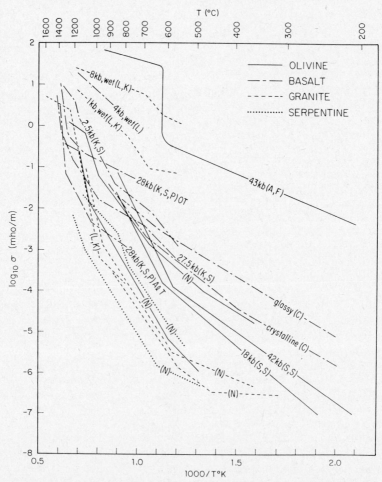

Fig. 1. Electrical conductivity of basic and granitic rocks illustrating the influence of water and pressure. Modified from *Richards* [1970]. Pressure is noted in kilobars. Data are from *Akimoto and Fujisawa* [1965] *(A, F)*; *Coster* [1948] *(C)*; *Khitarov and Slutsky* [1966] *(K, S)*; *Khitarov et al.* [1970] *(K, S, P)*; *Lebedev and Khitarov* [1964] *(L, K)*; and *Noritomi* [1961] *(N)*. On curves *K, S, P, OT* indicates olivine tholeiite, and *Al T* indicates aluminum tholeiite.

229

length of the plate is

$$F = \Delta \rho \; g \; w \; L_s H / L$$

There is no certain information on $\Delta \rho$. *Press* [1970] suggests that the density of oceanic lithosphere, 3.5 to 3.6 g/cm³ at 100 km, decreases to 3.3 to 3.5 g/cm³ at 300 km within the asthenosphere. The resulting value

$$\Delta \rho = 0.15 \pm 0.15 \quad \text{g/cm}^3$$

is positive but uncertain in magnitude. We assume the same contrast is maintained between foundered lithosphere and deeper mantle materials. Let $g = 10^3$ cm/sec², $w = 3$ cm/yr, $L_s = 400$ km, $H = 100$ km, and $L = 5000$ km. Then $F = 15$ ergs cm⁻² sec⁻¹, or one-quarter the average heat flow.

In order to continue the creation of the plate, there must be an equal but opposite conversion from heat to potential energy during creation. Is an efficiency of 25% possible in such a thermodynamic engine?

If the process of lithospheric formation and dissolution is cyclic, we can estimate the thermal efficiency by identifying the working substance, its cyclic changes, and departures from thermodynamic equilibrium. In the cycle illustrated in Figure 2, we assume that basalt is the working substance and that it undergoes the phase changes illustrated. These phases are in accord with the studies of quenched samples by *Ringwood and Green* [1966] and *Ito and Kennedy* [1970], but studies by the electrical conductivity method [*Khitarov et al.*, 1970] suggest a more complicated transition from eclogite to molten basalt. Deviations from thermodynamic equilibrium are impractical to estimate. In particular, solid-solid transitions at low temperature may be slow even on a geologic time scale. Consequently, a detailed study of efficiency is not possible. It is clear, however, that the maximum efficiency $(T_2 - T_1)/T_2$ is much greater than 25%, because the minimum temperature T_1 is representative of the fully cooled lithosphere, say 800°K, and the maximum T_2 is the temperature of the high pressure melt, about 1600°K.

This speculation indicates that friction may well be one of the major sources of heat at the base of the lithosphere. If the asthenosphere is granular, then variations of friction and corresponding variations of frictional heating are possible. It is attractive to associate linear vol-

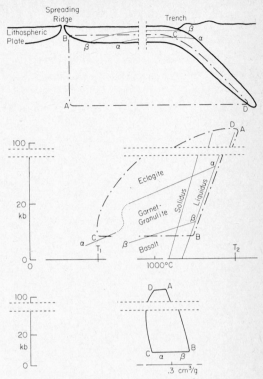

Fig. 2. Thermodynamic cycle of basalt-like substance in the lithosphere and asthenosphere. The upper panel shows assumed motions, the middle panel shows phase diagrams according to *Ito and Kennedy* [1970], and the lower panel shows a pressure volume diagram for basalt. The phase changes are along lines marked α_1, α_2 and β_1, β_2.

canic features such as the Hawaiian chain with such a mechanism. According to this view, the chain would be the wake of an asperity in the asthenosphere as the lithosphere is dragged across it. The dates at which the islands broke through the sea surface (E. L. Winterer, personal communication, 1970) are consistent with wake formation at the rate of 8.5 cm/year, with Midway being created first and Hawaii created last.

The implied production of magma can be estimated from the dimensions of the Hawaiian ridge and its rate of extension. In the vicinity of the populated islands, the breadth of the ridge is about 200 km at the sea floor and its height is variable from 5 to 10 km. If this is lengthened 8.5 cm/year by magma which cools 1200°, the volume extruded and the heat released are on the average, respectively, 4×10^5 cm³/sec and $4 \times$

10^8 cal/sec. The latter figure is equal to the average conductive heat flow through the old ocean crust over an area of 3×10^4 km^2.

The production and migration of magma, whatever the cause of melting, are a feature of the lithosphere that will tend to make the temperature and composition horizontally nonuniform.

With due regard for these uncertainties, we estimate the distribution of electrical conductivity roughly as follows. Unconsolidated oceanic sediments have a measured conductivity about one-half that of sea water. The upper layers of the consolidated crust are sediments and basalts or interbedded basalt and sediment. The conductivity of water-saturated basalts from three ocean floor sites has been measured. Values are listed in Table 1. At greater depths the conductivity probably decreases because of closing of pores and cracks [Brace and Orange, 1968], but the conductivity is very sensitive to moisture content of rocks (Figure 1). If we assume a basalt-eclogite composition, the conductivity will rise to 0.1 mho/m at a temperature of 900°C (if dry), or it will rise higher if appreciable water is present. Accordingly, at the mid-ocean ridge system and at volcanic vents, highly conducting material probably rises very close to the sea floor.

2. ELECTROMAGNETIC METHODS OF CONDUCTIVITY ESTIMATION

Analysis of magnetic fluctuations has been used since the time of Schuster [1889] for inferring the structure of electrical conductivity within the earth. Recent works by Lahiri and Price [1939], Rikitake [1966], and Banks [1969]

are based on an assumed spherical symmetry of the earth. Separation of observed magnetic fields into parts of external and internal origin leads to estimates of internal conductivity. Since this separation requires use of the vertical magnetic component as a diagnostic and the observations of the vertical force at frequencies of 1 cpd and higher show much scatter (presumably owing to the inhomogeneity of the conductivity in the upper mantle and crust coupled with wide and nonuniform spacing of magnetic observatories), we consider that the results are unreliable with respect to details of the upper mantle. Clearly, no structure based on spherical symmetry is capable of yielding data on ocean-continent differences.

Another method, based on magnetic observation in a dense net of stations, has been pioneered by Rikitake [1959], Schmucker et al. [1964], and Schmucker [1970]. This method also uses the vertical component as a diagnostic tool. It can yield information on the internal structure of the earth only if the conductivity as a function of depth is known outside the edges of the net; otherwise, the interpretation of the magnetic signature within the net is not unique. In particular, if the vertical force does not vary throughout the net, one gets no information on the depth of highly conducting structures.

Under some conditions, combined electric and magnetic measurements are useful to resolve this ambiguity. Such combined measurements at sea are reported by Filloux [1967], Cox et al. [1970], and Richards [1970]. The first two papers report a series of electric and magnetic measurements extending offshore from central

TABLE 1.   Conductivity of Water-Saturated Basalts from Three Ocean Floor Sites
(Also see the note added in proof.)

Sample	Water Depth, m	Depth of Sample,* m	Conductivity, (ohm m)$^{-1}$
Pre-Mohole drilling, Guadaloupe site†	3820	186	2×10^{-3} at 22°C
Deep-sea drilling, western south Atlantic‡	4800	142	1.9×10^{-2} at 6°C
Deep-sea drilling, western north Pacific§	3300	335	1.4×10^{-3} at 24°C

* Depth of sample below water sediment interface.

† Sample EM 7, run 5, from Somerton [1961]; fine-grained basalt with calcite in the fractures.

‡ Leg 3, site 19, core 12, section 1; highly weathered basalt, sample soaked in sea water for 30 days before measurement.

§ Leg 6, site 57.0, core 2; doleritic olivine basalt emplaced before Oligocene.

California. At the California shore line there is an anomaly of the magnetic fluctuations of the type discovered by *Parkinson* [1962] to be characteristic of many shore lines. Parkinson's effect pertains to magnetic field fluctuations with periods near one-half hour. Near the shore, he finds that the downward component is correlated and in phase with the onshore component of the field. As a result, the vector fluctuations lie in a plane that dips shoreward. *Cox et al.* [1970] reason that this effect could be the magnetic expression of a band of enhanced electric currents flowing in the ocean parallel to and near the shore or of a similar band within the earth beneath the ocean near the shore. The former condition requires that all conducting matter in the solid earth be insulated from and deeply buried under ocean and land; the latter condition requires that the conducting matter would be shallow under the ocean but deeply buried beneath the land.

These two possibilities can be distinguished from each other by measurement of the electric current (or field) in the ocean. Offshore from California, about half of Parkinson's effect for 1-cph fluctuation is caused by oceanic current flow. Therefore, the remainder must be caused by currents within the earth. The inferred conductivity structure within the earth requires that rocks with electrical conductivity about 1/10 that of sea water rise to within 30 km of the sea bottom offshore from California. This result tends to be confirmed by measurements as far as 700 km offshore, where a simple magnetotelluric method of interpretation is believed to be appropriate.

The ubiquity of Parkinson's effect along coast lines raises the possibility that the oceanic lithosphere is commonly highly conductive at a shallower depth than the sub-continental mantle. This speculation is attractive when it is considered in relation to the well established mobility of ocean lithospheric plates and the shallowness of the seismic low velocity layer beneath the oceans.

On the other hand, there is an exception to Parkinson's effect which has been observed in Peru. At the near-coastal station of Huancayo (100 km inland) the effect is absent or reversed, whereas on the coast the effect is weakly present. *Schmucker et al.* [1964] have investigated the coastal effect at large numbers of magnetic sta-

tions in Peru. They conclude that there is a buried ridge of highly conducting matter beneath the Andes. *Richards* [1970] has analyzed these and new electric observations at sea in the Peru trench and finds that the observed conductivity structure within the earth again requires highly conducting matter close (0 to 160 km) to the sea bottom, but in contrast to the situation in California, this does not lead to a normal Parkinson effect, owing to the electric current flow in the shallow conducting matter under the Andes.

In contrast to these studies with relatively high-frequency sources, the analysis of low-frequency effects by *Banks* [1969] requires that highly conducting parts of the mantle be below 300 km depth. It would appear that there is a discrepancy between the interpretation based on high- and low-frequency fields (Figure 3). Possibilities for resolution of this discrepancy are of three types.

1. The high conductivity found by Cox, Filloux, and Richards at shallow depths is in a thin layer that would scarcely influence the low-frequency fields studied by Banks; see for example *Price* [1970] and the re-evaluation of Bank's data by *Parker* [1971]. According to Parker there is a 'trade-off' between the best possible depth resolution and the accuracy of

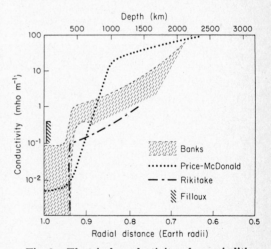

Fig. 3. Electrical conductivity of oceanic lithosphere according to *Filloux* [1967] compared with spherically symmetric models derived from study of low-frequency magnetic fields according to *Price* [1970], *McDonald* [1957], *Rikitake* [1966], and *Banks* [1969]. Price's model also requires a thin surficial layer of high conductivity (not shown).

conductivity estimation. The optimum resolution possible with Banks' data is insufficient to detect, within an accuracy of a factor of two, the conductivity of materials within the upper 80 km of the earth's lithosphere. Nevertheless, the optimum model does suggest conductivity rising toward the earth's surface, in contrast to Banks' estimates.

The implied reduction in conductivity below the effective depth of measurements made by Cox et al. is consistent with the known decrease of conductivity of basalt with increasing pressure (Figure 1). This can be understood on a physical basis if the conductivity is dominated by the mobility of large ions, since these ions would tend to be locked into place by increasing pressure in much the same way that increasing pressure inhibits melting. Alternatively, changes of composition and especially water content may dominate the conductivity distribution.

2. The analysis of Banks is based on a small number of stations mostly near shore lines. If there is in the deep mantle a counterpart of the ocean-continent dichotomy, the location of stations may lead to systematic errors of inter-

Rock Type	Source
Peridotite	Tonga trench
Basalt	Indian Ocean ridge
Gabbro	Indian Ocean ridge
Serpentinized gabbro	Indian Ocean ridge
Serpentinite	Indian Ocean ridge

pretation.

3. The oceanic observations by Cox, Filloux, and Richards are not representative of typical subocean conductivity structures. At the present time, it is not possible to choose between these alternatives.

All electromagnetic methods that use ionospheric signal sources are severely limited in their ability to detect low conductivity materials of the ocean crust. The high-frequency fields necessary for such detection are shielded from the crust by the ocean and underlying water-saturated sediments and basaltic layers.

3. DIRECT METHODS

Direct use of drilled cores or direct measurement through drill holes avoids this difficulty. At present, there are only a few samples of the oceanic second layer drilled from below the sediments, and these samples are saturated with

sea water. The conductivity is of the order of 10^{-2} mho/m. Recent technological improvements in accurate positioning of the drill bit permit changing of the bit and subsequent reentry of a hole. Even with the present limited capacity for supporting a long drill string from the drilling ship, it should be possible to drill more than 2 km into the igneous rocks. Two possibilities exist. One can collect core samples for laboratory studies, and one can use an array of deep drilled holes for artificial geoelectric study of the rocks in situ, free from the shorting effect of the overlying sea.

Note added in proof. The conductivities listed in Table 1 were measured at atmospheric pressure. R. Stresky and W. F. Brace have recently measured conductivities of sea floor rocks at pressures ranging up to 4 kb. The rocks were dredged from the walls of fractures cutting the Indian Ocean ridge and from the Tonga trench and were kindly made available by C. Engel and R. Fisher. During measurement the rocks were saturated with sea water and measured at room temperature. Results are summarized as follows:

Porosity	Conductivity, $(ohm\ m)^{-1}$
0.002	2×10^{-5}
0.03	4×10^{-4}
0.025	1×10^{-3}
0.025	5×10^{-2}
0.15	3×10^{-1}

The high conductivity of the serpentenized rocks is striking. According to R. Stresky and W. F. Brace (personal communication, 1971), 'The presence of magnetite plus the shape of the resistivity-pressure curve indicates mineral conduction for [the serpentinized gabbro] as opposed to conduction through pore fluids which characterize all other samples.' They note that magnetite or other oxides are present as a fine dust in the hydrothermally altered rocks. The characteristic presence of oxide dust and associated high mineral conductivity in such altered ultramafic rocks taken in conjunction with the observations by Cox, Filloux, and Richards of high conductivity at shallow depths in the oceanic lithosphere suggests an alternate hypothesis for the nature of electrical conductivity in the lithosphere as follows: At the top of the lithosphere where basalt predominates, the conductivity is controlled by pore fluids and decreases with increasing pressure because of gradual closing of cracks and pores. At greater depths and where an appreciable fraction of rocks are hydrothermally altered, the conductivity rises because of mineral conduction. At still greater depths, where altered ultramafics are unstable because of high temperature the conductivity will decrease because of the low conductivity of silicates even at these moderately high temperatures.

Acknowledgment. I acknowledge with gratitude the hospitality of the Department of Earth and Planetary Sciences of MIT while this report was being prepared. Deep sea drilling samples were provided through the assistance of the National Science Foundation.

The research was supported by the Office of Naval Research.

REFERENCES

Akimoto, S., and H. Fujisawa, Demonstration of the electrical conductivity jump produced by the olivine-spinel transition, *J. Geophys. Res., 70,* 443–449, 1965.

Banks, R. J., Geomagnetic variations and the electrical conductivity of the upper mantle, *Geophys. J., 17*(5), 457–487, 1969.

Brace, W. F., and A. S. Orange, Further studies of the effect of pressure on electrical resistivity of rocks, *J. Geophys. Res., 73,* 5407–5420, 1968.

Coster, H. P., The electrical conductivity of rocks at high temperatures, *Monthly Notices Roy. Astron. Soc., Geophys. Suppl., 5,* 193–199, 1948.

Cox, C. S., J. Filloux, and J. Larsen, Electromagnetic studies of ocean currents and the electrical conductivity below the sea floor, Chap. 17 in *The Sea,* vol. 4, part 1, edited by E. C. Bullard and A. E. Maxwell, 1970.

Elsasser, W. M., Convection and stress propagation in the upper mantle, *Tech. Rep. 5,* Princeton Univ., Princeton, New Jersey, June 15, 1967.

Filloux, J., Oceanic electric currents, geomagnetic variations and the deep electrical conductivity structure of the ocean-continent transition of central California, Ph.D. thesis, Univ. of California, San Diego, 1967.

Ito, K., and G. C. Kennedy, Fine structure of the basalt-eclogite transition, *Mineral. Soc. Amer. Spec. Pap. 3,* 77–83, 1970.

Khitarov, N. I., and A. V. Slutsky, Influence of temperature and pressure on electrical conductivity of albite and basalt, *J. Chem. Phys.,* in French, *64,* 1085–1091, 1966.

Khitarov, N. I., A. V. Slutsky, and V. A. Pugin, Electrical conductivity of basalts at high T-P and phase transitions under upper mantle conditions, *Phys. Earth Planet. Interiors, 3,* 334–342, 1970.

Lahiri, B. N., and A. T. Price, Electromagnetic induction in nonuniform conductors and the determination of the conductivity of the earth from terrestrial magnetic variations, *Phil. Trans. Roy. Soc., A, 237,* 509–540, 1939.

Lebedev, E. B., and N. I. Khitarov, The dependence of onset of melting of granite and the electrical conductivity of granite melt upon high water vapor pressure, *Geokhimia,* in Russian, *3,* 195–201, 1964.

McDonald, K. L., Penetration of the geomagnetic secular variation field through a mantle with variable conductivity, *J. Geophys. Res., 62,* 117–141, 1957.

Noritomi, K., The electrical conductivity of rocks and the determination of the electrical conductivity of the earth's interior, *J. Mineral. Coll. Akita Univ. A, 1,* 27–59, 1961.

Parker, R. L., The inverse problem of electrical conductivity in the mantle, *Geophys. J. Roy. Astron. Soc., 22,* 121–138, 1971.

Parkinson, W. D., The influence of continents and oceans on geomagnetic variations, *Geophys. J. Roy. Astron. Soc., 6,* 441–449, 1962.

Press, F., The suboceanic mantle, *Science, 165,* 174–176, 1969.

Press, F., Earth models consistent with geophysical data, *Phys. Earth Planet. Interiors, 3,* 3–22, 1970.

Price, A. T., The electrical conductivity of the earth, *Quart. J. Roy. Astron. Soc., 11,* 23–42, 1970.

Richards, M. L., Study of electrical conductivity in the earth near Peru, Ph.D. thesis, Univ. of California, San Diego, 1970.

Rikitake, T., Anomaly of geomagnetic variations in Japan, *Geophys. J. Roy. Astron. Soc., 2,* 276–287, 1959.

Rikitake, T., *Electromagnetism and the Earth's Interior,* 308 pp., Elsevier, N. Y., 1966.

Ringwood, A. E., and D. H. Green, An experimental investigation of the gabbro-eclogite transformation and some geophysical implications, *Tectonophysics, 3,* 383–427, 1966.

Schmucker, U., Anomalies of the geomagnetic variations in the southwestern United States, in *Bull. 13, Scripps Inst. Ocean.,* pp. 1–165, Univ. of Calif., San Diego, 1970.

Schmucker, U., O. Hartmann, A. A. Giesecke, M. Casaverde, and S. E. Forbush, Electrical conductivity anomalies in the earth's crust in Peru, *Carnegie Inst. Washington Yearbook, 63,* 354–362, 1964.

Schuster, A., The diurnal variation of terrestrial magnetism, *Phil. Trans. Roy. Soc. London A, 180,* 467–518, 1889.

Sclater, J. G., and J. Francheteau, Implications of terrestrial heat flow observation on current tectonic and geochemical models of the crust and upper mantle of the earth, *Geophys. J. Roy. Astron. Soc., 20,* 509–542, 1970.

Somerton, W. H., Physical properties of the basalt, in Experimental Drilling in the Deep Water at the La Jolla and Guadaloupe Sites, *Nat. Acad. Sci. Nat. Res. Counc. Publ., 914,* 146–151, 1961.

DISCUSSION

Madden: I do not think any of these 'world-wide' pictures are characteristic of more than the continents because the data come only from the continents. There must be a tremendous difference between the electrical properties of the upper mantle under oceans and under continents.

Cox: If the conductivity is as high as the 0.4 mho/m suggested by our work, then it certainly cannot be the same material. Such a high conductivity at such a shallow depth is certainly not usual.

Porath: Have you recomputed the Andes anomaly with your resistivity section for the ocean?

Cox: Richards has used old electric and magnetic daily variation data recorded at Huancayo, Peru, to make a magnetotelluric analysis. He finds evidence for a conductivity of 0.3 (ohm m)$^{-1}$ at a depth of 40 to 120 km below the observatory. This evidence, together with the magnetic fluctuation anomalies near shore, supports the model of Schmucker and others.

Landisman: Dziewonski has shown there is no need to have the high density in the upper mantle that Press has suggested. It may well be there, but it is not required by the data.

Cox: Press told me this result was controversial.

Whitcomb: If you pull the lithosphere over a partially melted zone and generate heat at the base, then this heat will go mainly toward melting more rock and will have little effect on temperature or heat flow.

Cox: No, this has been going on for a long time, say 100 million years, and if it were continuing to melt rock you would have an enormous melt zone down there.

Hales: It is possible that the melted mantle comes out as volcanic material.

Cox: Present estimates put the amount of heat carried up by volcanic rocks at a very small fraction of the average heat flow, whereas on this model, 25% of the heat flow is put into this potential energy form and later is reliberated.

Hales: I do not think you can say this unless you are prepared to defend the density differences you are assuming, which are extreme. You may just be melting material which later flows out of the mid-Atlantic and other ridges.

Cox: Then you are requiring this hot, frictionally melted rock to flow underneath the crust where it is created, namely in the older parts of the crust, laterally for some thousands of miles to the ridge. That would redistribute the heat very effectively.

Hales: If you squeeze the bottom of a toothpaste tube, the toothpaste that comes out comes from the top, but you squeezed the bottom.

Cox: We are discussing the heat, which is generated where the friction is, and not the toothpaste.

4. LABORATORY GEOPHYSICS

Imperfect Elasticity of Rock:
Its Influence on the Velocity of Stress Waves

ROBERT B. GORDON

Department of Geology and Geophysics, Yale University
New Haven, Connecticut   06520

DENNIS RADER

Department of Engineering and Applied Science, Yale University
New Haven, Connecticut   06520

Abstract. Motion on interfaces within rock results in a decrease in stress-wave velocity and the presence of internal friction. When there is a fluid phase in the interfaces, these effects can persist deep within the earth's crust. Comparison of results obtained by stress-pulse propagation, driven resonance, and cyclic loading experiments shows that, although the imperfect elasticity due to interface motion is frequency and amplitude insensitive in the range characteristic of seismic signals, properties observed in the millihertz (seismic) frequency range are significantly different from those observed by the ultrasonic methods commonly used in the laboratory. Introduction of intergranular fluid, for example, increases the ultrasonically measured sound velocity but decreases the velocity measured with millihertz-frequency signals. The experimental results show that if the measured velocity in a zone of the crust falls below that expected for the known (or estimated) composition and temperature in that zone, the presence of continuous intergranular fluid phase is indicated.

If a rock were perfectly elastic, the velocity of stress waves through it would be entirely determined by the elastic moduli and densities of the constituent minerals. However, the elasticity of real rock is sometimes very imperfect; there may be significant internal friction, resulting in attenuation of stress waves, and large departures of the velocity of propagation from that calculated by use of the measured velocities in the minerals present. In the presence of such imperfect elasticity, the values of the stress-wave propagation constants may be more indicative of the microstructure and physical environment of the rock than of its composition. Seismic data can then be used, in conjunction with laboratory results on imperfect elasticity, to establish the environmental conditions at depth. For example, the presence of intergranular fluid phase may be detected.

Imperfect elasticity in rock can arise from many sources, most of which are sufficiently complex that they must be evaluated through laboratory experiments with only general guidance from theory. The application of laboratory observations to the interpretation of field data is complicated in experiments in which mechanical properties are measured; first, difficulty is encountered in attempting to reproduce the true rock environment, including temperature, large lithostatic and small shear stress, and

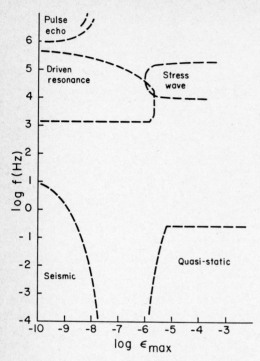

Fig. 1. Domains of frequency and strain amplitude (ϵ_{max}) in which various laboratory techniques used for the measurement of mechanical properties of solids are effective. No laboratory method duplicates the conditions of a stress wave in a teleseism.

1. Small velocity change
   thermally activated atom movements
   dislocation damping
   solid-solid phase transformations
   viscous grain boundary damping
2. Large velocity change
   dislocation breakaway
   interface sliding
   partial melting

These sources of internal friction are arranged in two classes. First are those sources in which large internal friction is accompanied by only a small change in wave velocity. These are damping mechanisms that cause the rock to behave as a viscoelastic material, and they are important principally at high temperature ($T/T_m > 0.5$, where T_m is the solidus temperature). They are probably more significant sources of imperfect elasticity for upper mantle than for crustal rock.

The second group contains those processes that can give rise to large changes in sound velocity as well as internal friction. When dislocation breakaway or interface sliding occurs in a material it cannot be described as viscoelastic; under cyclic loading the stress-strain curve is a hysteresis loop even under quasi-static conditions, and there is no longer a fixed relation between internal friction and the change in modulus (or velocity). Significant, nonlinear dislocation damping has not yet been observed in rock-forming minerals. Hence, in interpreting wave velocities measured in the earth, interface sliding and partial melting are of greatest interest; it is the presence of these mechanisms that can make the sound velocity through rock substantially different from that predicted from the elastic properties of the constituent minerals. This paper deals with the imperfect elasticity of rock at temperatures below T_m, which can result in large velocity changes, i.e., the imperfect elasticity due to interface sliding. Many of the characteristics of this type of imperfect elasticity have been described in an earlier paper [*Gordon and Davis*, 1968], in which it was shown that interface damping in rock is insensitive to frequency, strain amplitude, and temperature and that it is suppressed by a clamping pressure of about 500 bars. This clamping pressure was shown to be the difference between the externally applied hydrostatic pressure and the pore-fluid pressure. This paper presents additional data on interface internal

fluid pressure; second, there is the impossibility of duplicating in the laboratory seismic-wave conditions of low frequency and low strain amplitude. The second point is illustrated in Figure 1; in this figure are plotted the domains over which the presently available laboratory techniques are effective. None penetrate the frequency and amplitude domain of the stress waves usually observed in seismology. Finally, the theory of many of the standard laboratory methods of measurement of stress-wave propagation constants assumes that internal friction effects are small; this may not be true of rock samples. For all these reasons, rather extensive laboratory data on departures from perfect elasticity obtained over a wide range of experimental conditions are required if field observations of imperfect elasticity are to be used in evaluating rock structure and environment.

Listed below are the known sources of imperfect elasticity in rock:

friction that were obtained with new experimental methods (and more sensitive equipment).

STRESS-WAVE PROPAGATION EXPERIMENTS

Stress-wave experiments offer the opportunity of making measurements in the kilohertz range of frequencies at strain amplitudes that are as large as those where low frequency, quasi-static observations of modulus and attenuation are possible. These experiments are also of interest because the imperfect elasticity that results from interface friction may be large and non-linear. Because these conditions are excluded in the usual analysis of mechanical-resonance and pulse-echo measurements, there may be doubt that properties measured by forced oscillation of a sample in the laboratory will be found by the stress-wave propagation experiments, which utilize a technique closer to that used in the field. An experimental test to resolve this uncertainty has been made by doing successively, on one bar of rock, wave propagation, resonance, and quasi-static experiments. A single bar must be used because the imperfect elasticity due to interface sliding is highly structure-sensitive; there are sometimes wide variations in properties between samples of the same rock, and even careful handling of a specimen in the laboratory may result in a change in its properties.

The rock used was a bar of Chester (Rhode Island) granite. For the stress-pulse experiments, semiconductor strain gages were mounted in pairs at two stations along the bar, and each pair was connected in series to cancel bending strains. Pulse amplitudes were controlled by varying the impact speed of a steel ball at one end of the bar. The contact made between the ball and the bar (the impact end of the bar was coated with electrically conducting paint) was used to trigger a time-delay circuit, which in turn triggered the oscilloscope sweep after a predetermined interval. The speed of the compression pulse was measured during the first pass of the pulse down the bar, and the speed of the tension pulse was measured after reflection of the pulse at the free end.

Quasi-static experiments in tension and compression at a strain rate of about 10^{-5} sec^{-1} were performed to obtain the stress-strain curve for the granite in uniaxial loading. The technique is generally similar to that described in *Gordon and Davis* [1968]. The specimen used in the static tests was the same one used in the pulse-propagation experiments, in which the output of the two pairs of gages was recorded separately. The tangent modulus was measured at various strain levels for the outputs of each gage pair in both loading and unloading. Because the properties of the granite are different in tension and compression, it could not be assumed that complete cancellation of bending strains was obtained by wiring each gage pair in series. The testing machine was therefore carefully aligned and the experiments were repeated after a quarter turn rotation of the specimen about its longitudinal axis. The stress-strain curves obtained for the two positions matched very closely; therefore, it can be safely assumed that the observed stress-strain curves were not influenced by bending strains.

The central portion of the test specimen was also used in longitudinal-resonance experiments to measure its thin-rod velocity. Quartz crystal transducers were used, and the resonant frequency was measured for the first and third modes. Another specimen (number 2) cut from the same block of material as the original test specimen was used in a set of resonance measurements over a lower range of frequencies. For the resonance study, the relatively long specimen was cut in half for each successive test.

For the stress-pulse propagation method the wave speed c is measured for three pulse amplitudes in both tension and compression. The results are shown in Table 1. The tabulated

TABLE 1. Pulse Propagation Speed in Chester Granite

ϵ_1*	ϵ_2*	c, km/sec	Y, Mbar
Compression			
123×10^{-6}	93×10^{-6}	3.33	0.295
51	38	3.33	0.295
7.7	6.3	3.33	0.295
Tension			
100×10^{-6}	85×10^{-6}	3.08	0.252
46	37	3.33	0.295
6.8	5.6	3.63	0.349

* Amplitudes of stress pulse passing successive strain gages.

TABLE 2.   Longitudinal Bar Resonance of Chester
Granite

Specimen	Frequency, kHz	c, km/sec
1	25	3.640
	75	3.620
2	6	3.37
2a	12	3.45
2b	12	3.53
2c	25	3.66
2d	25	3.68

Young's modulus Y is obtained from the thin-rod equation, $Y = c^2\rho$, using the measured density of the rock, $\rho = 2.65$ g/cm³. Wave speeds calculated from the measured longitudinal-resonance frequencies are recorded in Table 2.

The principal frequency components of the pulses used were in the 10 to 50-kHz range, but the lowest strain amplitude of the pulse, about 10^{-6}, was greater than the 10^{-7} used in the resonance experiment. The tensile-pulse data show a decrease in c with increasing amplitude. This amplitude dependence of the velocity probably results in lower velocities being observed in the pulse experiments than in the resonance experiments on the same rock bar.

A decrease in Young's modulus with increase in tensile strain amplitude is also found in the quasi-static experiments, the results of which are shown in Figure 2. Under compression the modulus remains constant at strain amplitudes ranging up to 100×10^{-6}. The presence of some inhomogeneity in the test bar is shown by the data in both Table 1 and Figure 2. The rock is stiffer near gage 2. Comparison of the dynamic and quasi-static data taken at the same amplitude shows that there is a frequency dependence in the modulus; it is lower at low frequency. The resonance measurements on bar 2 also show a decrease of wave speed when the frequency drops below 25 kHz.

If the imperfect elasticity resulting from sliding at the interfaces between rock grains is due to macroscopic friction, it is expected to be independent of frequency and insensitive to strain amplitude except at some small amplitude where normal forces on the interfaces due to residual stresses become important. The experiments reported on in *Gordon and Davis* [1968] show that amplitude dependence of the internal friction and the dynamic modulus are not found in crystalline rock at strain amplitudes as low as 5×10^{-10}. When the stress-pulse method is used to extend the amplitude range above that which can be attained with piezoelectric resonators driven with high voltage, amplitude dependence is found in tension but not in compression. A significant frequency dependence of the modulus is found between the kilohertz and millihertz range. These results, then, cast doubt on the validity of the interpretation that interface damping is due simply to macroscopic friction. Better data at millihertz frequencies, particularly at low-strain amplitudes, are required for a further attempt to identify the active mechanism.

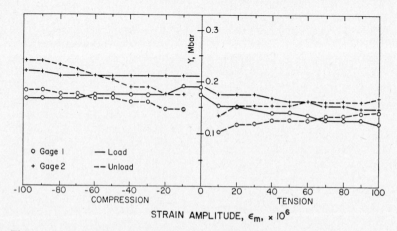

Fig. 2.   The tangent modulus read from the stress-strain curve of a test bar of dry Chester granite subjected to cyclic loading.

LOW-FREQUENCY EXPERIMENTS

The basic difficulty encountered in extending the quasi-static experiments into the seismic domain is the measurement of small strains with sufficient precision. In the experiments reported on in *Gordon and Davis* [1968], the lowest strain amplitude attained was 10^{-4}. In new experiments, an attempt has been made to lower this limit and to improve the sensitivity with which internal friction is detected. Rock bars fitted with strain gages were subjected to sinusoidally varying stress developed by a spring compressed by a rotating crank of variable throw. Stress amplitude was controlled by changing the crank throw while the alignment of the test bar in the compression fixture remained undisturbed. As before, the stress-strain curve was recorded while the stress was cycled. Typical results, (see Figure 3) show that the internal friction decreases with decreasing strain amplitude. This pattern was not detected in the earlier experiments because of the relatively low sensitivity of the equipment then in use. Somewhat better than a factor of ten has been gained in the lowest amplitude at which measurements can be made, but, unfortunately, this is still not enough to allow laboratory measurements to duplicate seismic wave characteristics.

Extrapolation of the data in Figure 3 indicates the presence of interface internal friction at the lowest strain amplitudes. There is no available evidence to show that such internal friction does not exist at low amplitude. However, there also must be a significant frequency dependence, since the internal friction measured at the lowest amplitudes in the low frequency stress-cycling experiments is approximately an order of magnitude less than that observed in the high frequency, low-amplitude resonator measurements on the same rock.

Laboratory experience shows the imperfect elasticity of rock that is due to interface sliding to be very structure sensitive, owing to the fact that the number of active interfaces in the rock is easily altered by applied stress, thermal strains, and the presence of fluids. Internal friction and significant changes in modulus result primarily from interface sliding. Laboratory evidence indicates that this relationship will continue to exist in the seismic domain of frequency and amplitude.

EFFECT OF FLUID PHASE

When a rock contains mechanically active interfaces, its properties, as mentioned above, are sensitive to the presence of intergranular

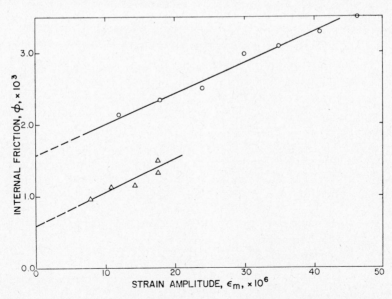

Fig. 3. Internal friction ϕ measured in two samples of Chester granite in cyclic compression as a function of strain amplitude. An attempt was made to attain the lowest possible amplitude.

fluids. It was reported in *Gordon and Davis* [1968] that the presence of water causes a large increase in internal friction at kilohertz frequencies but only a small change at low frequencies. A much more extensive series of measurements on this has been made using the improved, more sensitive cyclic-loading equipment. The earlier results were confirmed; it was found that marked changes in modulus occurred as a result of the presence of water. The method of measurement used was the observation of the change in modulus and internal friction that results when a thoroughly dry rock sample is saturated with water. (Precautions were taken to remove air from the interstices before water was added.) The experiments were performed with both longitudinal and torsional deformation; the changes observed were generally the same in both modes. The results of experiments in the longitudinal mode are shown in Figure 4. At low frequencies, observations were made at the lowest possible strain amplitude (about 10^{-5}). Relative changes in the velocity of stress waves were calculated from measured changes in modulus, δY, by the relation $\delta c/c = \delta Y/2Y$.

The results show clearly that at low frequencies the addition of water results in a decrease in velocity and that at high frequencies it results in an increase. Very small changes are observed in the pyroxenite, because this particular rock sample is so dense that very few active interfaces are present in laboratory specimens. Larger effects are observed in rocks with a more open structure. *Nur and Simmons* [1969] have reported a set of measurements of the velocity changes that result from saturating the intergranular spaces of granite with glycerine or water. Introduction of these fluids results in an increase in the shear and compressional velocities measured at 500 kHz. A velocity dependence is found as the viscosity is varied over fourteen powers of ten by changing the temperature of the glycerine saturated rock. Nur and Simmons assert that the frequency dependence of the fluid phase contribution to the velocity and attenuation remains fixed when the product of frequency and viscosity ($f\eta$) is held constant. However, since the introduction of intergranular fluid results in a decrease in velocity when it is measured at low frequency, it is evident that physical processes other than those controlled by the fluid viscosity are active in the millihertz range. The velocity decrease is greater for introduced water than for pentane, kerosene, or glycerine. Evidently, water facilitates grain boundary sliding. These results illustrate the difficulty of using laboratory data obtained by ultrasonic methods for the interpretation of seismic ob-

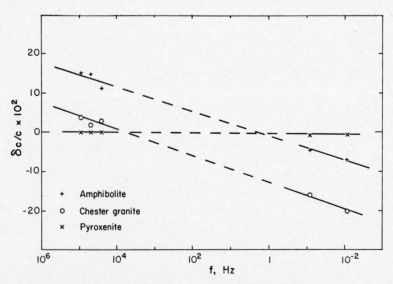

Fig. 4.   Relative change $\delta c/c$ in wave propagation velocity which results from saturating samples of three different rock types with water.

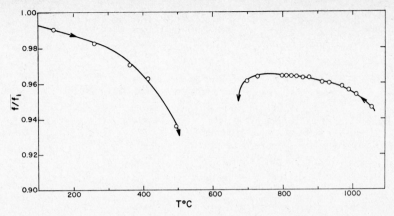

Fig. 5. Change in the velocity of sound, as measured by the change of the frequency of longitudinal resonance f of a bar of basalt during heating and subsequent cooling. Between about 1000° and 700°C the grain interfaces remain adherent; there is no interface damping. The resonant frequency at room temperature is f_1.

servations made with signals of much lower frequency.

Since it is probable that the water-induced reduction in stiffness will also occur in the seismic strain amplitude range, it follows that the presence of intergranular fluids will cause a marked decrease in the velocity of stress waves through rock containing active interfaces.

OBSERVATIONS AT HIGH TEMPERATURE

Since large changes in modulus and, hence, in velocity, result from interface sliding, it is important to know the domain of temperature and pressure in which this type of imperfect elasticity can occur. The pressure domain has been established by quasi-static internal-friction experiments done under hydrostatic pressure; interfaces in typical crystalline rock are clamped when the confining pressure exceeds the fluid (or neutral) pressure by about 500 bars. The temperature domain is established by the experiment illustrated in Figure 5; a rock bar is mounted on a composite piezoelectric resonator fitted with a buffer rod so that the sample can be heated. Changes in the frequency of longitudinal resonance, f, are observed during heating and cooling, as in Figure 5, in which data obtained by *Stocker* [1971] are shown for a sample of basalt (East Rock sill, Connecticut). Upon heating, there is an initial, linear decrease in f due to the decrease of the elastic constants of the constituent minerals with temperature. Above approximately 300°C,

the decrease in f becomes rapid. This pattern occurs because internal strains resulting from unequal thermal expansion unclamp additional grain interfaces in the rock. Measurements cannot be continued above 500°C because the mismatch of sample and resonator frequencies is too great. However, heating is continued and at a temperature of about 1100°C the damping of the system is observed to become low enough for measurements to commence again. After holding at high temperature, cooling is commenced and a regular increase in f is observed. This is found to correspond to the average increase in elastic stiffness of the constituent minerals. At about 700°C there is again a strong decrease in f.

The experiment may be interpreted as follows: At about 1100°C, spontaneous bonding of the grain boundaries occurs. Because the rock grains are cemented together, interface damping is suppressed. The grains remain cemented during cooling until accumulating thermal stresses are great enough to again break open the grain interfaces. This phenomenon occurs at about 700°C. The same experiment has been done with samples of dacite with generally similar results. Below about 700°C, the grain boundaries no longer remain bonded and a large decrease in sound velocity is found.

The temperature to which interfaces would remain bonded during cooling of rock in the earth's crust would be somewhat lower than that indicated in this experiment for two reasons: First, at slower cooling rates there will be greater relaxation of stress concentrations at grain boundaries by means of plastic flow within the grains. Second, presence of fluid phases may promote boundary welding. It is expected that the activation energy that controls relaxation of stress concentrations will be large. Then the boundary break-apart temperature will be relatively insensitive to cooling rate. A reduction of about 100°C may be expected for conditions in the crust. Laboratory evidence for the effect of fluid phases on boundary welding is lacking.

These high-pressure and high-temperature experiments indicate that imperfect elasticity due to interface sliding will be present in crystalline rock whenever $T < \sim 600°C$ and the clamping pressure is less than 500 bars. Since high fluid pressure may occur at all depths in the crust, zones of low velocity can result from interface sliding without changes in composition, changes in phase, or partial melting. If seismic data reveal a region where the stress wave velocity is below that level associated with the expected temperature and composition of that region, there is evidence for the presence of a continuous film of intergranular fluid. Clearly, the success of this method of fluid detection depends on the accuracy with which an 'expected' velocity can be determined, i.e., the accuracy with which temperature and composition can be inferred from other evidence. If it could be shown that the zone of low velocity is also a zone of low Q, the validity of the interpretation in terms of fluid films would be greatly strengthened.

Acknowledgment. We are indebted to R. L. Stocker for permission to quote some of his data before publication.

The research reported in this paper was supported by the U. S. Office of Naval Research.

REFERENCES

Gordon, R. B., and L. A. Davis, Velocity and attenuation of seismic waves in imperfectly elastic rock, *J. Geophys. Res., 73,* 3917–3935, 1968.

Nur, A., and G. Simmons, The effect of viscosity of a fluid phase on velocity in low porosity rocks, *Earth Planet. Sci. Lett., 1,* 99–108, 1969.

Stocker, R. L., The acoustic properties of partial melts, Ph.D. thesis, Yale University, New Haven, Connecticut, 1971.

Resistivity of Saturated Crustal Rocks to 40 km based on Laboratory Measurements

W. F. BRACE

Department of Earth and Planetary Sciences, Massachusetts Institute of Technology
Cambridge, Massachusetts 02139

Abstract. Recent studies in solution chemistry, together with our earlier high-pressure work, provide the information needed to construct resistivity-depth profiles for typical crustal rocks saturated with aqueous solutions. Profiles are presented for three heat flow provinces: the Sierra Nevada, the Basin and Range, and the eastern United States, to a depth of 40 km. The effects of high pressure and temperature and the variation of pore pressure from hydrostatic to lithostatic are discussed. For aqueous solutions alone, resistivity typically decreases with the conditions encountered at depth. However, this decrease is nearly counteracted by the effect of decreasing porosity of typical rocks at depth, so that below a few kilometers, resistivity of solution-saturated rocks should vary only slightly until temperatures sufficient for mineral conduction are reached. Mineral conduction should become significant at a depth of 10 to 40 km, depending on heat flow province. At no depth should resistivity of solution-saturated crustal rocks be greater than 10^6 Ω-m.

Much is known about electrical resistivity of rocks of the sort thought to make up the earth's crust, for pressures and temperatures appropriate to depths of 50 km or more [see *Keller,* 1966; *Parkhomenko,* 1967]. In most of the laboratory studies upon which these data are based, the rocks were nominally dry or, if they were water-bearing, they were at atmospheric pressure [see *Scott et al.,* 1967]. In the earth, rocks below the water table are saturated with aqueous solutions to at least 4 to 5 km, based on observations in deep wells, and possibly to even greater depths. What is the resistivity of crustal rocks if these conducting solutions are taken into account? No estimates appear to be available in which all the important factors such as temperature, salinity, and pore pressure are taken into account. The purpose of this paper is to attempt such an estimate based on laboratory studies. Two assumptions are made: that laboratory samples are truly representative of intact rock in the crust and that crustal rocks are saturated. Actual field measurements of resistivity of deeply buried rocks may themselves provide a test of these assumptions.

What factors determine resistivity of rocks in the crust? From earlier studies at room tem-perature [*Keller,* 1966; *Brace and Orange,* 1968b], porosity is the sole rock property that determines the resistivity of water-saturated rocks composed of nonconducting minerals; grain size and mineralogy have almost no effect. Resistivity also depends strongly on the detailed chemistry of the pore water. Pressure and temperature are important because they affect both porosity of the rock and conductivity of the pore solution.

Of these factors, porosity, salinity, and pressure have been studied in some detail for saturated rocks [*Brace et al.,* 1965; *Brace and Orange,* 1968a, b; *Greenberg and Brace,* 1969]. Temperature, particularly in its effect on solution resistivity, is the principal remaining factor whose effect is poorly understood. The complete experiment, in which both pressure and temperature are varied for saturated rock, does not seem to be feasible at present. Instead, we will use in this paper some new measurements of resistivity of various dilute solutions to 4 kb and 800°. Fortunately, these solutions, which were studied for quite a different purpose, happen to be of geologic interest, and the pressures and temperatures are quite appropriate for the crust. We combine these data with our earlier room-temperature and high-pressure results for rocks and thereby estimate the resistivity of a solution-saturated rock at different depths in the crust. All the factors that determine resistivity vary not only vertically in the crust but also horizontally. In this paper we narrow the choice somewhat by selecting three heat flow provinces [*Roy et al.*, 1968; *Lachenbruch*, 1970]. Resistivity-depth profiles are constructed for each province, by using the appropriate temperature-depth profiles. Each resistivity profile gives the range of resistivities that is shown by a group of typical crustal rocks and the variation with depth (which is the combined effect of pressure on the rock and of pressure and temperature on the solution). Variations in resistivity as pore pressure ranges from hydrostatic to lithostatic are discussed; the effect of mineral conduction at high temperatures is also considered.

Porosity. Porosity must be considered carefully, since it is the prime factor controlling conduction in saturated rocks at lower temperatures. Natural rocks have three kinds of porosity: fracture, crack, and pore. We adopt the term fracture porosity here for porosity that is associated with joints, faults, bedding planes, cleavage, and other features typically not included in the small samples studied in the laboratory. Laboratory samples contain cracks and pores [*Brace*, 1965; *Walsh and Brace*, 1966]; the cracks close at low pressure as, presumably, does fracture porosity. Pores are somewhat reduced by pressure but still appear to provide a network of conduction paths to very high pressure [*Brace et al.*, 1965].

To what extent is the porosity of laboratory samples identical with that of rock in place? Obviously, fracture porosity is lacking in laboratory samples; there may be differences with regard to crack and pore porosity as well [*Brace and Byerlee*, 1967]. Cracks may be introduced during sampling, particularly if the rock in place is under stress [*Nur and Simmons*, 1970]. However, it seems likely that most crack and pore porosity already exists prior to sampling, inasmuch as it probably results from changes in pressure and temperature which the rock experiences during its geologic history. For example, crack porosity of laboratory samples of granites is particularly high relative to either a pure quartz rocks or to rocks with no quartz [*Nur*

and Simmons, 1970]. This relation would be expected if the porosity resulted from temperature changes; the thermal expansion of quartz is quite different from that of feldspar and other common minerals. Thus, values for crack and pore porosity of laboratory and intact rock in situ are probably not very different. The most sensitive test of this would be a detailed comparison of laboratory and in situ resistivities. Although one such comparison was attempted [*Simmons and Nur*, 1968], the results were quite inconclusive [*Orange*, 1969]. As noted above, we will have to assume here that laboratory resistivity is the same as that of rock in place as long as factors such as pressure and salinity are the same.

The effects of temperature on porosity are not well understood and need to be explored, for thermal stresses may in certain cases cause cracks and increase porosity. Probably, cracking due to temperature change is minor as long as effective pressure is high. One guide in this respect is *Birch's* [1943] study of shear velocity at high temperature and pressure. None of the pronounced dropoff in velocity that might be expected from new cracks was observed up to 600° for pressures of 3 to 4 kb; evidently any cracking was suppressed by the pressure. We will assume that this is generally true and that porosity is unaffected by temperature change as long as pressure and temperature increase together. The changes in porosity due simply to thermal expansion are probably negligible.

Pressure. Pressure also must be carefully considered, for it plays a varied role. On the one hand, resistivity of a saturated rock depends on effective or grain-to-grain pressure, $\bar{P}$, which is total pressure less the pore pressure [*Brace and Orange*, 1968a]. On the other hand, resistivity of the pore fluid itself depends only on pore pressure, P_{H_2O}. We will consider two situations that may be of interest geologically: when pore pressure is lithostatic $(P_{H_2O})_{LITH}$ and when it is hydrostatic $(P_{H_2O})_{HYD}$, where the corresponding effective pressures are $(\bar{P})_{LITH}$ and $(\bar{P})_{HYD}$, respectively. Not enough is known about water in deeply buried rocks to determine whether other situations (such as zero or sub hydrostatic pore pressure) are more relevant. When pore pressure is lithostatic, it balances the weight of overlying rock; effective pressure $(\bar{P})_{LITH}$ is therefore zero. When pore pressure is hydrostatic, effective pressure $(\bar{P})_{\mathrm{HYD}}$ is total less hydrostatic pore pressure, or $(\rho_R - \rho_W)gz$, where ρ_R and ρ_W are average densities of overlying rock and water, respectively, z is depth, and g is the acceleration of gravity. Pore pressure might be hydrostatic if all pore space were continuously connected to the surface.

Salinity. Resistivity of a solution-saturated rock depends strongly on the concentration of salts in the pore solutions. This dependence was explored [*Bracc et al.*, 1965] for two pore solutions that ranged in resistivity from 50 Ω-m (tap water) to 0.25 Ω-m (NaCl solution). Data from this study and from *Brace and Orange* [1968b] are given in Figure 1 for a wide range of crystalline igneous and metamorphic rocks, at the same effective confining pressure of 4 kb. Surface conduction [*Madden and Marshall*, 1959] lowered the resistivity of the samples with tap water by a factor of 10 to 20, so that resistivity for the two pore solutions differed only by a factor of 10 (rather than the factor of 200 to be expected from the ratio of solution resistivities).

What is the resistivity of pore fluids likely to be present in rocks in the crust? This is, of course, not known and can only be surmised from the resistivity of waters found in igneous and metamorphic rocks near the surface. Data from numerous wells [*White et al.*, 1963; *Keller*, 1966] reveal that resistivity is almost always within the limits 0.5 to 100 Ω-m, with average values for wells of 1 to 30 Ω-m. Recent sediments, young sedimentary rocks, and some volcanic environments contain waters of lower resistivity, ranging down to 0.1 Ω-m.

If we adopt 1 to 30 Ω-m as typical of natural pore solutions, then, from Figure 1, resistivity of rocks of 0.001 to 0.01 porosity will be about 10^5 to 10^6 Ω-m at $\bar{P}$ equals 4 kb. This narrow range is a helpful result of surface conduction which, in effect, requires that a rock behave as though pore fluid resistivity were about 2 Ω-m, although in reality the fluid might be much more resistive pure water.

Heat flow provinces. Basing their study on regional similarity of heat flow and surface heat production, *Roy et al.* [1968] defined three heat flow provinces: the eastern United States, the Sierra Nevada, and a zone that includes the Basin and Range tectonic province. Within each province, crustal temperature has been estimated [*Roy et al.*, 1968] with greater than normal precision (Table 1, Figure 2). For each

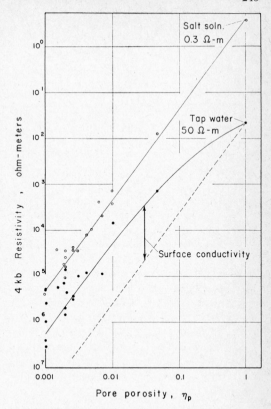

Fig. 1. Measured resistivity of igneous and metamorphic rocks at 4-kb effective pressure as a function of pore porosity [from *Brace et al.*, 1965; *Brace and Orange*, 1968b].

province, overburden or lithostatic pressure was determined as a function of depth based on appropriate seismic profiles. For Basin and Range and Sierra Nevada, data of *Bateman and Eaton* [1967] were used, and for eastern U. S., data of *James et al.* [1968] were used. The density-velocity relation of *Bateman and Eaton* [1967] was used. Table 1 includes lithostatic pressure P_{LITH} and effective pressure $(\bar{P})_{\mathrm{HYD}}$ for the case when pore pressure $P_{\mathrm{H_2O}}$ is hydrostatic. For the other case, that $P_{\mathrm{H_2O}}$ equals P_{LITH}, then $(\bar{P})_{\mathrm{LITH}}$ is zero. P_{LITH} is simply $\rho_R gz$, where rock density ρ_R is derived from the seismic profiles noted above, z is depth, and g is the acceleration of gravity.

Mineral conduction. At high temperature, conduction even through normally insulating minerals like quartz and feldspar becomes important. Conduction varies appreciably among mineral species and even within members of a solid solution series. However, some general

TABLE 1. Pressures and Temperatures of Heat Flow Provinces

(P_{LITH} is overburden pressure and $(\bar{P})_{\text{HYD}}$ is effective pressure when pore pressure is hydrostatic.)

	East U. S.			Sierra Nevada			Basin and Range		
Z, km	T, °C	P_{LITH}, kb	$(\bar{P})_{\text{HYD}}$, kb	T, °C	P_{LITH}, kb	$(\bar{P})_{\text{HYD}}$, kb	T, °C	P_{LITH}, kb	$(\bar{P})_{\text{HYD}}$, kb
5	100	1.35	0.85	80	1.35	0.85	150	1.35	0.85
10	170	2.80	1.80	120	2.70	1.70	250	2.70	1.75
15	240	4.25	2.75	160	4.10	2.60	380	4.05	2.70
20	310	5.70	3.70	200	5.50	3.50	510	5.55	3.82
25	370	7.15	4.65	235	6.90	4.40	630	7.05	5.0
30	420	8.6	5.6	270	8.4	5.4	750	8.65	6.3
35	460	10.0	6.5	300	9.9	6.4	860	10.3	7.7
40	500	11.7	7.7	340	11.4	7.4	970	11.9	9.1

trends are clear, based on extensive laboratory studies both at room pressure and at high pressure. Two curves of conductivity versus temperature between which most reported measurements fall are shown in Figure 3. Resistivities of a wide range of dry granitic, intermediate, mafic, and ultra mafic rocks [see *Parkhomenko*, 1967; *Keller*, 1966] fall between these limits, as does the olivine (0 to 15 per cent fayalite) studied by *Hamilton* [1965] to 42 kb. Partially melted rocks, particularly in the presence of water, fall outside these limits; in this case, resistivity can be many orders of magnitude lower [*Lebedev and Khitarov*, 1964].

Detailed mineralogy of the actual rocks in the crust is not known, and even if it were known, high temperature resistivity might not be exactly predictable without knowledge of minor elements and impurities [*Hamilton*, 1965]. In what follows we assume that the two curves in Figure 3 limit the actual resistivity of subsolidus rocks in the crust. As will become clearer later, these curves which were obtained for dry materials probably also apply to water-saturated rocks at high temperature, for, above a few hundred degrees, mineral conduction dominates conduction through pore fluids.

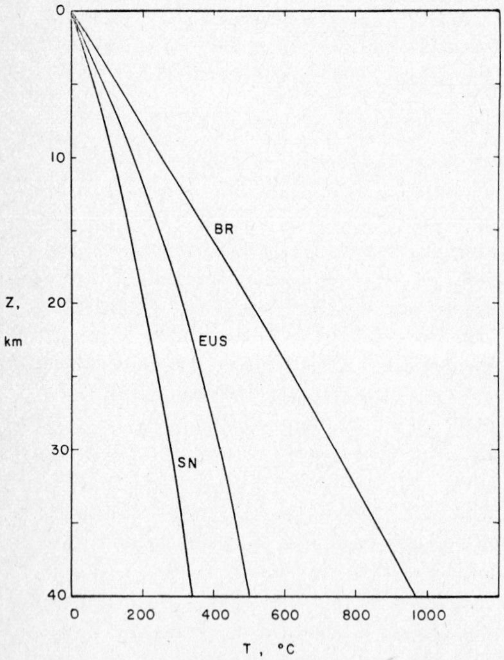

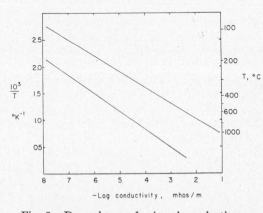

Fig. 2. Geotherms after *Roy et al.* [1968]. SN refers to Sierra Nevada, BR to Basin and Range, and EUS to eastern United States.

Fig. 3. Dependence of mineral conduction on temperature. The two lines give the limits between which most measurements for dry silicate rocks fall [from *Keller*, 1966; *Parkhomenko*, 1967; *Hamilton*, 1965].

CONDUCTIVITY OF SOLUTIONS AT HIGH PRESSURE
AND TEMPERATURE

Before an examination of the role of solute ions, it is of interest to follow the changes in pure water that occur when it is subjected to the pressures and temperatures encountered at depth in the three heat flow provinces. Density is shown in Figure 4 for the three areas; data are based on *Kennedy and Holser* [1966]. Two curves are given for each area, termed 'HYD' when P_{H_2O} is hydrostatic and 'LITH' when P_{H_2O} is lithostatic. It is seen that density generally decreases with depth, except for Sierra Nevada, the abnormally 'cold' province. When pore pressure is lithostatic, density change nowhere exceeds 15 per cent. Even in Basin and Range, the abnormally 'hot' region, density never falls below 0.5 g/cm³.

The conductivity of a variety of dilute aqueous solutions has been measured by *Quist et al.* [1963] and *Quist and Marshall* [1966, 1968,

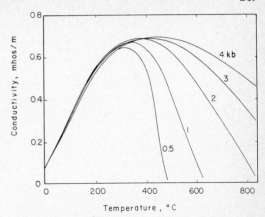

Fig. 5. Isobaric conductivity of a 0.01-m NaCl solution as a function of temperature. The number on each curve gives P_{H_2O} in kilobars [from *Quist and Marshall*, 1968].

1969, 1970]. Typical results are shown in Figure 5 for a 0.01-molar NaCl solution as a function of pressure and temperature. Conductivity is seen to go through a pronounced maximum with temperature; the value of maximum conductivity varies little with pressure, although the maximum is shifted to higher temperature with increasing pressure. These somewhat curious changes in conductivity may be partly understood by considering the changes in the properties of water with increasing temperature. Viscosity decreases with increasing temperature, as does the dielectric constant and, to a lesser extent, density [*Quist and Marshall*, 1968]. The combination of these effects makes it initially easier for ions to move in the solution in response to an applied electric field; at higher temperature, however, motion is impeded as the effect particularly of lowered dielectric constant overtakes that of lowered viscosity.

Merely from inspection of Figure 5 it is clear that, initially at least, conductivity of a solution such as 0.01-molar NaCl would increase by a factor of about 6, irrespective of the detailed way in which pressure changes. Beyond approximately 400°, however, there is a more complicated interaction of pressure and temperature effects. For the problem at hand, we must consider the actual pressures and temperatures of the three areas of interest. This has been done; the results are shown in Figures 6 and 7. Relative conductivity was determined from the data of Quist and co-workers from figures such as

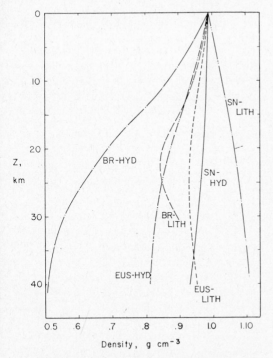

Fig. 4. Density of water as a function of depth for the three heat flow provinces, when pore pressure is hydrostatic (HYD) and lithostatic (LITH). Data are from *Kennedy and Holser* [1966]. SN refers to Sierra Nevada, BR to Basin and Range, and EUS to eastern United States.

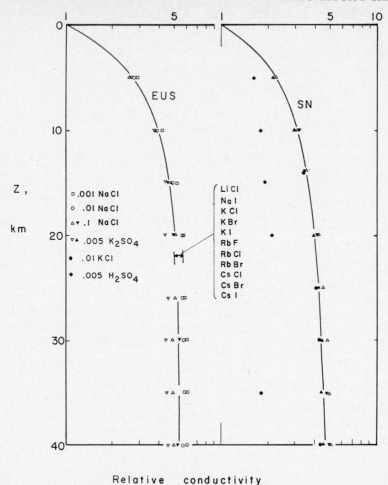

Fig. 6. Relative conductivity of various solutions for the eastern United States (EUS) and the Sierra Nevada (SN) as a function of depth. White symbols are hydrostatic pore pressure and black symbols are lithostatic pore pressure. Curves give average values.

Figure 5 for a variety of solutions using the pore pressures and temperatures listed in Table 1. The relative conductivity is obtained by dividing conductivity by the value at room pressure and temperature.

The effect of solute concentration is clearly indicated by results for the eastern United States (Figure 6). Data are shown for 0.001-, 0.01-, and 0.1-molar NaCl solutions. For this range of concentration, for which resistivity ranges from about 1 to 100 Ω-m at standard conditions, variation with depth is seen to be nearly the same. Variation with depth is also very similar for Sierra Nevada (Figure 6) and for the Basin and Range (Figure 7) when pore

pressure is lithostatic. For the hydrostatic Basin and Range, variation in conductivity is markedly different; first it increases and then rapidly decreases with depth.

Variation of solute ion, for the species so far investigated by Quist and co-workers, also seems to have little effect on the variation of conductivity with depth. With the exception of H_2SO_4, all the different solutions group rather closely, as shown in Figure 6.

Curiously enough, differences between hydrostatic and lithostatic pore pressure are also minor, except for the Basin and Range. Apparently, conductivity for the three geotherms rather quickly reaches a value close to the maximum such as shown in Figure 5. As seen from this figure, the value of the maximum is not too sensitive to pressure as long as temperature is increasing slowly. For the Basin and Range hydrostatic case, temperature increases so rapidly that conductivity falls away from the maximum with increasing depth; see Figure 7.

APPLICATION TO CRUSTAL ROCKS

Porosity changes with depth. Not only fluid conductivity but also porosity of typical crustal rocks changes with depth. As noted above, only the effects of increasing pressure are believed to be significant. As depth increases, both lithostatic and pore pressures typically change; as

long as the difference in these pressures, $\bar{P}$, increases with depth, then porosity decreases as well. Of the many possibilities, two cases are considered: when P_{H_2O} is lithostatic and when it is hydrostatic. For the first case, $\bar{P}$ equals zero and porosity remains unchanged with depth. For the second case, $\bar{P}$ increases and porosity decreases continuously with depth.

We next consider how $\bar{P}$ affects pore, crack, and fracture porosity and, therefore, conduction through pores, cracks, and fractures. We will attempt to define average behavior so that in later sections we can discuss characteristics of an average crustal profile.

Porosity of a wide variety of rocks decreases

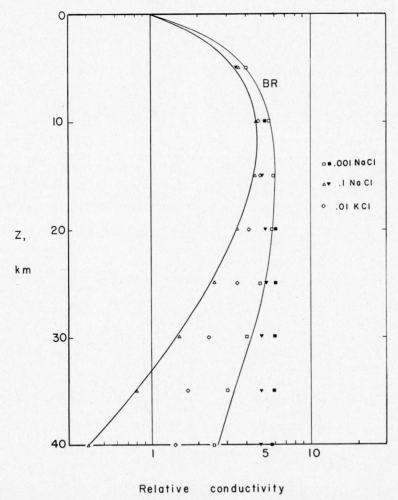

Fig. 7. Conductivity changes of various solutions for the Basin and Range (BR) as a function of depth. Black symbols are hydrostatic and white symbols are lithostatic pore pressure. Curves bound hydrostatic values.

with $\bar{P}$ in very nearly the same way between 2 and 10 kb [*Brace and Orange*, 1968*b*]. At these effective pressures, most cracks and, in the earth, most fractures would be closed, so that the porosity affected here is pore porosity. To a good approximation in this pressure region, conductivity change due to decreasing pore porosity is given [*Brace and Orange*, 1968*b*] by

$$1/\sigma \, d\sigma/d\bar{P} = -0.10 \text{ kb}^{-1} \qquad (1)$$

where σ is conductivity. In other words, conductivity decreases by a factor of 10 per 10 kilobars increase in $\bar{P}$.

Conductivity changes due to elimination of crack porosity under pressure vary rather widely among different rocks. Decrease of conductivity in the first 1 to 2 kb of a factor of 5 to 10 is typical [*Brace et al.*, 1965; *Brace and Orange*, 1968*b*].

The response of fracture porosity to increasing pressure is not amenable to study in the laboratory, but it can be estimated on the basis of certain field observations. *Snow* [1968] used measured flow rates at ten-foot intervals in drill holes at some 35 dam sites in granite, gneiss, and meta volcanics to calculate fracture porosity as a function of depth. Porosity ranged from 5×10^{-4} at 10 m or so below the surface to 5×10^{-6} at about 100 m. This porosity is appreciably lower than typical pore and crack porosity; for example, pore porosity is typically 10^{-2} to 10^{-3}, and crack porosity is approximately 1 to 2×10^{-3}. However, conduction through fracture porosity may still be important. Fractures, in the sense used here, are probably more nearly straight and therefore have lower tortuosity than cracks. Earlier measurements indicate that conduction through fractures may therefore vary more nearly as (porosity)^1 than as (porosity)^2 as observed for cracks and pores [*Brace et al.*, 1965; *Brace and Orange*, 1968*a*]. Conduction through fractures will therefore be a factor of 5 to 10 greater than crack conduction. To judge from Snow's observations, fractures are nearly closed at around 100 m, equivalent to $\bar{P}$ of about 20 bars.

Contributions from fracture, crack, and pore conduction for saturated rocks can now be combined schematically. In a plot of log conductivity versus effective pressure $\bar{P}$, this

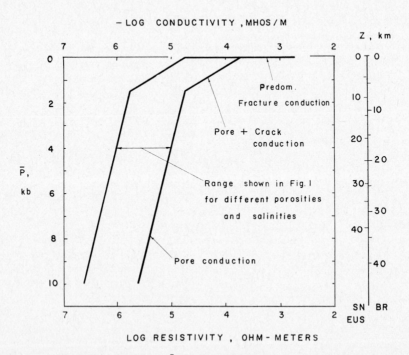

Fig. 8.   Generalized effect of pressure $\bar{P}$ on conductivity of rocks saturated with aqueous solutions which remain at room temperature and pressure. Depth Z for the three heat flow provinces is shown, based on Table 1.

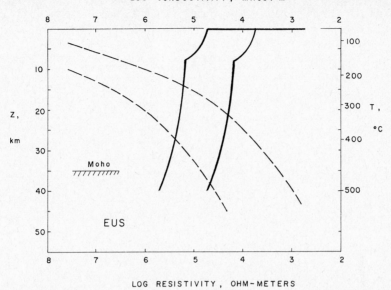

Fig. 9. Resistivity versus depth for the eastern United States (EUS) for hydrostatic pore pressure. Dry mineral conduction falls between the dotted line bounds; conduction through pore solutions falls between the continuous line bounds.

characterization (Figure 8) consists of three line segments. Close to zero pressure, conduction through fractures dominates. From zero to 1.5 kb, pore and crack conduction combined change by a factor of 10. Above 1.5 kb, pore conduction according to (1) is assumed to be dominant. The two lines in Figure 8 reflect variation in resistivity due to a porosity variation of 0.001 to 0.01, as summarized in Figure 1, for a pressure of 4 kb. Thus, the two lines bound values to be expected for typical saturated crustal rocks of these porosities.

Combined effects. The curves in Figure 8 bound the conductivity of average crust with the rock at high pressure and temperature and the pore solutions at room pressure and temperature. The effects of elevated pressure and temperature on the pore solutions can now be added. The lines drawn through the data in Figures 6 and 7 were taken to be the typical response of solutions to the pressure and temperature at depth in the three provinces. These effects were combined by multiplying rock conductivity at a particular depth (from Figure 8) by the relative conductivity of pore fluid appropriate to that depth (from Figures 6 and 7). In this way, the pairs of continuous curves shown in Figures 9, 10, and 11 were generated. For the

Basin and Range, the data in Figure 7 vary, depending on salinity and pore pressure; the curves shown in that figure bound the hydrostatic case. The lower bound from Figure 7 was used for the lower bound in Figure 11, et cetera.

The continuous curves in Figure 9 to 11 show conductivity versus depth for the hydrostatic case. If pore pressure becomes equal to lithostatic pressure below some depth, then effective pressure $\bar{P}$ remains zero below that depth. Further changes in conductivity below that depth would be simply those appropriate to the pore solutions.

The broken lines in Figures 9 to 11 bound the conductivity of dry rocks; they are based on the data summarized in Figure 3. These broken lines are seen to intersect the continuous lines; they represent conduction through pore solutions, at depths which vary considerably with the particular province, from a high of around 10 km for the Basin and Range to as deep as 40 km for the Sierra Nevada. For comparison, the approximate location of the base of the crust is indicated on the left-hand side of each figure. A few kilometers above the depth of intersection, conduction through pore solutions should dominate, whereas a few kilometers below, mineral conduction should be the more significant.

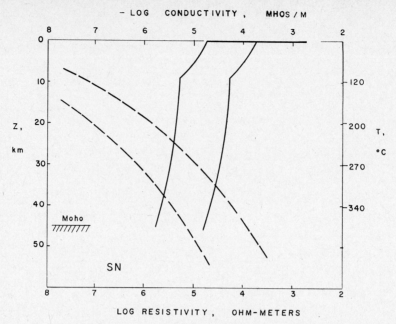

Fig. 10.   Same as Figure 9, for the Sierra Nevada.

DISCUSSION

Before discussion of the three profiles above, we restate the assumptions on which they are based:

1. Crustal rocks are typical igneous or metamorphic varieties.
2. Laboratory samples have the same crack and pore porosity as rock in situ when both are at the same effective pressure.
3. Crustal rocks are saturated with dilute aqueous solutions similar to those found in surface igneous and metamorphic rocks.
4. Natural dilute aqueous solutions respond to elevated pressure and temperature in the same way as the solutions studied by Quist and co-workers.
5. Surface conduction (Figure 1) is independent of pressure and temperature.
6. Temperature remains below the solidus.

The resistivity profiles for the three regions are combined in Figure 12; for each heat flow province, a curve midway between the bounds in Figure 9, 10, or 11 has been drawn. Interesting similarities and differences in the curves are apparent.

The upper parts of all three profiles are nearly identical. Resistivity rises rapidly to a maximum of nearly the same value (10^5 Ω-m), which is reached at nearly the same depth (10 km). The abrupt corners in the profiles are, of course, fictitious. In contrast, the lower parts of the three profiles differ markedly. Rapid decrease in resistivity occurs at a depth that ranges from 15 km for the Basin and Range to 40 km for the Sierra Nevada. This variation reflects the marked differences in geothermal gradient for the different provinces.

Maximum resistivity, even if we allow for the variation suggested by the sets of bounds in Figures 9, 10, and 11, is less than 10^6 Ω-m. This resistivity is too low for efficient transmission of electromagnetic radiation [*Levin,* 1968].

Field measurements of crustal resistivity in continental areas appear to give values of the same order, as suggested by the profiles above. Crystalline rocks near the surface range in resistivity from 10^2 to 10^3 Ω-m [*Keller et al.,* 1966]. For highly resistive rocks at depth in, for example, the New England and New York area, resistivities as high as 10^4 to 10^5 Ω-m are reported [*Anderson and Keller,* 1966]. These values are similar to scattered measurements made in Europe [*Keller et al.,* 1966].

Based on the evidence presented here, a typical resistivity profile is nearly independent of mineralogy and, therefore, of rock type, and nearly independent of pore water pressure, as

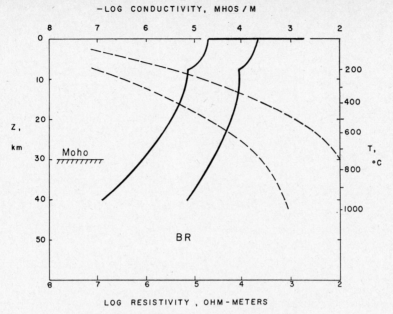

Fig. 11.   Same as Figure 9, for the Basin and Range.

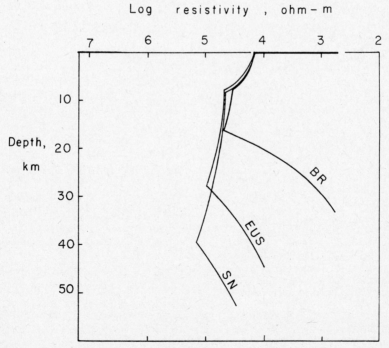

Fig. 12.   Comparison of resistivity-depth profiles for the three heat flow provinces.

long as this pressure is at least hydrostatic. Thus, we should expect little correlation between profile of velocity (which depends strongly on mineralogy) and resistivity (which depends primarily on porosity). Similarly, regions of high pore pressure will not be revealed by resistivity measurements.

Of the assumptions listed above, 2 and 3 are most critical and if invalid, could cause major departures of the natural profiles from those derived above. Unfortunately, both have the same general effect and are probably indistinguishable by field measurement. If either pore porosity does not exist in natural rocks or if the rocks are water-free, then resistivity in the 10- to 40-km region of the profiles could be much higher than the maximum suggested above. Neither possibility can be ruled out on the basis of present laboratory or field studies. Either deep drilling or additional deep resistivity soundings are needed to settle this question.

Acknowledgments. I am particularly indebted to Isadore Amdur, who brought the new studies of Quist and co-workers to my attention. T. R. Madden was a constant source of ideas and criticism, and discussion of various aspects of the problem with C. Cox and D. Blackwell was particularly helpful.

The study was supported by the National Science Foundation as grant GA 18342.

REFERENCES

Anderson, L. A., and G. V. Keller, Experimental deep resistivity probes in the central and eastern United States, *Geophysics, 31,* 1105–1122, 1966.

Bateman, P. C., and J. P. Eaton, Sierra Nevada Batholith, *Science, 158,* 1407–1417, 1967.

Birch, F., Elasticity of igneous rocks at high temperatures and pressures, *Bull. Geol. Soc. Amer., 54,* 263–287, 1943.

Brace, W. F., Some new measurements of linear compressibility of rocks, *J. Geophys. Res., 70*(2), 391–398, 1965.

Brace, W. F., and J. D. Byerlee, Recent experimental studies of brittle fracture of rocks, paper presented at 8th Symposium on Rock Mechanics, Univ. of Minnesota, Minneapolis, 1967.

Brace, W. F., and A. S. Orange, Electrical resistivity changes in saturated rocks during fracture and frictional sliding, *J. Geophys. Res., 73*(4), 1433–1445, 1968a.

Brace, W. F., and A. S. Orange, Further studies of the effect of pressure on electrical resistivity of rocks, *J. Geophys. Res., 73*(16), 5407–5420, 1968b.

Brace, W. F., A. S. Orange, and T. M. Madden, The effect of pressure on the electrical resistivity of water-saturated crystalline rocks, *J. Geophys. Res., 70*(22), 5669–5678, 1965.

Greenberg, R. J., and W. F. Brace, Archie's law for rocks modelled by simple networks of resistors, *J. Geophys. Res., 74*(8), 2099–2102, 1969.

Hamilton, R. M., Temperature variation at constant pressure of the electrical conductivity of periclase and olivine, *J. Geophys. Res., 70,* 5679–5692, 1965.

James, D. E., J. T. Smith, and J. S. Steinhart, Crustal structure of the middle Atlantic States, *J. Geophys. Res., 73,* 1983–2008, 1968.

Keller, G. V., Electrical properties of rocks and minerals, in *Handbook of Physical Constants,* edited by S. P. Clark, *Geol. Soc. Amer. Mem., 97,* 553–577, 1966.

Keller, G. V., L. A. Anderson, and J. I. Pritchard, Geological Survey investigations of the electrical properties of the crust and upper mantle, *Geophysics, 31,* 1078–1087, 1966.

Kennedy, G. C., and W. T. Holser, Pressure, volume, temperature, and phase relations of water and carbon dioxide, in *Handbook of Physical Constants,* edited by S. P. Clark, *Geol. Soc. Amer. Mem., 97,* 371–383, 1966.

Lachenbruch, A. H., Crustal temperature and heat production: Implications of the linear heat-flow relation, *J. Geophys. Res., 75*(17), 3291–3300, 1970.

Lebedev, E. B., and N. I. Khitarov, Beginning of melting in granite and the electrical conductivity of its melts in relation to pressure of pore water, *Geokhimiya, 3,* 195–201, 1964.

Levin, S. B., Model for electromagnetic propagation in lithosphere, *Proc. IEEE, 56,* 799–804, 1968.

Madden, T. R., and D. J. Marshall, Induced polarization, *U.S. At. Energy Comm., RME-3160,* 80 pp., 1959.

Nur, A., and G. Simmons, The origin of small cracks in igneous rocks, *Int. J. Rock Mech. Mineral. Sci., 7,* 307–314, 1970.

Orange, A., Granitic rock, properties in situ, *Science, 165,* 202–203, 1969.

Parkhomenko, E. I., *Electrical Properties of Rocks,* 314 pp., Plenum, New York, 1967.

Quist, A. S., and W. L. Marshall, Electrical conductances of aqueous solutions at high temperatures and pressures, 3, The conductances of potassium bisulfate solutions from 0 to 700° and at pressures to 4000 bars, *J. Phys. Chem., 70,* 3714, 1966.

Quist, A. S., and W. L. Marshall, Electrical conductances of aqueous sodium chloride solutions from 0 to 800° and at pressures to 4000 bars, *J. Phys. Chem., 72,* 684, 1968.

Quist, A. S., and W. L. Marshall, The electrical conductances of some alkali metal halides in aqueous solutions from 0 to 800° and at pressures to 4000 bars, *J. Phys. Chem., 73,* 978, 1969.

Quist, A. S., and W. L. Marshall, A reference solution for electrical measurements to 800° and 12,000 bars, *J. Phys. Chem., 74,* 1970.

Quist, A. S., E. U. Franck, H. R. Jolley, and W. L. Marshall, Electrical conductances of aqueous solutions at high temperature and pressure, 1, The conductances of potassium sulfate-water solutions from 25 to 800° and at pressures up to 4000 bars, *J. Phys. Chem., 67,* 2453, 1963.

Roy, R. F., D. D. Blackwell, and F. Birch, Heat generation of plutonic rocks and continental heat flow provinces, *Earth Planet. Sci. Lett., 5,* 1–12, 1968.

Scott, J. H., R. D. Carroll, and D. R. Cunningham, Dielectric constant and electrical conductivity measurements of moist rock: A new laboratory method, *J. Geophys. Res., 72,* 5101–5116, 1967.

Simmons, G., and A. Nur., Granites: Relation of properties in situ to laboratory measurements, *Science, 162,* 789–791, 1968.

Snow, David T., Rock fracture spacings, openings, and porosities, *J. Soil Mech. Foundations Div. Proc. Amer. Soc. Civil Eng., 94*(SM1), Pap. 5736, 73–91, 1968.

Walsh, J. B., and W. F. Brace, Cracks and pores in rocks, *Proc. Int. Congr. Rock Mech., Lisbon,* 643–646, 1966.

White, D. E., J. D. Hem, and G. A. Waring, Chemical composition of subsurface waters, *U.S. Geol. Surv. Prof. Pap. 440-F,* 67 pp., 1963.

DISCUSSION

Ward: I want to repeat my warning about the sampling problem. When we look at rocks at the surface we see the more competent ones exposed. Those hidden beneath swamps and muskeg are the more common, and these often include the pyritic and graphitic rocks. They do not show up in surface mapping, but we see them in aerial electromagnetic surveys. I think you are giving us a crust biased toward the more competent rocks.

Brace: Yes, I have only considered the relatively resistive rocks, because the objective of this conference was to see if a highly resistive layer is likely in the crust. If it is unlikely, based on laboratory measurements on typical, resistive silicate rocks, then I think we can rule it out. For this reason, I have not looked at graphitic rocks, although in certain regions they may well be common and may dominate the situation electrically.

Alldredge: You showed resistivities up to 10^6 Ω-m, but no field investigation has yielded numbers higher than 40,000. Is this because your laboratory assumptions are way out of line?

Brace: Yes, this could well indicate that certain of our assumptions are not correct, although my greatest resistivity (5×10^5 Ω-m) does not seem to be way out of line with field measurements reported by Anderson and Keller (10^4–10^5 Ω-m).

H. W. Smith: I don't think you have a way in the world of measuring resistivities of the order 10^7 in the field. You cannot measure more than 5000 Ω-m in a borehole by any known technique because of the drilling mud. We must think very carefully about this if we actually drill a hole.

A Discussion of Water in the Crust

PETER J. WYLLIE

Department of Geophysical Sciences, University of Chicago, Chicago, Illinois 60637

This comment was stimulated by the suggestion that a geochemist discuss the possibility that water is present in the crust. This material is relevant to the papers of T. Madden and W. F. Brace.

Wyllie: I do not know what actually happens to water in the crust, but I can outline the conditions and variables that we are concerned with and the types of theoretical and experimental systems that provide limits for the natural conditions.

Significant variables in the crust are composition, mineralogy, temperature (T), pressure as a function of depth ($P_t = P_{total}$), the pressure on the interstitial pore fluid if present (P_f), and the partial pressures of the pore fluid components. A rock in the crust may consist of any combination of anhydrous minerals, hydrous minerals, and a pore fluid generally considered to be dominantly aqueous. Let us arbitrarily omit carbonates for simplicity. Changes in depth and temperature cause reactions to occur among the rock components. For example, increasing temperature may cause dehydration of a hydrous mineral such as brucite:

$$Mg(OH)_2 = MgO + H_2O$$

releasing H_2O to the pore fluid.

I have listed on the board five possible conditions for a rock in the crust. First,

$$P_f = P_{H_2O} = P_t$$

This assumes that the pore fluid is composed essentially of pure H_2O. Second,

$$P_f = P_{H_2O} < P_t$$

This refers to a rock in which the pore fluid is composed of pure H_2O, but part of the lithostatic pressure is supported by grain-to-grain boundary contacts, and there just isn't enough

fluid in the rock to fill the available pore space and reach the confining pressure. Third,

$$P_{H_2O} < P_f = P_t$$

This occurs in a rock in which P_{H_2O} is reduced compared with P_f because the fluid phase is diluted by other components in solution, such as CO_2 and alkali salts. Fourth,

$$P_f = 0; \qquad P_{H_2O} < P_t$$

This condition occurs if there is no pore fluid present in the rock at all. P_{H_2O} may still have a finite, fictive value if the vapor-absent rock consists of both hydrous and anhydrous minerals. Fifth,

$$P_f > P_t$$

These conditions can be illustrated using phase diagrams for the system MgO-H_2O, which is about the simplest system involving examples of anhydrous minerals, hydrous minerals, and a vapor phase (= pore fluid). In Figure 1a, I have sketched a P-T diagram for the univariant dissociation reaction under condition 1, with $P_f = P_{H_2O} = P_t$; in Figure 1b, I have sketched an isobaric section through the system which shows the phase assemblage $Mg(OH)_2$ + vapor reacting to yield MgO + vapor under the same condition 1 and another field for the vapor-absent assemblage $MgO + Mg(OH)_2$, corresponding to condition 4.

There are two types of systems that can be treated with thermodynamic rigor—a closed system and an open system. In a closed system, no material is lost or gained during a reaction. In an open system, certain components may migrate into or from the reacting system, in response to defined controls outside of the system: these are called perfectly mobile components. Rock systems in the crust may be

257

treated as if they correspond to one or the other of these ideal thermodynamic systems, but in reality the natural rock systems probably behave in some intermediate fashion. Kinetic factors such as the rates of reaction and the rate of migration of pore fluids are involved in the crust, but thermodynamics tells us only about equilibrium conditions once they have been achieved.

The concept of a closed system as applied to a rock in the crust is a familiar one, but it is doubtful whether any rock is completely closed to the movement of pore fluids. The concept of an open system as applied to the kind of rock that we are using as an example requires that H_2O be able to migrate in or out of the rock, through the pore fluid; the chemical potential of H_2O in the open rock system is controlled by the regional chemical potential gradient appropriate for the crustal megasystem enclosing the rock system being considered. P_{H_2O} is thus determined independently of the reactions occurring in the system, in contrast to the closed system where P_{H_2O} is a dependent variable. In an open system, H_2O is termed a mobile component. It is doubtful whether any rock system comprising a substantial volume in the crust

behaves as an ideal open system, with H_2O as a perfectly mobile component.

We have already considered the dissociation of brucite in a closed system, using the phase diagrams, Figure 1. The univariant curve on the P-T projection gives the maximum stability temperature for brucite, for condition 1 where $P_f = P_{H_2O} = P_t$. The reaction temperature becomes lower at constant P_t if P_{H_2O} is reduced compared with P_t. This is achieved in a closed system if there are components additional to H_2O in the vapor phase, which corresponds to condition 3, $P_{H_2O} < P_f = P_t$. It is also achieved physically in a closed system with condition 2, $P_f < P_t$. A third way for the dissociation temperature to be reduced at constant P_f is in an open system, where P_{H_2O} can be lowered independently of P_t by factors external to the system. So we see that the products of a particular dissociation reaction at a given depth in the crust are not produced simply because the appropriate temperature is reached; they may be produced at various temperatures depending on the pore fluid composition and behavior. For this reason, it is very difficult to deduce the nature of the pore fluid in a rock from the study of its mineralogy. Notice again the water-deficient

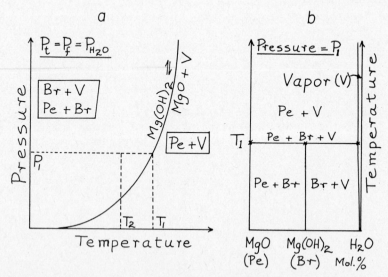

Fig. 1. Phase relationships in the system MgO–H_2O. 1a shows a univariant curve for the dissociation of brucite (Br) in the presence of a vapor phase (V) composed of pure water, to yield periclase (Pe) and vapor. The phase assemblages stable at temperatures above and below the conditions for reaction are shown. These can be seen on Figure 1b. Isobaric section through the system at pressure P_1 from Figure 1a. Notice the vapor-absent field for periclase plus brucite. Brucite becomes unstable at temperature T_1 (see Figure 1a).

region in the isobar for the system MgO-H₂O; if no vapor phase is present, the assemblage brucite + periclase exists at temperatures well below the brucite dissociation temperature. If there is insufficient H₂O in the system to cause the maximum hydration of MgO, then no vapor phase exists.

Healy: Are you saying that in a normal crust we'd be in a non-vapor phase region?

Wyllie: No. In the crust we could have a free vapor phase, but the phase diagrams show also that a vapor-absent phase assemblage is theoretically possible.

We can extend the simple MgO-H₂O system to provide an illustration of processes in open systems. If we add a non-reacting volatile component X, we can use a triangle MgO-H₂O-X, Figure 2, to show the phase relationships at constant P_t and constant temperature. We can lower P_{H_2O} from its maximum value in MgO-H₂O by diluting the vapor phase with the additional material X. At a given P_t and T, the two minerals brucite + periclase coexist with a vapor phase of fixed composition. The assemblage can be hydrated to yield brucite or dehydrated to yield periclase just by changing the vapor phase composition at constant P_t and T. This is what can happen in an open system, when the regional chemical potential gradient for H₂O causes H₂O to migrate into or out of the rock. (This topic was discussed in more detail, following *Wyllie* [1962a, b]).

I want to call your attention in particular to the assemblage brucite + periclase + vapor in the ternary system. What we have done is to add a mixed vapor phase, H₂O-X, to the vapor-absent region of the binary system MgO-H₂O. The assemblage is still water-deficient, because it is not saturated with H₂O. Imagine this assemblage in an open system, and assume that water starts leaking into the system. It will not change the vapor phase composition because it will be used up by reaction, producing brucite, if equilibrium is maintained, so that the water will not be able to migrate freely through the system. It has to pass through the system as a front. It converts all of the periclase to brucite in each small volume before whatever remains is able to pass through the system, changing the vapor phase composition. There was a question this morning about how water coming into a rock would get soaked up. If the mineralogy of the rock is unsaturated with H₂O, then H₂O cannot

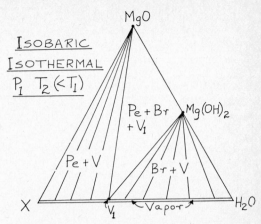

Fig. 2. Phase relations in the system MgO-H₂O-X where components X are non-reacting components soluble in the aqueous fluid (vapor = V) phase. An isobaric isothermal section through the system at pressure P_1 and temperature T_2 from Figure 1a. Figure 1b shows that in the presence of H₂O at constant pressure, periclase and brucite coexist at only one temperature (T_1). If the vapor phase is diluted with components X, as in this figure, then these two minerals coexist with a specific vapor phase V_1 at a lower temperature T_2. The reaction Mg(OH)₂ ⇌ MgO + H₂O can be made to occur at constant total pressure (P_t) and constant temperature (T_2) by diluting the vapor phase with X.

pass through it without first reacting to saturate the minerals, to the extent required by the conditions of P_t and T. This assumes, of course, that chemical equilibrium is maintained.

Madden: For a given temperature, is there no water pressure at which a fluid phase would be present without being absorbed into the minerals?

Wyllie: If the bulk composition of a rock is unsaturated with H₂O for the given conditions of T and P_t, then any additional H₂O would have to react with the minerals until saturation were complete, and during this process P_{H_2O} would be buffered by the mineral assemblage.

Madden: No matter what the water pressure is, even if it is at zero pressure?

Wyllie: Pore fluid with zero water pressure; that would be composed of CO₂, for example, and certainly you could pass CO₂ through the rock without its reacting (unless carbonation reactions become involved).

Madden: No, I mean supposing it was a pore fluid, but with very low water pressure.

Wyllie: The phase diagram and other phase diagrams for more complex systems indicate that a pore fluid with very low water pressure could exist with an anhydrous mineral assemblage. If H_2O were added to the system, the pore fluid composition would change and P_{H_2O} would increase until the hydrous minerals began to form, and thereafter additional H_2O would react, with P_{H_2O} buffered by the minerals, until the rock became saturated with H_2O and no more hydrous minerals could be produced. Only then could more H_2O be added to the rock without being absorbed by the minerals.

The phase diagram for the system MgO-H_2O shows that if you have a vapor-absent H_2O-undersaturated rock, then you can add H_2O to the system without producing any pore fluid at all until all of the anhydrous minerals are fully hydrated.

Madden: No matter what the pressure is?

Wyllie: Yes, as long as you start off with a vapor-absent rock and H_2O-undersaturated mineral assemblage, and this condition is theoretically possible. Whether it ever occurs in the crust is another question.

Madden: Well, you're telling us then that you can look at a rock and tell us that actually there could have been no water whatsoever no matter what the fluid pressure was?

Wyllie: No, I don't think you can do this. The phase diagrams show that a mineral assemblage that is H_2O-unsaturated can exist either without any pore fluid at all or with a pore fluid present.

Madden: Another question is Are the pore fluids common in certain crustal environments?

Wyllie: I started out by saying that I don't know what the actual regime of water is in the crust. The phase diagrams do not tell us this; they only show conditions that are possible in the crust, if chemical equilibrium is achieved. I would guess that there is a pore fluid present in most parts of the crust.

Healy: You can't rule out free water.

Wyllie: The phase diagrams show that the same mineral assemblage can exist either with or without a pore fluid, so study of mineralogy is not sufficient to tell you whether or not a pore fluid containing H_2O is present. It seems to me that an aqueous pore fluid could exist in equilibrium with any known crustal rock, provided that conditions of P_t, P_f, and T are suitable.

REFERENCES

Wyllie, P. J., The petrogenetic model, an extension of Bowen's petrogenetic grid, *Geol. Mag.*, *99*, 558–569, 1962a.

Wyllie, P. J., The effect of 'impure' pore fluids on metamorphic dissociation reactions, *Mineral. Mag.*, *33*, 9–25, 1962b.

5. HIGH PRESSURE-HIGH TEMPERATURE GEOCHEMISTRY

The Nature of the Crust of the Earth,
with Special Emphasis on the Role of Plagioclase[1]

A. L. BOETTCHER

Department of Geosciences, The Pennsylvania State University
University Park, Pennsylvania 16802

Abstract. This paper is primarily a discussion of the evolution, composition, and structure of the crust-mantle system of the earth; we attempt to determine the nature of any layers, including high electrical-resistance and low seismic-velocity zones, within upper levels. Our concepts regarding the composition and structure of the crust and mantle are significantly influenced by our prejudiced view of the mechanisms by which the earth formed and evolved. A 'hot-earth' model in which the earth accreted from silicate-rich dust appears to be most nearly consonant with available geochemical data. This model implies that the crust evolved by crystal-liquid differentiation of a partially or wholly molten silicate mantle. Estimates of the composition of the lower crust range from intermediate to basic. On the basis of available experimental data for such compositions, we can say that melting cannot be responsible for continuous layering within at least the upper 30 km, but compositional changes and phase transformations could provide geophysically detectable layers within the crust and upper mantle. Within the oceanic crust, zones of acidic rocks and the transformation of basalts to rocks including those in zeolite, greenschist, and amphibolite facies may create continuous layers. In continental areas, layers possibly result from zones of amphibolite or even anorthosite, but the major seismically detectable layers probably occur at the first appearance of garnet at 0–40-km depth and at the disappearance of plagioclase at depths as great as 70–90 km, based on the behavior of plagioclase and selected basic and intermediate rock compositions in the laboratory at high temperatures and pressures.

The theory of the existence of a low-velocity layer [e.g., *Press,* 1970] within the upper mantle of the earth is now accepted by some as being well established. Furthermore, this proposed layer appears to be more than just a temperature effect on the rigidity and density, although the question remains open. Experimental evidence on melting and other phase transformations [*Spetzler and Anderson,* 1968; *Lambert and Wyllie,* 1968; *Hill and Boettcher,* 1970] strongly suggests that partial melting within zones of the upper mantle can be responsible for any attenuation of seismic waves within a low-velocity layer in the mantle. Less well-known, but perhaps of equal significance, is the theory of layers (including low-velocity as well as high electrical-resistivity channels) within the crust of the earth proposed by several research groups in the past five years [e.g., *Pálmason,* 1967; *Barrett and Aumento,* 1970; *Mueller and Landisman,* 1966; *Clowes and Kanasewich,* 1970]. Many earth scientists doubt the existence of extensive, continuous layers

---

[1] Contribution 70-12, College of Earth and Mineral Sciences, The Pennsylvania State University, University Park.

within the crust, and this symposium was held to deliberate this question. The purpose of this paper is to consider possible causes of such layers and the mechanisms by which they may originate.

EVOLUTION OF THE CRUST-MANTLE SYSTEM

Before considering the possible nature of any layered structure in the crust, we must first examine what is known of the over-all chemical and physical properties. There are currently two major schools of thought regarding the evolution of the earth and the crust-mantle system. One school thinks that the division of the earth into layers, including core, mantle, and crust, resulted from differentiation of an entirely or partially molten earth that originally was more nearly chemically homogeneous. The second school maintains that the layers are primary features, resulting from accretion and accumulation of matter in order of increasing vapor pressure.

The first and most widely accepted school maintains that the earth accreted from dust, either in a single stage as proposed by *Ringwood* [1966] or in several stages (accretion into lunar-size planetesimals, followed by collision, disintegration, degassing and desilication, and reaccumulation into the planets) as postulated by *Urey* [1963]. Within the framework of this first school, some suggest that the earth remained cold ($< \sim 1000°C$) during final accretion and heated slowly by radioactive decay until the core-forming alloy became molten. The significance of the contribution of radioactivity to the thermal budget of the deep interior of the earth is discussed in somewhat contrasting views by *Lubimova* [1967, 1969] and *Hanks and Anderson* [1969]. Others argue that accretional release of gravitational energy alone was sufficient to melt all or nearly all of the earth. In either situation, the combination of the release of original accretional energy [*Ringwood*, 1966], radioactive disintegration [e.g., *Lubimova*, 1969], and energy released upon accretion of the core [*Birch*, 1965a; *Elsasser*, 1963; *Urey*, 1952] would likely be sufficient to melt nearly all of the earth. The presence of a core in the earth, in itself, supports this melting concept, because accretion of the core through a solid, silicate-rich proto-earth is difficult to conceive. The absence of detectable cores in

the bodies Mars and the moon possibly is attributable to lack of melting resulting from their small mass and relatively low accretional energies. Additional support for melting in the early earth is given by the apparent upward concentration of lithophile elements, including K, U, and Th, in the upper mantle [e.g., *Birch*, 1965b; *Ringwood*, 1966; *Gast*, 1960, 1968], whereas the mantle is more nearly homogeneous with respect to the major oxides, such as FeO, MgO, and SiO_2. As pointed out by *Ringwood* [1966], these geochemical distributions of major and 'incompatible' elements could result from convection in a nearly liquid mantle, with fractionation of the lithophile minor and trace elements near the surface, or by zone melting [*Harris*, 1957; *Vinogradov*, 1961] in which all or most of the mantle was molten, but with the zone of melting beginning near the core and progressing toward the surface; such a pattern would require only a fraction of the mantle be melted at any given time.

The second major prevailing concept regarding primary accretion of the earth was published most recently by *Turekian and Clark* [1969]. Basing their theory on earlier work by *Eucken* [1944], *Larimer* [1967], *Cameron* [1962], and *Anders* [1968], they postulate that structuring of the earth into layers, including core, mantle, crust, and their subdivisions, is primary, and that the structure results from accretion of matter from primordial solar nebula in order of increasing vapor pressure. In other words, condensation begins with the core-forming alloy and is followed by refractory mantle-forming silicates and by alkali silicates and the volatile elements and compounds in the crust, hydrosphere, and atmosphere. Although this model has several attractive features, it fails to explain the major-element homogeneity of the mantle (neglecting such possibilities as a higher FeO/MgO ratio in the lower mantle [e.g., *Anderson and Jordan*, 1970; *Press*, 1968] and the heterogeneity of the minor-element distribution in the mantle, for example, the upward concentration of the non-volatile elements U and Th and the rare earths. Such element distributions are best interpreted on the basis of a once molten or partially molten mantle (the hot-earth hypothesis); this theory is consonant with the recent results of many investigators [e.g., *Hanks and Anderson*, 1969; *Lee*, 1968].

In the second model, in which layers are primary and result from accretion, the relationship between the over-all bulk chemistry of the crust and that of the mantle is obscure, owing to the state of our present knowledge; it is controlled by such unknown quantities as rate of condensation, degree of equilibration, nature of the condensed species, and temperature [*Larimer*, 1967]. In the first model, in which the mantle of the earth was molten, or nearly so, and in which the crust evolved by crystal-liquid differentiation, followed by upward migration of the more felsic components, the chemical relationships between crust and mantle are easier to conceive. For example, unless they were disturbed by a mechanism such as convective overturn throughout the mantle, the composition of the crust and the composition of the underlying upper mantle from which the crust evolved together would average to be what we consider the bulk composition of the mantle (peridotitic). This model is valid regardless of whether the Moho is a phase change or a compositional change. For this reason, the crust-mantle system to depths of about a hundred km could have approximately the same composition for continental and oceanic regions alike.

COMPOSITION OF THE CRUST

Various methods have been used in the past 50 years to obtain the bulk composition of the continental crust. *Clarke and Washington* [1924] averaged the results of over 5000 igneous rocks without regard for distribution or abundances of the rocks, hoping that statistics and good fortune would operate to yield a reliable answer. This list included many scarce, unusual rocks, and the potential error is, therefore, large. *Goldschmidt* [1933] prudently allowed Mother Nature to do the work; he analyzed several scores of samples of till and other glacial debris in Scandinavia, arguing that glaciers would objectively sample the crystalline basement rocks exposed in that region. More recently, *Poldervaart* [1955], *Vinogradov* [1962], *Taylor* [1964], *Ronov and Yaroshevsky* [1969], and *Tan and Chi-lung* [1970] have weighed the compositions against proposed crustal abundances of rocks to obtain a better estimate of

the entire crust. In a different approach, *Pakiser and Robinson* [1966] used seismic velocities of compressional (P) waves from different parts and layers of the crust in the United States, *Birch's* [1960, 1961] compilations of seismic velocity versus rock type, and *Nockolds'* [1954] averages for the chemical compositions of these rocks to arrive at a bulk chemical analysis of the continental crust. The results of some of these estimates appear in Table 1. Additional data and comments on the compositions of the upper crust appear in *Wedepohl* [1969] and *Turekian* [1968].

These results are in surprising agreement, but it must be recalled that they are only estimates subject to considerable modification when additional data become available. These compositions are intermediate between gabbroic and granitic, but they correspond to no known common igneous rock and, if correct, probably represent a mixture of basic and acidic rocks. However, *Taylor and White* [1965] have suggested that these values may be somewhat in error and that they reflect a crust consisting largely of andesitic composition. The most commonly generated model of crustal structure has been that of a granitic sial overlying a gabbroic lower crust and separated from it by a seismic discontinuity, the 'Conrad discontinuity.' This model has been the object of increasing criticism in recent years, [see *James and Steinhart*, 1966; *Prodehl*, 1970]. We will consider its validity later in this paper.

LAYERS IN THE CRUST

Compositional changes. With this introduction, let us examine some of the possible causes of layers within the crust. The concept that the major divisions of the earth (mantle, crust, and core) developed during the primary accretion of the earth [*Turekian and Clark*, 1969] has something to recommend it, but the possibility that extensive continuous layers have persisted since the beginning of geologic time in a crust-mantle system subjected to metamorphism, melting, and tectonism must be discarded. However, the possibility must be entertained that, since its formation, layers have developed in the crust from compositional changes or phase changes at various levels. For example, *Pakiser* [1965] and *Pakiser and Zietz* [1965] have suggested that the

TABLE 1.   Estimates of the composition of the earth's crust calculated H_2O-free
and normalized to 100 wt %*
(The table is modified from *Ronov and Yaroshevsky* [1969].)

Oxide	Continental Crust							Entire Crust	
	a	b	c	d	e	f	g	h	i
SiO_2	60.3	60.5	63.4	60.4	59.4	57.9	61.9	55.2	59.3
TiO_2	1.0	0.7	0.7	1.0	1.2	1.2	0.8	1.6	0.9
Al_2O_3	15.6	15.7	15.3	15.7	15.5	15.2	15.6	15.3	15.9
Fe_2O_3	3.2	3.3	2.5		2.3	2.3	2.6	2.8	2.5
FeO	3.8	3.5	3.7	7.2	5.0	5.5	3.9	5.8	4.5
MnO	0.1	0.1	0.1	0.1	0.1	0.2	0.1	0.2	0.1
MgO	3.5	3.6	3.1	3.9	4.2	5.3	3.1	5.2	4.0
CaO	5.2	5.2	4.6	5.8	6.7	7.1	5.7	8.8	7.2
Na_2O	3.8	3.9	3.4	3.2	3.1	3.0	3.1	2.9	3.0
K_2O	3.2	3.2	3.0	2.5	2.3	2.1	2.9	1.9	2.4
P_2O_5	0.3	0.3	0.2	0.2	0.2	0.3	0.3	0.3	0.2

* In the table the sources are as follows: a is *Clarke and Washington* [1924]; b is *Goldschmidt* [1933]; c is *Vinogradov* [1962]; d is *Taylor* [1964], where total Fe is listed as FeO; e is *Poldervaart* [1955]; f is *Pakiser and Robinson* [1966]; g is *Ronov and Yaroshevsky* [1969]; h is *Poldervaart* [1955]; and i is *Ronov and Yaroshevsky* [1969].

continental crust was originally more silicic than it is now and that it became more basic by addition of mantle material throughout its evolution. This proposal is in accord with the observations of *Pakiser and Robinson* [1966] that the crust in the tectonically inactive eastern United States is compositionally more basic as the result of previous orogenic cycles than is the crust in the western United States, where it is still evolving. Nevertheless, it is improbable that the addition or intrusion of basic rocks into the crust would result in continuous layers spanning various geologic provinces.

It has often been suggested that there is an anorthosite layer within the crust, [e.g., *Herz*, 1969; *Buddington*, 1943], and the discovery of anorthosites among the lunar samples [*Wood et al.*, 1970] and among material dredged from the mid-Indian Ocean ridge [*Engel and Fisher*, 1969] may revive this concept. Such material could conceivably result in seismically or electrically definable layers, but *Middlemost* [1970] recently suggested that a crustal anorthosite layer is currently evolving and is, at present, discontinuous and irregular. *Birch*'s [1960] and *Christensen*'s [1965] values of seismic velocity versus rock type indicate that anorthosites have P-wave velocities similar to common mafic rocks. Also, a crustal anorthosite layer would provide a first-order discontinuity suggested by

Mueller and Landisman [1966] as necessary to explain the low-velocity layer in the crust. Beyond this, however, there is little geologic or geophysical evidence for such a layer. Similar arguments can be proposed for serpentinite layers.

The possibility of other compositional layers must also be considered. For example, the oceanic crust has usually been considered as a basaltic horizon overlying a periodotitic mantle and veneered with marine sediments. However, recent dredgings in the ocean have disclosed samples ranging from lherzolite, anorthosite, and gabbro [*Engel and Fisher*, 1969] to nepheline gabbro [J. Honorez, personal communication, 1970; *Honorez and Bonatti*, 1970]. Other examples, including nepheline-bearing rocks, representing advanced differentiation, are known from the earlier works of *LaCroix* [1924] and *Daly* [1925].

The presence of granitic xenoliths in the basaltic lavas of Surtsey on the mid-Atlantic ridge [*Sigurdsson*, 1967] makes it tempting to postulate granitic and other compositional layers in some regions of the ocean crust. One such inclusion that I observed on Surtsey had a pronounced gneissic texture, suggesting metamorphism under deep burial. Such a layer may be the source of the approximately 10 per cent acidic rocks [*Walker*, 1959; *Thorarinsson*,

1967] associated with the Icelandic basalts.

Molecular H_2O, as opposed to chemically bound water or $(OH)^-$, may be distributed along certain horizons within the crust. It is well known that many parameters of rocks, such as electrical resistivity [e.g. *Parkhomenko*, 1967], are a function of water content, but there is no geologic evidence to support the concept of water-rich or water-poor layers in the intermediate to deep crust.

Phase changes. Phase transformations would explain the presence of at least some of any layers that may exist within the crust, and metamorphic petrology can furnish insight into the nature of such transformations. *Den Tex* [1965, p. 123] considered the question of layers in the continental crust, and he speculated that the lower crust consists of intermediate rocks of the granulite facies, together with dry migmatites and rocks of the charnockite anorthosite norite pyroxenite clan. His conclusion was based largely on the lack of first-order velocity changes associated with any seismic (Conrad) discontinuity that may exist at the base of the upper, granitic crust. *Ringwood and Green* [1966a] have also postulated that the lower continental crust is of intermediate composition, but they propose that it consists of rocks in the eclogite facies or, where sufficient water is available, in the amphibolite facies. I find this model attractive, after some modification.

Basalt-eclogite transformation. The basalt (gabbro)-eclogite transformation has been suggested as the mechanism responsible for the Mohorovicic discontinuity [*Lovering*, 1958; *Kennedy*, 1959]. However, *Green and Ringwood* [1967a] and *Ringwood and Green* [1966b] interpreted their own experimental data to indicate that eclogite is the stable assemblage for basaltic compositions throughout most, if not all, of the crust. I find the possibility untenable that eclogite-facies rocks form continuous layers within the upper crust. As stated by *O'Hara and Mercy* [1963, p. 301], 'No case of conversion of basic rock to eclogite within the crust can be regarded as established.' In addition, it is now certain that amphibolites of basaltic composition are stable throughout the range of pressure-temperature conditions in the crust [e.g., *Yoder and Tilley*, 1962; *Hill and Boettcher*, 1970]. Also, experiments over a time span of more than six months (Boettcher and Modreski, unpublished data, 1970) on an olivine tholiite and an alkali basalt at temperatures between 400° and 700°C and pressures from 1 to 3 kb, both with excess H_2O and 50/50 mole per cent H_2O-CO_2, produced an amphibolite mineralogy with no indication of change toward eclogite.

Experiments covering a longer time span that are now in progress on intermediate and basic compositions in the presence of less than 1 wt% H_2O will add insight to the possibility of eclogites being stable in the upper crust, but available geologic evidence suggests that eclogites will not occur in important quantities under such low-pressure conditions. For example, most eclogites in the strict sense [type A of *Coleman et al.*, 1965; see also *White*, 1964] occur only as high-pressure xenoliths in kimberlites and basalts. Also, in a recent field study of eclogite-facies rocks metamorphosed in situ within rocks of obvious sedimentary derivation in Venezuela, *Morgan* [1970] noted that eclogites occur only within pockets that he interprets as having been 'dry.' Most of the rocks of similar composition are amphibolites, garnet amphibolites, and eclogite amphibolites. Using available experimental data for assemblages in the surrounding pelitic and calcareous rocks, Morgan suggests that the conditions under which the eclogites were formed were 525 ± 50°C and 5–10 (probably 7) kb, which are conditions certainly in excess of those at depths of 10 km in the earth's crust.

In another recent study in France, *Velde et al.* [1970] describe garnet amphibolites and amphibole eclogites; the latter were interpreted as forming from the amphibolite at pressures in excess of about 6.5 kb at 650°C. The presence of the hydrous phases amphibole and zoisite in the eclogites suggests that the activity of water was high in the eclogites as well as in the amphibolites but that eclogites required the high total pressures. In yet another example, *den Tex* [1965] presents a scheme in which eclogite-facies rocks occur at pressures greater than those of the amphibolite and granulite facies. I will return to this subject at the end of this paper, but it appears that eclogites cannot be called upon to explain layers in at least the upper 20 km of the crust.

Other metamorphic phase transformations.

There are no documented, indubitable cases of a seismic discontinuity occurring at a metamorphic-facies transition. Nevertheless, some such discontinuities may exist. For example, in Iceland, *Pálmason* [1967; unpublished data, 1970], from seismic refraction studies, has deduced layers that transcend geologic provinces. These layers extend nearly horizontally under the Quaternary Neovolcanic lavas along the projected strike of the mid-Atlantic ridge in central Iceland with the Tertiary flood basalts flanking them on each side. Layer 0 occurs everywhere in the Neovolcanic zone and consists of porous, vesicular basalts with P-wave velocities (V_p) of up to 3.4 km/sec. Layers 1 and 2, V_p of about 4.1 and 5.1, respectively, consist principally of the Tertiary flood basalts and usually extend to depths of about 3 km but may extend to depths of as great as 9 km. Layer 3, V_p of about 6.5, is found nowhere at the surface and extends to depths of about 7 to 15 km. Layer 4 has an anomalously low value of about 7.2 km/sec. The abundance of zeolites in some of the Tertiary Icelandic lavas [*Walker*, 1960] makes it tempting to speculate that the boundary between layers 1 and 2 is related to the formation of rocks in the zeolite facies, as was first suggested by *Einarsson* [1965]. Layer 3, the 'oceanic layer,' may consist of metamorphosed basalts (G. Pálmason, unpublished data, 1970), possibly amphibolites [*Cann*, 1968]. Compelling evidence that metamorphism of basalts and gabbros accounts for the observed crustal layering is found in the correlation between the depth to layer 3 and the temperature as obtained from bore holes (G. Pálmason, unpublished data, 1970).

Similar layers are revealed in an interesting study of the mid Atlantic ridge, near 45°N, by *Barrett and Aumento* [1970]. They propose that the layering results from fresh basalt (layer 2) about 0.5 km thick overlain by vesicular and, in some cases, zeolitic basalts and underlain by altered basalts and gabbros in the greenschist facies. They also measured P-wave velocities at elevated pressures of dredged samples of these rocks and equated their results with those obtained from their seismic measurements in the crust. Although V_p for the altered gabbros and basalts was less than 6 km/sec, Barrett and Aumento correlate it with layer 3 (the oceanic layer), with the assumption that

V_p will increase to about 6 km/sec at depths greater than several kilometers. Serpentinites were recovered from isolated pods at all elevations in the study area, and they do not appear to contribute to the layering.

It should be noted here that *Melson and Thompson* [1970] have discovered a layered, basic intrusion on the Romanche fracture in the Atlantic Ocean. Although no other examples in oceanic regions are known, some oceanic crustal layers possibly result from such features.

Melting. Unquestionably, melting of 'wet' silicic and intermediate rocks occurs in the crust and upper mantle, and the melting and differentiation of mantle-derived crustal material, such as andesite [*Harris*, 1967], has been proposed as the mechanism leading toward a layered crust. *Clowes and Kanasewich* [1970] have revealed in deep-reflection seismograms the presence of a fine structure for the basal crust in Alberta, Canada, consisting of layers less than 0.2 km thick and totalling less than 1 km; they suggest that partial melting may be responsible for such layering. *Wyllie* [1971] considers the conditions under which melting can occur for specific conditions of composition and water content, primarily under mantle conditions. Here, I will simply examine the minimum temperature conditions under which melting could occur in extensive crustal and upper mantle layers. Emphasis is placed on the feldspars or their high-pressure breakdown products, because they are the low-melting fraction in nearly every rock proposed for the crust.

It is now well known that water is an efficacious flux, capable of dissolving in silicate liquids under pressure and greatly lowering the temperatures of the beginning of melting. Melting curves for the major end-member feldspars under water-saturated conditions are now known to pressures of 20 kb and higher. The first study made above 10 kb was that for the system $NaAlSiO_4$(nepheline)- SiO_2-H_2O by *Boettcher and Wyllie* [1967, 1969]; some of the results of this study appear in Figure 1. The important point to note is that the melting curves of these assemblages at pressures below 10 to 15 kb decrease in temperature with increasing vapor pressure, but they have positive slopes (dp/dT), similar to dry fusion curves, at pressures above which they intersect solid-solid phase transformations. This change in slope results in temperature minimums on the curves representing the beginning of melting, thus indicating minimum temperatures that are necessary to cause melting. For example, the curve representing the beginning of melting for the reaction albite (Ab) + vapor (V) $\rightleftharpoons$ liquid (L) decreases in temperature, reaching a minimum at 620°C at the invariant point I_4, where it intersects the reaction albite $\rightleftharpoons$ jadeite (Jd) + quartz (Qz). At pressures above I_4 (17.3 kb), albite + water compositions melt according to the reaction Jd + Qz + V $\rightleftharpoons$ L, which has a positive dP/dT. The melting of albite +

quartz + H_2O compositions (Ab + Qz + V $\rightleftharpoons$ L) provided an estimate of the minimum melting temperatures likely to be encountered in the terrestrial crust.

Melting relationships of the other plagioclase end-member, anorthite, have recently become available to pressures of 35 kb in a study of the system $CaO-Al_2O_3-SiO_2-H_2O$ [Boettcher, 1970]. Lambert et al. [1969] have determined the curves representing the beginning of melting for the potassic feldspar, orthoclase, in the presence of water and quartz + water to pressures of 18.5 kb. Some of the results of these

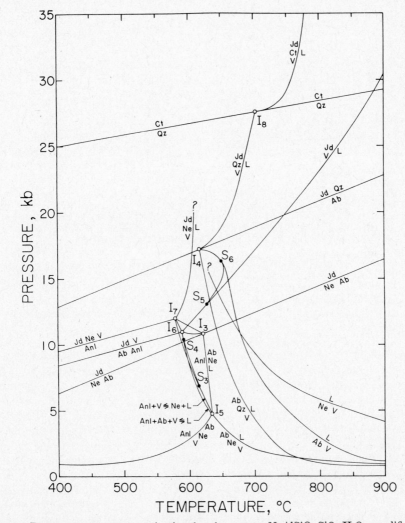

Fig. 1. Pressure-temperature projection for the system $NaAlSiO_4-SiO_2-H_2O$; modified after *Boettcher and Wyllie* [1969, p. 897]. Abbreviations: Ab is albite, Anl is analcite, Ct is coesite, Jd is jadeite, L is liquid, Ne is nepheline, Qz is quartz, and V is vapor.

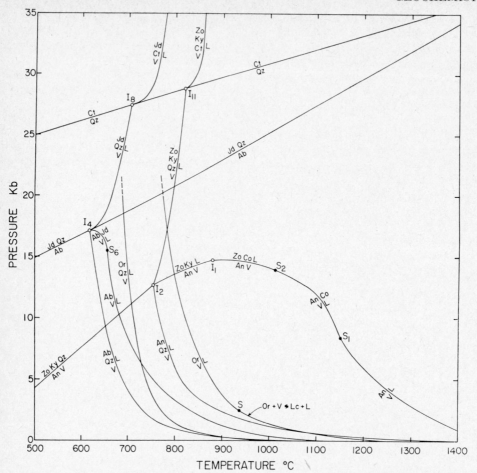

Fig. 2. Pressure-temperature projection of the beginning of melting of feldspars and feld-spar-quartz compositions in the presence of excess water; it is based on results of *Boettcher* [1970], *Boettcher and Wyllie* [1969], and *Lambert et al.* [1969]. Abbreviations: Ab is albite, An is anorthite, Co is corundum, Ct is coesite, I is invariant point, Jd is jadeite, Ky is kyanite, L is liquid, Or is orthoclase (potassic feldspar), Qz is quartz, S is singular point, V is vapor, and Zo is zoisite.

studies appear in Figure 2, together with the water-saturated melting curves of albite and albite + quartz from Figure 1. Potassic feldspar undergoes no phase transformations in the pressure interval studied, but the profound effects of phase transformations involving anorthite on the melting curves is apparent, as was also shown to be true for albite in Figure 1. In Figure 2, anorthite (An) + vapor (V) compositions melt along the curve passing through S_1, S_2, I_1, I_2, and I_{11}. Quartz is much more effective in lowering the melting temperature of anorthite than it is in the case of albite; the beginning of melting for anorthite + quartz +

water compositions is the same, within experimental error [*Boettcher*, 1970, p. 354], as the curve through I_2 and I_{11}.

As emphasized in Figure 3, melting in these model, synthetic systems provided the basis for investigations of the melting behavior of water-saturated rocks, as shown by the minimum melting curves for granite-H_2O and basalt-H_2O, which are compared with those of basalt, basalt-H_2O-CO_2, peridotite, and peridotite-H_2O. The factors limiting our knowledge of the levels and regions within the earth where melting occurs are, first, limitations on our knowledge of the chemical compositions of the earth's interior and, second, perhaps even less well known, our knowledge of the geothermal gradient. Nevertheless, best estimates of these factors, as shown in Figure 3, indicate with a fair degree of reliability that partial melting cannot be responsible for continuous layers at depths above at least 30 km (corresponding to pressures of about 10 kb).

Stability of plagioclase. A knowledge of the stability of plagioclase and its breakdown products at high pressures is essential to understanding the nature of the lower crust, regardless of whether it is gabbroic, andesitic, or eclogitic, for example. The stability of plagioclase merits special consideration, particularly in regard to the formation of eclogites from basic and intermediate rocks. The subsolidus phase relationships of one of the plagioclase endmembers, albite, are relatively simple and fairly well known [*Boettcher and Wyllie,* 1968*b*], as shown in Figure 4. The presence of nepheline reduces the upper pressure stability limit of

albite according to the reaction albite + nepheline = jadeite, but this reaction is not as significant petrologically as that for the silica-saturated reaction, albite $\rightleftharpoons$ jadeite + quartz. One complication that arises is that the position of the equilibrium curve for these reactions is a function of the structural state of the albite, as suggested by *Hlabse and Kleppa* [1968]. For the latter reaction, the equilibrium curves of *Boettcher and Wyllie* [1968*b*] were obtained by using synthetic high albite. (High albite is the high-temperature, high-entropy structural form of albite (NaAlSi$_3$O$_8$) in which the Al and Si ions are completely disordered in the tetrahedral sites. Low albite is the low-temperature modification in which the Al and Si ions are ordered.) *Newton and Smith* [1967] used natural low albite. There is good agreement between these two curves in the temperature interval over which experiments were conducted (500°–800°C), probably because this is the approximate temperature at which the low $\rightleftharpoons$

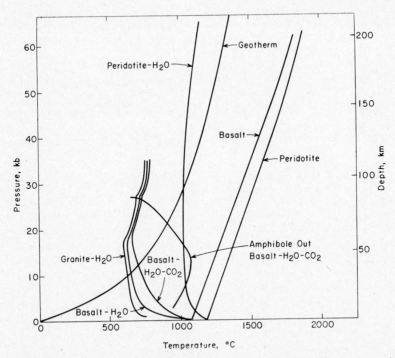

Fig. 3. Pressure-temperature projection for the beginning of melting of granite-water [*Boettcher and Wyllie,* 1968*a*], basalt-water [*Lambert and Wyllie,* 1968; *Hill and Boettcher,* 1970], basalt-water-carbon dioxide [*Hill and Boettcher,* 1970], peridotite and peridotite-water [*Kushiro et al.,* 1968], and basalt [*Cohen et al.,* 1967]. The geotherm is that calculated by *Clark and Ringwood* [1964], assuming a heat flow of 1.5 hfu. The curve for the stability of amphibole is from *Hill and Boettcher* [1970].

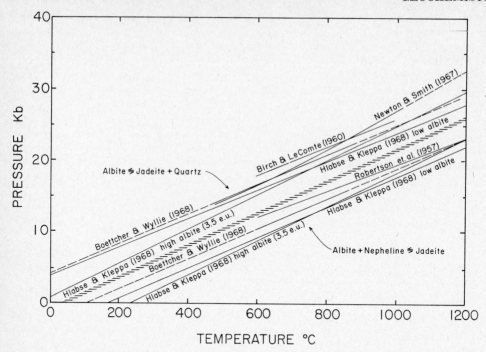

Fig. 4. Pressure-temperature projection for the stability of jadeite and jadeite + quartz. Jadeite + quartz is the high-pressure assemblage for the upper set of curves; jadeite is the high-pressure assemblage for the lower set. Abbreviation: e.u. is entropy units (cal/mole-deg).

high transformation occurs [e.g., *Holm and Kleppa*, 1968]. The effect of this Al-Si order-disorder is best seen by comparing the calcu-lated curves for low and high albite [*Hlabse and Kleppa*, 1968], assuming an entropy change (ΔS) of ordering of 3.5 cal/mole-deg (Figure 4).

Subsolidus phase relationships for the other end-member plagioclase, anorthite, are more complicated, in part, the result of the hydrous mineral zoisite (Figure 5). The maximum high-pressure stability of anorthite is governed by the breakdown to grossularite + kyanite + quartz. However, in the presence of water, anorthite reacts at a pressure about 5 kb lower to form zoisite + kyanite + quartz. The Al and Si ions in anorthite are completely ordered, and the effects of other structural modifications, including position disorder of the Ca ions [*Barth*, 1969; *Marfunin*, 1966], is probably in-significant.

Armed with these data on the end-number plagioclases, our task is to establish the stability of plagioclases of intermediate compositions, for it is these compositions that occur in rocks that

are candidates for the crust. In the more com-plex compositions found in these rocks, there are, of course, numerous reactions other than those shown in Figures 4 and 5 that conceivably could reduce the stability of plagioclase. For example, in rocks of basaltic composition, the anorthite component of plagioclase reacts with mafic minerals such as olivine or pyroxenes [*Cohen et al.*, 1967; *Green and Ringwood*, 1967a] to yield garnet-quartz assemblages at pressures below those for the breakdown of pure anorthite (Figure 5). Higher pressures are re-quired for the incorporation of the albite com-ponent by clinopyroxenes, and *Cohen et al.* [1967, p. 510] and *Green and Ringwood*, [1967a, p. 802] report that increasing the Ab/An ratio in their basalts extends the stabil-ity of the plagioclase to pressures that are higher but that are still below the reaction for pure albite (Figure 4). Nevertheless, experi-ments at high pressures and temperatures on some compositions including basalt-H_2O [*Lam-bert and Wyllie*, 1968; *Hill and Boettcher*, 1970], basalt-H_2O-CO_2 [*Hill and Boettcher*, 1970], granite [*Green and Lambert*, 1965], gabbroic anorthosite and diorite [*T. Green*, 1970] suggest that, under certain conditions, intermediate-composition plagioclase is stable to pressures similar to or higher than either end-member, albite or anorthite.

I am currently experimentally investigating the phase relationships of these intermediate-composition, plagioclase-bearing assemblages, but this research is experimentally and conceptually difficult, and final results will not be forthcoming for several years. However, based on preliminary results from this study, it is possible to indicate the general pattern of plagioclase breakdown at high pressures. For this purpose, let us consider the case in which albite reacts to jadeite + quartz and anorthite reacts to grossularite + kyanite + quartz.

If we assume an ideal solution between albite and anorthite components in the plagioclase and assume that no other crystalline solution occurs, it is possible to calculate the stability of plagioclase of intermediate composition [e.g., *Lewis and Randall*, 1961, p. 249] using the following relationships.

$$RT \ln \frac{a_{An}{}^{XP1^\circ}}{a_{An}{}^{XP1^1}} = (P^\circ - P^1)\, \Delta V^{3An \rightleftharpoons Gr + 2Ky + Qz}$$

$$RT \ln \frac{a_{Ab}{}^{XP1^\circ}}{a_{Ab}{}^{XP1^1}} = (P^\circ - P^1)\, \Delta V^{Ab \rightleftharpoons Jd + Qz}$$

where

R is gas constant.

T is absolute temperature in °K.

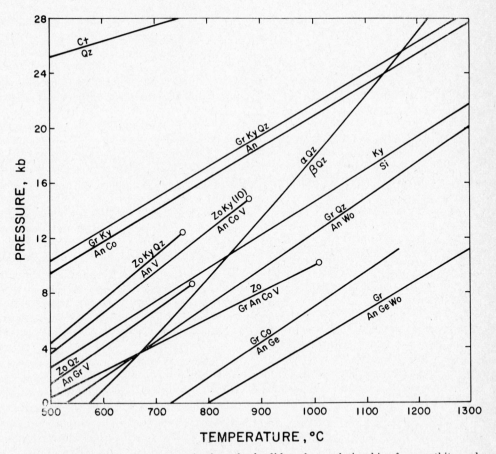

Fig. 5. Pressure-temperature projection of subsolidus phase relationships for anorthite and anorthite-bearing compositions in the system CaO-Al$_2$O$_3$-SiO$_2$-H$_2$O, modified after *Boettcher* [1970]. Abbreviations: An is anorthite, Ct is coesite, Co is corundum, Ge is gehlenite, Gr is grossularite, Ky is kyanite, Qz is quartz, Si is sillimanite, Wo is wollastonite, and Zo is zoisite.

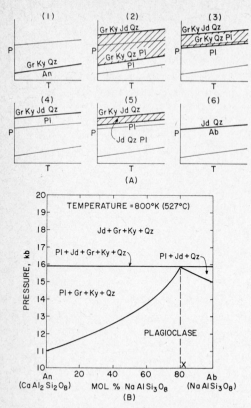

Fig. 6. Section A shows schematic pressure-temperature projections for the stability of plagioclase. The albite (NaAlSi$_3$O$_8$) component increases progressively from 0 in (1) to 100 in (6). For reference, the curves for the stability of albite and anorthite are shown in all six diagrams as light lines. Section B is a calculated isothermal, pressure-composition diagram for the stability of plagioclase. See the text for details. Abbreviations: Ab is albite, An is anorthite, Gr is grossularite, Jd is jadeite, Ky is kyanite, Pl is plagioclase, and Qz is quartz.

$P°$ is equilibrium pressure for the reaction involving plagioclase $Pl°$.

P^1 is equilibrium pressure for the reaction involving plagioclase Pl^1.

ΔV is equilibrium volume change.

X is number of moles of Pl in balanced reaction.

$a_{Ab}{}^{P1}$ is activity of albite component in plagioclase.

For pure albite and anorthite,

$$\ln a_{An}{}^{XP1°} = 0 \quad \text{and} \quad \ln a_{Ab}{}^{XP1°} = 0$$

Molar volume data are tabulated by *Robie and*

Waldbaum [1968]. The results of these calculations at a temperature of 800°K (527°C) are shown graphically in Figure 6. It should be stated that the plagioclase crystalline solution is unlikely to be ideal, particularly in the low range of temperatures in the crust, and that it may be necessary to revise the model when additional data become available.

In this model, for plagioclase of any composition, assuming that crystalline solution occurs in none of the other phases, plagioclase will persist up to the pressure of the 5-phase element, at which point it will have the composition X (Figure 6). At this pressure, jadeite appears and the system is isothermally invariant until all the plagioclase is consumed. In the actual case, where the CaAl$_2$SiO$_6$ component dissolves in the jadeite, the pressure at which jadeite appears at any given temperature is a function of the composition of the initial plagioclase. Nevertheless, on the basis of this model, for the restricted range of plagioclase compositions encountered in acid and intermediate-composition rocks, the maximum-pressure stability of plagioclase should be only slightly above that for pure albite. This theory is in accord with experimental results on granite-H$_2$O [*Boettcher and Wyllie*, 1968a], granite [*Green and Lambert*, 1965], and diorite and gabbroic anorthosite [*T. Green*, 1970], where the plagioclase disappears at about the pressure of the reaction albite ⇌ jadeite + quartz. Thus, these relationships provide reliable limits to the maximum pressures (depths) for the stable existence of plagioclase-bearing rocks, such as basalt (gabbro), andesite (diorite), and granulite. As shown in Figure 7 by the intersection of the continental geotherm with the curves representing the upper-pressure stability limit of these rocks, the disappearance of plagioclase could provide a seismic discontinuity at depths as great as 90 km.

Stability of plagioclase-bearing rocks. For more basic compositions, as occur for some basalts, where the normative albite/anorthite ratio is lower and the pyroxene/plagioclase ratio is higher, all the plagioclase components may be consumed by the pyroxene at a lower pressure, as shown in experiments of *Cohen et al.* [1967] and *Green and Ringwood* [1967a] on basalts. This effect can be seen in the results of the experiments on the quartz tholeiite composition

shown in Figure 7. Also, the high-pressure stability of plagioclase is reduced by reaction with
olivine, as was demonstrated experimentally by
Kushiro and Yoder [1966] and *Green and
Hibberson* [1970].

In complex natural compositions, plagioclase
begins to break down at pressures lower than
the reaction anorthite ⇌ grossularite + kyanite
+ quartz. This results from complex reactions
between the plagioclase and mafic minerals, including olivine and orthopyroxene, producing
garnet at lower pressures [*Cohen et al.*, 1967;
Green and Ringwood, 1967a, b]. The range of
conditions under which garnet first appears and
plagioclase disappears, as determined from experiments on several rock compositions, is
shown in Figure 7. The uncertainty in the extrapolation of these curves to the lower temperatures is evident from the crossing of the
curves for the quartz tholeiite, which is an improbable configuration.

Thus, although plagioclase breaks down over
a single wide multivariant pressure interval at a
given temperature, the garnet-producing reactions involving the anorthite component at the

lower pressure boundary of the interval and
the disappearance of the albite-rich plagioclase
at the higher pressure boundary provide two
relatively narrow reaction intervals (Figure 7)
that should be seismically detectable within
the crust and mantle. This is in concert with a
recent experimental investigation of the basalt-
eclogite transition determined by *Ito and Kennedy* [1970], where basalt transforms to eclogite via two density discontinuities. As shown
by the position of the geotherm in Figure 7,
basalt, gabbro, and garnet-free rocks of intermediate composition are unstable throughout
most, if not all, of the crust. In other words, the
geotherm everywhere is at pressures higher
than are required for the stable appearance of
garnet for all compositions shown. However,
such garnet-free rocks would persist metastably and not transform to the high-pressure
garnetiferous assemblages until at depths where
temperatures would be sufficient to allow
the garnet-producing reactions to proceed at
geologically significant rates. Thus, the actual first appearance of garnet would occur at
pressures along the geotherm and above those

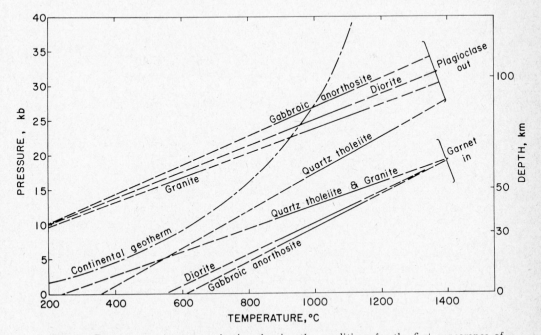

Fig. 7. Pressure-temperature projection showing the conditions for the first appearance of
garnet and the disappearance of plagioclase for diorite and gabbroic anorthosite [*Green*, 1970],
quartz tholeiite [*Green and Ringwood*, 1967a], and granite [*Boettcher and Wyllie*, 1968a;
Green and Lambert, 1965]. The geotherm was calculated with the assumption of a heat flow of
1.5 hfu [*Clark and Ringwood*, 1964].

shown by the curves representing the stable appearance of garnet in Figure 7. If the transformation to garnet-bearing assemblages began at 500°C (a reasonable estimate for near the base of the crust), then the position of the continental geothermal gradient indicates that the pressure at which garnet would appear is about 7 kb, corresponding to a depth of about 25 km. It is of interest to note that *Prodehl* [1970] determined the boundary between crust and mantle in much of the western United States to be a zone up to 10 km thick, not a sharp discontinuity as it is often conceived. This zone could be the result of the formation of garnet over a pressure (depth) interval, possibly beginning at a depth of 25 km.

Two final considerations arise regarding layers in the earth and their effect on seismic waves. First, the compositional or phase changes must significantly alter the compression- and shear-wave velocities or attenuation of the waves to allow detection. For example, the first appearance of garnet or the final disappearance of plagioclase at depth may escape detection. Little information is available on this subject, but *Vaišnys* [1968] presents an interesting discussion of the propagation of acoustic waves through a system undergoing phase transformations. The second consideration is that resolution of the layer requires a minimum thickness of the order of the wavelength of the seismic wave. *Clowes and Kanasewich* [1970] have proposed layers of less than 0.2-km thickness, based on their reflections studies of the crust in Alberta, and this represents the approximate minimum thickness detectable with standard techniques. Most, or all, of the phase transformations considered in this section, if present, would likely occur as layers exceeding these minimum detection values.

CONCLUSIONS

Layering in the crust and upper mantle can conceivably occur from compositional changes and-or phase transformations. Partial melting on regional scales at depths above at least 30 km is ruled out as a mechanism by which such layering can result.

In the continental crust, most estimates of the bulk composition suggest that the lower levels are of intermediate composition in either the granulite or eclogite facies. More basic compositions would be present as amphibolites if sufficient water is present and possibly as eclogites in the deepest crustal levels. Under either wet or dry conditions, gabbro and basalt are unstable throughout the crust. These rocks probably invert to relatively stable garnet-bearing assemblages under the higher temperature conditions in the lower crust.

The nature of the oceanic crust is equally uncertain, but the concept of a basaltic layer overlying a peridotitic mantle and veneered by sediments is undoubtedly too simple. Compositional stratification, possibly including acidic and intermediate layers, and phase changes to zeolites, greenschists, amphibolites, and other rocks are very likely to occur in the oceanic crust.

Plagioclase or its breakdown products constitutes a significant part of most rocks that are candidates for the lower crust. A knowledge of the stability of intermediate composition plagioclase provides valuable restrictions on most proposed lower-crust models, and available experimental results indicate that the disappearance of albitic plagioclase could provide a seismically detectable density discontinuity at depths as great as 70 to 90 km for intermediate and some basic rocks. Consumption of plagioclase by reaction with olivine and pyroxene would occur at much shallower depths for peridotitic compositions. Another generally shallower discontinuity will likely occur at the first appearance of garnet at depth, but its position may range from about 0 to 40 km, depending on the depths to which the garnet-free rocks persist metastably.

Our knowledge of the structure of the lower crust and upper mantle is still incomplete. When additional data on heat flow, geothermal gradients, and the compositions of the various levels of the interior of the earth become available, the results of experimental petrologic investigations at elevated temperatures and pressures will be applied with greater confidence.

Acknowledgments. This paper has benefited from critical reviews of the manuscript by William C. Luth, of Stanford University, and James F. Hays, of Harvard University. I thank Gudmundur Pálmason, of the University of Iceland, for the use of his unpublished seismic data.

This research is supported by National Science Foundation grants NSF-GA-1364 and NSF-GA-12737.

REFERENCES

Anders, E., Chemical processes in the early solar system, as inferred from meteorites, *Accounts Chem. Res., 1,* 289–298, 1968.

Anderson, D. L., and T. Jordan, The composition of the lower mantle, *Phys. Earth Planet. Interiors, 3,* 23–35, 1970.

Barrett, D. L., and F. Aumento, The Mid-Atlantic Ridge near 45°N, 6, Seismic velocity, density, and layering of the crust, *Can. J. Earth Sci., 7,* 1117–1124, 1970.

Barth, T. F. W., *Feldspars,* 261 pp., Interscience, New York, 1969.

Birch, F., The velocity of compressional waves in rocks to 10 kilobars, 1, *J. Geophys. Res., 65,* 1083–1102, 1960.

Birch, F., The velocity of compressional waves in rocks to 10 kilobars, 2, *J. Geophys. Res., 66,* 2199–2224, 1961.

Birch, F., Energenics of core formation, *J. Geophys. Res., 70,* 6217–6221, 1965a.

Birch, F., Speculations on the Earth's thermal history, *Geol. Soc. Amer. Bull., 76,* 133–154, 1965b.

Birch, F., and P. LeComte, Temperature-pressure plane for albite composition, *Amer. J. Sci., 258,* 209–217, 1960.

Boettcher, A. L., The system CaO-Al₂O₃-SiO₂-H₂O at high pressures and temperatures, *J. Petrol., 11,* 337–379, 1970.

Boettcher, A. L., and P. J. Wyllie, Hydrothermal melting curves in silicate-water systems at pressures greater than 10 kilobars, *Nature, 216,* 572–573, 1967.

Boettcher, A. L., and P. J. Wyllie, Melting of granite with excess water to 30 kilobars pressure, *J. Geol., 76,* 235–244, 1968a.

Boettcher, A. L., and P. J. Wyllie, Jadeite stability measured in the presence of silicate liquids in the system NaAlSiO₄-SiO₂-H₂O, *Geochim. Cosmochim. Acta., 32,* 999–1012, 1968b.

Boettcher, A. L., and P. J. Wyllie, The system NaAlSiO₄-SiO₂-H₂O to 35 kilobars pressure, *Amer. J. Sci., 267,* 875–909, 1969.

Buddington, A. F., Some petrological concepts and the interior of the Earth, *Amer. Mineral., 28,* 119–140, 1943.

Cameron, A. G. W., The formation of the Sun and planets, *Icarus, 1,* 13–69, 1962.

Cann, J. R., Geological processes at mid-ocean ridge crests, *Geophys. J. Roy. Astron. Soc., 15,* 331–341, 1968.

Christensen, N., Compressional wave velocities in metamorphic rocks at pressures to 10 kilobars, *J. Geophys. Res., 70,* 6147–6164, 1965.

Clark, S. P., Jr., and A. E. Ringwood, Density distribution and constitution of the mantle, *Rev. Geophys. Space Phys., 2,* 35–88, 1964.

Clarke, F. W., and H. S. Washington, The composition of the earth's crust, *U.S. Geol. Surv. Prof. Pap. 127,* 117 pp., 1924.

Clowes, R. M., and E. R. Kanasewich, Seismic attenuation and the nature of reflecting hori-zons within the crust, *J. Geophys. Res., 75,* 6693–6705, 1970.

Cohen, L. H., K. Ito, and G. C. Kennedy, Melting and phase relations in an anhydrous basalt to 40 kilobars, *Amer. J. Sci., 265,* 475–518, 1967.

Coleman, R. G., D. E. Lee, L. B. Beatty, and W. W. Brannock, Eclogites and eclogites: Their differences and similarities, *Geol. Soc. Amer. Bull., 76,* 483–508, 1965.

Daly, R. A., The geology of Ascension Island, *Daedalus, 60,* 1–80, 1925.

den Tex, E., Metamorphic lineages of orogenic plutonism, *Geol. Mijnbouw, 44,* 105–132, 1965.

Einarsson, Tr., Remarks on crustal structure in Iceland, *Geophys. J. Roy. Astron. Soc., 10,* 283–288, 1965.

Elsasser, W. M., Early history of the earth, in *Earth Science and Meteoritics,* edited by J. Geiss and E. D. Goldberg, p. 1, North-Holland, Amsterdam, 1963.

Engel, C. G., and Fisher, R. L., Lherzolite, anorthosite, gabbro, and basalt dredged from the Mid-Indian Ocean Ridge, *Science, 166,* 1136–1141, 1969.

Eucken, A., Physikalisch-chemisch Betrachtungen über die früheste Entwicklungsgeschicte der Erde, *Nachr. Akad. Wiss. Goettingen Math. Phys. Kl. 1,* 1-25, 1944.

Gast, P. W., Limitations on the composition of the upper mantle, *J. Geophys. Res., 65,* 1287–1297, 1960.

Gast, P. W., Upper mantle chemistry and evolution of the Earth's crust, in *History of the Earth's Crust,* edited by R. Phinney, p. 15, Princeton University Press, Princeton, New Jersey, 1968.

Goldschmidt, V. M., Grundlagen der quantitativen Geochemie, *Fortschr. Mineral. Kristallogr. Petrogr., 17,* 112–156, 1933.

Green, D. H., and W. Hibberson, The instability of plagioclase in peridotite at high pressure, *Lithos, 3,* 209–221, 1970.

Green, D. H., and I. B. Lambert, Experimental crystallization of anhydrous granite at high pressures and temperatures, *J. Geophys. Res., 70,* 5259–5268, 1965.

Green, D. H., and A. E. Ringwood, An experimental investigation of the gabbro to eclogite transformation and its petrological applications, *Geochim. Cosmochim. Acta., 31,* 767–833, 1967a.

Green, D. H., and A. E. Ringwood, The genesis of basaltic magmas, *Contrib. Mineral. and Petrol., 15,* 103–190, 1967b.

Green, T. H., High pressure experimental studies on the mineralogical constitution of the lower crust, *Phys. Earth Planet. Interiors, 3,* 441–450, 1970.

Hanks, T. C., and D. L. Anderson, The early thermal history of the Earth, *Phys. Earth Planet. Interiors, 2,* 19–29, 1969.

Harris, P. G., Zone refining and the origin of potassic basalts, *Geochim. Cosmochim. Acta., 12,* 195–208, 1957.

Harris, P. G., Segregation processes in the upper mantle, in *Mantles of the Earth and Terrestrial Planets*, edited by S. K. Runcorn, p. 305, Interscience, New York, 1967.

Herz, N., Anorthosite belts, continental drift, and the anorthosite event, *Science, 164*, 944–947, 1969.

Hill, R. E. T., and A. L. Boettcher, Water in the Earth's mantle: Melting curves of basalt-water and basalt-water-carbon dioxide, *Science, 167*, 980–982, 1970.

Hlabse, T., and O. J. Kleppa, The thermochemistry of jadeite (NaAlSiO₄), *Amer. Mineral., 53*, 1281–1292, 1968.

Holm, J. L., and O. J. Kleppa, Thermodynamics of the disordering process in albite, *Amer. Mineral., 53*, 123–133, 1968.

Honorez, J., and E. Bonatti, Nepheline gabbro from the Mid-Atlantic ridge, *Nature, 228*, 850–852, 1970.

Ito, K., and G. C. Kennedy, The fine structure of the basalt-eclogite transition, *Mineral. Soc. Amer. Spec. Pap. 3*, 77–83, 1970.

James, D. E., and J. S. Steinhart, Structure beneath continents: A review of explosion studies 1960–1965, in *The Earth Beneath the Continents, Geophys. Monogr. Ser.*, vol. 10, edited by J. S. Steinhart and T. J. Smith, p. 293, AGU, Washington, D.C., 1966.

Kennedy, G. C., The origin of continents, mountain ranges, and ocean basins, *Amer. Sci., 42*, 491–504, 1959.

Kushiro, I., and H. S. Yoder, Jr., Anorthite-forsterite and anorthite-enstatite reactions and their bearing on the basalt-eclogite transformation, *J. Petrol., 7*, 337–362, 1966.

Kushiro, I., Y. Syono, and S. Akimoto, Melting of a peridotite nodule at high pressures and high water pressures, *J. Geophys. Res., 73*, 6023–6029, 1968.

LaCroix. A.. Les roches eruptives de l'Archipel de Kerguelan, *Compt. Rend., 179*, 113–119, 1924.

Lambert, I. B., and P. J. Wyllie, Stability of hornblende and a model for the low velocity zone, *Nature, 219*, 1240–1241, 1968.

Lambert, I. B., J. K. Robertson, and P. J. Wyllie, Melting reactions in the system KAlSi₃O₈-SiO₂-H₂O to 18.5 kilobars, *Amer. J. Sci., 267*, 609–626, 1969.

Larimer, J. W., Chemical fractionations in meteorites, 1, Condensation of the elements, *Geochim. Cosmochim. Acta., 31*, 1215–1238, 1967.

Lee, W. H. K., Effects of selective fusion on the thermal history of the Earth's mantle, *Earth Planet. Sci. Lett., 4*, 270–276, 1968.

Lewis, G. N., and M. Randall, *Thermodynamics*, revised by K. S. Pitzer and L. Brewer, McGraw-Hill, New York, 1961.

Lovering, J. F., The nature of the Mohorovicic discontinuity, *Eos, Trans. AGU, 39*, 947–955, 1958.

Lubimova, E. A., Theory of thermal state of the Earth's mantle, in *The Earth's Mantle*, edited by T. F. Gaskell, p. 231, Academic, New York, 1967.

Lubimova, E. A., Thermal history of the Earth, in *The Earth's Crust and Upper Mantle, Geophys. Monogr. Ser.*, vol. 13, edited by P. J. Hart, p. 63, AGU, Washington, D.C., 1969.

Marfunin, A. S., *The Feldspars*, 317 pp., Translated from Russian by J. Kolodny, Israel Program for Scientific Translations, Jerusalem, 1966.

Melson, W. G., and G. Thompson, Layered basic complex in oceanic crust, Romanche fracture, equatorial Atlantic Ocean, *Science, 168*, 817–820, 1970.

Middlemost, E. A. K., Anorthosites: a graduated series, *Earth Sci. Rev., 6*, 257–265, 1970.

Morgan, B. A., Petrology and mineralogy of eclogite and garnet amphibolite from Puerto Cabello, Venezuela, *J. Petrol., 11*, 101–145, 1970.

Mueller, S., and M. Landisman, Seismic studies of the Earth's crust in continents, 1, Evidence for a low-velocity zone in the upper part of the lithosphere, *Geophys. J. Roy. Astron. Soc., 10*, 525–538, 1966.

Newton, R. C., and J. V. Smith, Investigations concerning the breakdown of albite at depth in the earth, *J. Geol., 75*, 268–286, 1967.

Nockolds, S. R., Average chemical compositions of some igneous rocks, *Geol. Soc. Amer. Bull., 65*, 1007–1032, 1954.

O'Hara, M. J., and L. P. Mercy, Petrology and petrogenesis of some garnetiferous peridotites, *Trans. Roy. Soc. Edinburgh, 65*, 251–314, 1963.

Pakiser, L. C., The basalt-eclogite transformation and crustal structure in the western United States, *U.S. Geol. Surv. Prof. Pap. 525-B*, 8 pp., 1965.

Pakiser, L. C., and R. Robinson, Composition of the continental crust as estimated from seismic observations, in *The Earth Beneath the Continents, Geophys. Monogr. Ser.*, vol. 10, edited by J. S. Steinhart and T. J. Smith, p. 620, AGU, Washington, D.C., 1966.

Pakiser, L. C., and I. Zietz, Transcontinental crustal and upper-mantle structure, *Rev. Geophys. Space Phys., 3*, 505–520, 1965.

Pálmason, G., Upper crustal structure in Iceland, in *Iceland and Mid-Ocean Ridges*, edited by S. Björnsson, p. 67, Prentsmidjan Leiftur H. F., Reykjavik, Iceland, 1967.

Parkhomenko, E. I., *Electrical Properties of Rocks*, Plenum, New York, 1967.

Poldervaart, A., Chemistry of the earth's crust, *Geol. Soc. Amer. Spec. Pap. 62*, 119–144, 1955.

Press, F., Density distribution in Earth, *Science, 160*, 1218–1221, 1968.

Press, F., Earth models consistent with geophysical data, *Phys. Earth Planet. Interiors, 3*, 3–22, 1970.

Prodehl, C., Seismic refraction study of crustal structure in the western United States, *Geol. Soc. Amer. Bull., 81*, 2629–2646, 1970.

Ringwood, A. E., Chemical evolution of the terrestrial planets, *Geochim. Cosmochim. Acta., 30,* 41–104, 1966.

Ringwood, A. E., and D. H. Green, Petrological nature of the stable continental crust, in *The Earth Beneath the Continents, Geophys. Monogr. Ser.,* vol. 10, edited by J. S. Steinhart and T. J. Smith, p. 611, AGU, Washington, D.C., 1966*a*.

Ringwood, A. E., and D. H. Green, An experimental investigation of the gabbro-eclogite transformation and some geophysical implications, *Tectonophysics, 3,* 383–427, 1966*b*.

Robertson, E. C., F. Birch, and G. J. F. MacDonald, Experimental determination of jadeite stability to 25,000 bars, *Amer. J. Sci., 255,* 115–137, 1957.

Robie, R. A., and D. R. Waldbaum, Thermodynamic properties of minerals and related substances at 298.15°K (25.0°C) and one atmosphere (1.013 bars) pressure and at higher temperatures, *U.S. Geol. Surv. Bull. 1259,* 256 pp., 1968.

Ronov, A. B., and A. A. Yaroshevsky, Chemical composition of the earth's crust, in *The Earth's Crust and Upper Mantle, Geophys. Monogr. Ser.,* vol. 13, edited by P. J. Hart, p. 63, AGU, Washington, 1969.

Sigurdsson, H., Acid xenoliths from Surtsey, in *Proc. Surtsey Res. Conf.,* Surtsey Research Society, Reykjavik, Iceland, 65–68, 1967.

Spetzler, H., and D. L. Anderson, The effect of temperature and partial melting on velocity attenuation in a simple binary system, *J. Geophys. Res., 73,* 6051–6060, 1968.

Tan, L., and Y. Chi-lung, Abundance of chemical elements in the earth's crust and its major tectonic units, *Int. Geol. Rev., 12,* 778–786, 1970.

Taylor, S. R., Abundance of chemical elements in the continental crust: A new table, *Geochim. Cosmochim. Acta., 28,* 1273–1285, 1964.

Taylor, S. R., and A. J. R. White, Geochemistry of andesites and the growth of continents, *Nature, 208,* 271–273, 1965.

Thorarinsson, S., Hekla and Katla, the share of acid and intermediate lava and tephra in the volcanic products through the geological history of Iceland, in *Iceland and Mid-Ocean Ridges,* edited by S. Björnsson, p. 190, Prentsmidjan Leiftur H. F., Reykjavik, Iceland, 1967.

Turekian, K. K., The composition of the crust, in *Origin and Distribution of the Elements,* edited

by L. H. Ahrens, p. 549, Pergamon, Oxford, 1968.

Turekian, K. K., and S. P. Clark, Jr., Inhomogeneous accumulation of the Earth from the primitive solar nebula, *Earth Planet. Sci. Lett., 6,* 346–348, 1969.

Urey, H. C., *The Planets: Their Origin and Development,* 285 pp., Yale University Press, New Haven, Connecticut, 1952.

Urey, H. C., The origin and evolution of the Solar System, in *Space Science,* edited by D. P. LeGalley, p. 123, John Wiley, New York, 1963.

Vaišnys, J. R., Propagation of acoustic waves through a system undergoing phase transformations, *J. Geophys. Res., 73,* 7675–7683, 1968.

Velde, B., F. Hervé, and J. Kornprobst, The eclogite-amphibolite transition at 650°C and 6.5 Kbar pressure, as exemplified by basic rocks of the Uzerche area, central France, *Amer. Mineral., 55,* 953–974, 1970.

Vinogradov, A. P., The origin of the material of the earth's crust, Communication 1, *Geochemistry, USSR, 1–32,* 1961.

Vinogradov, A. P., Average content of chemical elements in main types of igneous rocks of the earth's crust, *Geokhimiya, (7),* 555–571, 1962.

Walker, G. P. L., Geology of the Reydarfjördur area, eastern Iceland, *Quart. J. Geol. Soc. London, 114,* 367–393, 1959.

Walker, G. P. L., Zeolite zones and dike distribution in relation to the structure of the basalts of eastern Iceland, *J. Geol., 68,* 515–527, 1960.

Wedepohl, K. H., Composition and abundance of common igneous rocks, in *Handbook of Geochemistry,* edited by K. H. Wedepohl, p. 227, Springer-Verlag, Berlin, 1969.

White, A. J. R., Clinopyroxenes from eclogites and basic granulites, *Amer. Mineral., 49,* 883–888, 1964.

Wood, J. A., J. S. Dickey, Jr., U. B. Marvin, and B. N. Powell, Lunar anorthosites, *Science, 167,* 602–604, 1970.

Wyllie, P. J., Experimental limits for melting in the earth's crust and upper mantle, in *The Structure and Physical Properties of the Earth's Crust, Geophys. Monogr. Ser.,* vol. 14, edited by J. G. Heacock, AGU, Washington, D.C., this volume, 1971.

Yoder, H. S., Jr., and C. E. Tilley, Origin of basalt magmas: An experimental study of natural and synthetic rock systems, *J. Petrol., 3,* 342–532, 1962.

Experimental Limits for Melting in the Earth's Crust and Upper Mantle

PETER J. WYLLIE

Department of Geophysical Sciences, University of Chicago, Chicago, Illinois   60637

Abstract. The conditions for melting in the crust and upper mantle are governed by the mineralogy (determined by bulk composition, depth, and temperature), the water content, the physical state of the water (available in pore fluid or bound in crystals), P_{cH_2O}, and the temperature distribution. The average composition of the crust is andesitic and its mineralogy is dominated by feldspars and quartz. Melting curves in the presence of excess water at pressures ranging to more than 10 kb (40-km depth) have now been determined for individual feldspars, for most feldspar-quartz combinations, and for many major rock types. In the presence of an aqueous vapor phase, the granitic components of many crustal rocks combine to produce water saturated liquid of granite composition. Starting assemblages for melting in rock-water systems consist of minerals with interstitial vapor, hydrous and anhydrous minerals with no vapor, or anhydrous minerals with no vapor. Models for magma generation must consider whether the liquids produced are water saturated or water deficient under the conditions of melting. From estimates of temperatures in the crust it becomes apparent that no granitic liquids can be produced at a depth shallower than 20 km. Results from water-excess experiments and interpolated water deficient conditions indicate that the normal product of partial fusion of many crustal rocks is a water undersaturated granite liquid in a crystal mush which persists through a wide temperature range. It is not likely that liquids of intermediate composition are generated directly, because temperatures are too high, but crystal-liquid assemblages of intermediate bulk composition may move to higher levels in the crust by diapiric rise. The generation of basaltic magmas in a dry mantle requires unusually high temperatures. Thus, most basaltic magmas must be produced under conditions where the local temperature greatly exceeds that of the average geothermal gradient. However, the presence of trace amounts of water in the mantle does permit incipient melting of eclogite or peridotite with a normal geothermal gradient. The depth interval within which such melting occurs in rock-water systems coincides with the seismic low-velocity zone; this fact may explain the presence of the zone as a continuous layer in the earth's mantle. Crustal melting, localized in orogenic belts, is not likely to produce continuous layers. However, removal of pore fluids from deep seated continental basement rocks by repeated melting may have produced regions with laterally extensive uniform properties, despite their variegated compositions.

This review was prepared for a symposium concerned particularly with evaluation of the extent of lateral continuity of physical properties, especially electrical properties, at depth. Evidence for the existence of extensive layers at about the 10-km level and deeper in the crust and upper mantle is outlined elsewhere in this volume [see *Landisman et al.,* 1971]. My responsibility was to consider the possibility that melting processes contributed to the formation of such layers.

The eruption of lavas shows that melting does occur in the earth's interior. The conditions for magma generation, the compositions of liquids generated, and the history of upward migration, intrusion, and extrusion of magmas can be evaluated from several approaches. Field geology, petrology, and geochemistry of the rocks provide the basic information. Experimentally determined phase relationships in silicate systems elucidate petrogenetic interpretations of the rocks.

Magmas can be generated by the partial fusion of three types of assemblages: anhydrous minerals; anhydrous and hydrous minerals; and minerals in the presence of a pore fluid usually considered to be aqueous, although carbon dioxide, fluorine, chlorine, and other components are likely to be present and influential.

Experimental conditions are considerably simpler than those in the complex natural magmatic systems. Experimental approaches include the following [*Wyllie*, 1963]: liquidus studies of dry systems at atmospheric pressure, with or without controlled atmosphere; solid-liquid studies of dry minerals and rocks at high pressures; solid-liquid-vapor studies of mineral-water and rock-water systems at high pressures, with oxygen fugacity controlled independently in some experiments; and solid-liquid-vapor studies involving water and other volatile components.

Most published experimental results have dealt either with the melting relationships of dry silicate systems or with samples in the presence of excess water. Most deep seated magmatic processes occur with water pressure less than the total pressure and with insufficient water present to saturate the liquid; these are water deficient conditions which have received little experimental attention. Terms such as water-deficient and vapor-absent have been widely used in the literature without adequate definitions. The terms 'vapor-present' and 'vapor-absent' are used simply to denote the presence or absence of vapor in phase assemblages; *Yoder and Kushiro* [1969] used the terms 'gas-present' and 'gas-absent' for the same conditions. *Robertson and Wyllie* [1971a, b] prepared a consistent set of definitions for four types of silicate-water systems ranging from dry to water-excess. These provide limits for the natural magmatic systems. Type I, 'water-absent,' denotes an assemblage of anhydrous silicate minerals with no vapor phase. Type II, 'water-deficient and vapor-absent,' denotes an assemblage of silicate minerals including hydrous minerals but with no vapor phase. Type III, 'water-deficient and vapor-present,' denotes an assemblage of silicate minerals with or without hydrous minerals and with vapor phase. There is insufficient water present to saturate the liquid when the crystalline assemblage is completely melted at the existing pressure. Type IV, 'water-excess,' denotes an assemblage of silicate minerals with or without hydrous minerals and with more than sufficient water to saturate the liquid when the crystalline assemblage is completely melted at the existing pressure. A vapor phase is present.

Yoder [1952] introduced the term 'water-deficient' for subsolidus assemblages with insufficient water to permit maximum hydration of the crystalline phases. This corresponds to

type II. When a liquid is involved, a vapor-present subsolidus assemblage can become a vapor-absent assemblage above the solidus, when the vapor dissolves in the liquid. Such a system is obviously deficient in water. Therefore, the definition of water-deficient systems is extended to include the solubility of water in the liquid by introducing type III [*Robertson and Wyllie*, 1971a, b].

Magma generation within the earth may be associated with dynamic processes such as the diapiric uprise of rock masses under adiabatic conditions [*Green and Ringwood*, 1967] or those processes involved in plate tectonics [*Isacks et al.*, 1968]. Such processes form no part of the static, closed-system laboratory experiments. However, the experimental results determined under specified and controlled conditions provide limits for melting within the earth, regardless of the processes that may be involved in transferring masses of rock or magma across melting or reaction boundaries.

In the following pages we will examine first the assemblages of silicate minerals that may be melted in the crust and mantle and then the effect of excess water under pressure on the temperature of beginning of melting (type IV). Next, the effects of volatile components additional to water are outlined. Then we examine the phase relationships through the melting intervals of rocks in the presence of excess water, as a guide to the products of partial melting. Excess water is not available for magma generation [see *Luth*, 1969; *Dodge and Ross*, 1971], so the next step is to examine melting relationships with water-deficient conditions. Experimental results for solid-liquid phase relationships in water-deficient systems (types II and III) are scarce, but generalized diagrams for silicate-water systems can be constructed by interpolating between the dry (type I) and the water-excess (type IV) experimental results.

This review of the available types of experimental data in silicate-water–volatile component systems that provide limits for the conditions of melting is followed by applications to the problem of magma generation in the crust and mantle and consideration of the question of whether melting processes contribute to the development of layers.

PETROLOGY AND MINERALOGY OF THE CRUST AND UPPER MANTLE

The crust is composed primarily of rocks ranging in composition from granite to gabbro, together with the sedimentary suites. The average crustal composition corresponds to andesite [*Ronov and Yaroshevsky*, 1969]. The upper mantle is composed of peridotite, residual dunite remaining after removal of a basaltic melt fraction, precipitated peridotite from magma fractionating during uprise, and eclogite or gabbro from magma that failed to reach the crust [*Wyllie*, 1970]. For conditions of melting in the crust we therefore consider the systems granite-water, granodiorite-water, and gabbro-water; for the mantle we consider the systems peridotite-gabbro, peridotite-water, and gabbro-water (the rock names are used as compositional terms in this context).

Crustal mineralogy is dominated by feldspars and quartz. Figure 1 shows the variations in composition of a series of calc-alkaline rocks from the Wallowa batholith, expressed in terms of their normative feldspars and quartz. Gabbro, containing neither quartz nor alkali feldspar, plots in the central region of the edge An-Ab, and granite plots on or very near to the base, Ab-Or-Qz. The points plotted for tonalites and granodiorites lie fairly close to a line extending through the tetrahedron connecting gabbro to granite. Thus, the phase relationships in feldspar-quartz-water systems provide models for the more complex rock-water systems.

The mantle rocks are composed of olivine, orthopyroxene, and clinopyroxene, possibly with accessory aluminous minerals changing successively from plagioclase to spinel to garnet with increasing depth. Small proportions of hydrous minerals such as phlogopite and amphibole may be present locally in the uppermost mantle. Phase relationships in the system CaO-MgO-Al_2O_3-SiO_2 are used as models for the more complex mantle rock systems.

BEGINNING OF MELTING OF ROCKS AND MINERALS WITH EXCESS WATER OR SOLUTIONS

Systematic experimental studies of the effect of water under pressure on the melting relationships in silicate systems began with the publication by *Bowen and Tuttle* [1950] of the phase relationships in the system $KAlSi_3O_8$-$NaAlSi_3O_8$-H_2O. They cited the few earlier studies. In the twenty years since 1950, solidus curves for the beginning of melting in most feldspar-quartz-water and rock-water systems with excess water (type IV) have been determined. The effect of water under pressure is to lower the melting temperature. At high pressures, where plagioclase feldspars break down to yield dense minerals such as jadeite, zoisite, and kyanite, the slope of the solidus reverses.

Feldspar-quartz-water systems. Merrill et al. [1970] reviewed the data for solidus curves in these systems, with the results shown in Figures 2 and 3. The melting curves for individual feldspars and for most feldspar-quartz combinations have been located through a wide range of pressures.

Figure 2a shows the univariant curves for the individual feldspars and quartz. The curve for albite [*Boettcher and Wyllie*, 1969] terminates at an invariant point at about 16.5 kb with the formation of jadeite, but the curve for orthoclase extends to pressures higher than 20 kb [*Lambert et al.*, 1969]. The curve for quartz terminates at a second critical endpoint, according to *Kennedy et al.* [1962], but *Stewart* [1967] stated that the endpoint has not been

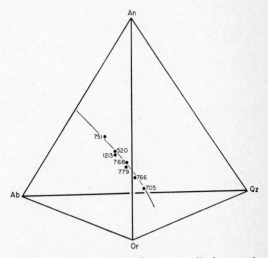

Fig. 1. Norms of tonalites, granodiorites, and granites from the Needle Point pluton, Wallowa batholith, recalculated to 100 per cent of quartz and feldspar in the granodiorite tetrahedron. The curved line through the numbered points representing specific rocks shows the general trend of mineralogical variation in the calc-alkaline rock series from gabbro to granite [after *Piwinskii and Wyllie*, 1968].

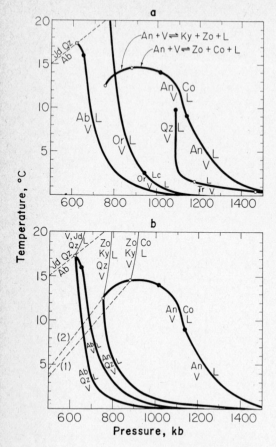

The univariant curve for the reaction between anorthite and water vapor passes through two singular points and two invariant points. The products of the reaction between anorthite and water change at each point, as depicted in Figure 2a. The albite curve passes through a singular point just below the jadeite reaction boundary, changing to an incongruent reaction $Ab + V \rightleftharpoons Jd + L$.

The melting curves for plagioclase feldspars and for plagioclase-feldspar-quartz mixtures are shown in Figure 2b. The curves for albite and albite-quartz [*Boettcher and Wyllie*, 1969] and for anorthite and anorthite-quartz [*Stewart*, 1967; *Boettcher*, 1970] terminate at high pressures at invariant points. These curves are terminated in this way because breakdown of the feldspar produces the same subsolidus phase assemblage from both feldspar and feldspar-quartz mixtures. Figure 2b shows in light broken lines the subsolidus reactions that limit the high pressure stability of the plagioclase feldspars. The light continuous lines indicate the hydrothermal melting curves which replace the feldspar and feldspar-quartz curves at higher pressures.

Figure 3 compares the melting curves for orthoclase and orthoclase-quartz with the two

Fig. 2. Curves for the beginning of melting in feldspar-quartz systems with excess water, compiled from various sources (see text) by *Merrill et al.* [1970]. Heavy lines are solidus curves; light lines are solidus curves at pressures above the stability limit of feldspars; light dashed lines are subsolidus reactions; white circles are invariant points; black circles are singular points. Ab is albite, Or is sanidine, An is anorthite, Qz is quartz, Lc is leucite, Tr is tridymite, Jd is jadeite, Co is corundum; Zo is zoisite, Ky is kyanite, L is liquid, V is vapor. 2a shows the system SiO_2-H_2O and reactions involving single feldspars. 2b shows univariant reactions in the system $CaAl_2Si_2O_8$-$NaAlSi_3O_8$-SiO_2-H_2O. Reaction 1 is $An + Co + V \rightleftharpoons Zo + Ky$. Reaction 2 is $An + V \rightleftharpoons Zo + Ky + Qz$.

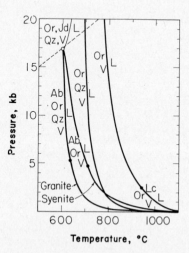

Fig. 3. Curves for the beginning of melting in the feldspar-quartz system with excess water [after *Merrill et al.*, 1970]. Line values and circles as for Figure 2. $Ab_{ss} + Or_{ss}$ approximates a syenite, and $Ab_{ss} + Or_{ss} + Qz$ approximates a granite in mineralogy; ss is solid solution. For other abbreviations, see Figure 2.

reached at 10 kb. *Kushiro* [1969a, Figure 7] presented a diagram indicating that the univariant reaction still persists at 20 kb, 1020° ± 15°C, but he referred to uncertainties of interpretation. *Stewart* [1967] determined the curve for anorthite to 10 kb, and this was extended to higher pressures by *Boettcher* [1970].

curves involving orthoclase solid solution and albite solid solution. The latter two curves meet at an invariant point where the last of the albite is replaced by jadeite. The light line shows the high-pressure reaction which replaces them. Details of the low-pressure portions of the curves were reviewed by *Scarfe et al.* [1966]. The assemblage $Ab_{ss} + Or_{ss} + Qz$ corresponds to a granite, and $Ab_{ss} + Or_{ss}$ corresponds to a syenite.

Rock-water systems. Figure 4 compares curves for the beginning of melting of major rock types in the presence of excess water [*Merrill et al.*, 1970]. The rock names are used as compositional terms. Curves for granite, tonalite, and gabbro, representing acid, intermediate, and basic members of the calc-alkaline suite, respectively, are roughly parallel. They are separated by only 30°C between 8 and 15 kb. Below about 15 kb, increased pressure causes progressive depression of the temperature, but this is reversed when the feldspars break down to dense phases such as jadeite, zoisite, and kyanite (compare Figure 2). At higher pressures, melting curves become similar in slope to those for dry silicates and dry rocks. The syenite curve transferred from Figure 3 crosses the curves for gabbro and tonalite and terminates at about 16.5 kb, where albite becomes unstable (compare Figure 2b). At higher pressures, the minerals present in syenite bulk compositions are the same as in granite, so both rock compositions have the same curve for the beginning of melting.

The melting behavior of these rocks (with the exception of peridotite) at the solidus is dominated by feldspars and quartz. The gabbros that have been studied at mantle pressures in the presence of excess water produce quartz when the plagioclase breaks down, so that their high pressure product is quartz eclogite. The peridotite in Figure 4 consists of olivine, two pyroxenes, and spinel; garnet appears at about 30 kb. *Kushiro et al.* [1968] show the peridotite-water curve passing through a temperature minimum in the region of 30 kb. This is related to the transformation of spinel peridotite into garnet peridotite and does not involve feldspars.

Experimental studies have also been published for melting relationships of various sedimentary and metamorphic rocks in the presence of excess water. *Wyllie and Tuttle* [1961a]

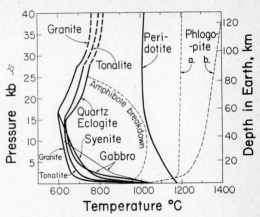

Fig. 4. Solidus curves for rocks in the presence of excess water compiled by *Merrill et al.* [1970]. Granite-H_2O is after *Piwinskii* [1968] and *Boettcher and Wyllie* [1968]. Tonalite-H_2O is after *Piwinskii* [1968] and *Lambert and Wyllie* [1970a and unpublished manuscript]. Gabbro-H_2O is after *Lambert and Wyllie* [1968, 1970a, and unpublished manuscript], *Hill and Boettcher* [1970], *Tuthill* [1969] at 5 kb, and *Holloway and Burnham* [1969] at 2, 5, and 8 kb. Light curve is after *Yoder and Tilley* [1962] for olivine tholeiite. Syenite-H_2O is based on Figure 3. Peridotite-H_2O is after preliminary curve of *Kushiro et al.* [1968]. Curves for the breakdown of amphibole and phlogopite are shown by dashed lines. Amphibole in gabbro with excess water is after *Lambert and Wyllie* [1968]. *Hill and Boettcher* [1970] have additional data. Amphibole in peridotite and in vapor-absent region is stable to higher pressures and temperatures: see *Wyllie* [1970]. Phlogopite *a* is in the presence of excess water (low temperature) and phlogopite *b* is in the vapor-absent region (higher temperature) [after *Yoder and Kushiro*, 1969].

studied shales; Winkler and his associates published a series of papers between 1957 and 1961. The results of these experimental studies on granitic rocks and on sedimentary and metamorphic rocks that produce granitic liquids on partial melting were reviewed by *Piwinskii and Wyllie* [1968]. The solidus curves for a wide variety of crustal rock types are situated close to, or just above (a few tens of degrees), the solidus for granite given in Figure 4.

Effects of other volatiles and salts in solution. Pore fluids that may be involved in magma generation contain ingredients additional to water. Carbon dioxide is the next most abundant volatile component in the crust, but small proportions of other components may be very influential. Early experiments by *Wyllie and Tuttle* [1959, 1960*b*] on the effect of CO₂ and CO₂-H₂O mixtures on the melting temperatures of albite and granite produced only qualitative results; they showed that addition of CO₂ at constant total pressure raises the temperature of beginning of melting. This conclusion has since been substantiated quantitatively by *Holloway and Burnham* [1969], *Hill and Boettcher* [1970] *and Millhollen* [1971]. *Wyllie and Tuttle* [1960*a*, 1961*b*, 1964] also studied the effect of other volatiles, in addition to water, on the melting temperatures of albite and granite. The temperature of beginning of melting of granite in the presence of pure water is lowered

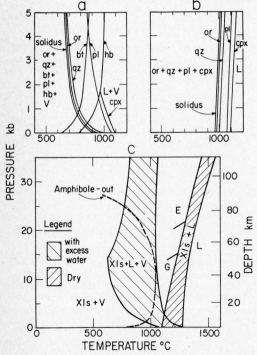

Fig. 5. Pressure-temperature projections contrasting the water-excess and dry melting relationships of rocks typical of the earth's crust and mantle [from *Robertson and Wyllie, 1971b*]. 5*a* shows water-excess melting relations of granodiorite 766 extrapolated to 5 kb [after *Piwinskii and Wyllie, 1968*]. 5*b* shows inferred melting relations for the anhydrous equivalent of granodiorite **766**. 5*c* shows water-excess and dry melting relations for gabbro [after *Lambert and Wyllie*, 1970*b*]. Abbreviations: bt is biotite, hb is amphibole, pl is plagioclase, cpx is pyroxene, G is gabbro, E is eclogite, xls is crystals. For others, see Figure 2.

by the addition of HF, P₂O₅, and SO₃ and is raised by the addition of HCl [see *von Platen*, 1965] and NH₃. The lowering of melting temperatures is appreciable (50°C or more) for small concentrations of HF and Li₂O, but higher concentrations of the other volatiles are required to raise the melting temperatures significantly.

In order to overcome the problems introduced in these studies by chemical reaction and breakdown of the feldspars by the acid solutions, the additional components were added in the form of salts. Melting relationships in the systems Ab-Na₂CO₃-H₂O, Ab-NaF-H₂O, and Ab-NaCl-H₂O have been published by *Koster van Groos and Wyllie* [1968*a*, *b*, 1969]. The NaF and NaCl systems confirmed the conclusions reached from the reconnaissance studies using HF and HCl; the presence of NaF in solution lowers the temperature of beginning of melting of silicates considerably, and NaCl lowers it by a few degrees only. *Burnham* [1967] reached similar conclusions from experiments with the Harding pegmatite and with the Spruce Pine pegmatite in the presence of a solution of NaCl+KCl.

The temperature of beginning of melting in the crust in the presence of impure pore fluids can thus be lowered or raised compared to the granite-water solidus curve in Figure 4, but the amount of melting produced at lower temperatures in the presence of the small proportion of pore fluid that is likely to be present is probably insignificant in terms of magma generation.

MELTING INTERVALS OF ROCKS: DRY AND WITH EXCESS WATER

Determination of the solidus curves for the beginning of melting of rocks is only the beginning of elucidation of the problem of magma generation. The first detailed experimental study of the phase relationships through the melting interval of a rock was by *Yoder and Tilley* [1962] on basalt compositions, with water excess (type IV) to 10 kb and water-absent (type I) at greater pressures. High-pressure studies on other dry rocks ranging in composition from acid through basic to ultrabasic have been presented and reviewed by *D. H. Green and Ringwood* [1967], *T. H. Green and Ringwood* [1968], *Cohen et al.* [1967], *Ito and Kennedy* [1967, 1968], *O'Hara* [1968], and *Kushiro et al.* [1968].

Phase relationships through the solid-liquid-vapor interval of crustal rock-water systems with excess water have been presented and reviewed by *Piwinskii and Wyllie* [1968, 1970], *Piwinskii* [1968], *Gibbon and Wyllie* [1969], *McDowell and Wyllie* [1971], and *Robertson and Wyllie* [1971a]. *Lambert and Wyllie* [1970a] studied tonalite-water and gabbro-water for the range between 10 and 25 kb. *Wyllie* [1970] reviewed the phase relationships for the systems gabbro-water and peridotite-water at mantle pressures, and *Kushiro* [1970] provided additional data for peridotite-water. *Hill and Boettcher* [1970] presented results for the system gabbro-water-carbon dioxide with excess vapor. Results for the rock-water systems are outlined below.

Granodiorite-water. Figures 5a and 5b show partly schematic phase relationships for a granodiorite that is representative of the dominant plutonic igneous rocks of the crust. Figure 5b shows the dry condition, in which the effect of pressure is to raise the melting temperature of silicates. For excess water, Figures 5a and 9a show that the effect of pressure is to lower the melting temperature. The phase relationships in Figures 5a and 9a, based on experimental studies by *Piwinskii and Wyllie* [1968], show a standard pattern (somewhat simplified) for calc-alkaline igneous rocks of intermediate composition. In some rocks of similar composition pyroxene is the liquidus mineral, and it may coexist with amphibole down to subsolidus temperatures [*Robertson and Wyllie*, 1971a]. The minerals quartz and potash-feldspar and the sodic portion of the plagioclase yield a granite liquid within a narrow temperature interval above the solidus, and biotite and hornblende persist to temperatures where they undergo reactive solution. This behavior probably occurs through a temperature interval; the curves show only the upper temperature limit for the minerals under the specific experimental conditions used by *Piwinskii and Wyllie* [1968, 1970]. Biotite and amphibole would react at lower temperatures with different oxygen fugacities [*Tuthill*, 1969; *Wones and Eugster*, 1965]. The lime-rich plagioclase coexists with the granite liquid through a considerable temperature interval.

The phase relationships for granodiorite composition under dry, water-absent conditions have

not been determined experimentally, but they must be similar to those depicted in Figure 5b. It is assumed that the water in the hydrous minerals biotite and hornblende has been removed by dehydration; clinopyroxene is shown as the major dehydration product [*Robertson and Wyllie*, 1971b].

If excess water is added to the dry granodiorite composition, the main differences in phase relationships produced at pressures of up to 10 kb are that: the solidus and liquidus temperatures are lowered; the temperature interval between liquidus and solidus is increased; and hydrous minerals, biotite and hornblende, are stabilized and become involved in reactions including liquid. These differences reported for granodiorite are standard for crustal rocks.

Gabbro-water and peridotite-water. Figure 5c compares the melting interval for dry gabbro with that for gabbro in the presence of excess water, as well as the stability range of amphibole (see also Figure 4 for amphibole and phlogopite stability ranges). Mineralogical details within the melting intervals are shown in Figure 10a. Figure 5c is based on experimental work by *Green and Ringwood* [1967] and *Cohen et al.* [1967] for dry gabbro and by *Lambert and Wyllie* [1968, 1970a, b] and *Hill and Boettcher* [1970] for gabbro-water.

The gabbro exhibits the same three differences listed for granodiorite, but two other factors are introduced at higher pressures corresponding to mantle conditions: first, the slope of the solidus for gabbro-water is reversed, so that with increasing pressure in the presence of excess water the temperature of beginning of melting increases (see Figures 2 and 4); second, at high pressures the breakdown temperature for amphibole decreases. At pressures greater than 25 kb the amphibole breaks down below the solidus, and there are no hydrous minerals involved in reactions including liquid.

Preliminary results by *Kushiro* [1970] show a similar pattern for the system peridotite-water; the curve for amphibole stability just overlaps the solidus curve [*Kushiro et al.*, 1968] between 8 and 16 kb (see Figure 11b). These pressure limits may vary considerably as a function of bulk composition and amphibole composition.

Rock series gabbro-granite-water. The earth's crust is composed primarily of the calc-alkaline rock series gabbro-diorite-granite and their metamorphosed equivalents, represented by the curved line in Figure 1. Experimental investigations of individual rock specimens have yielded much useful information, but we are usually concerned more with the origin of series of associated rocks. The phase relationships through the melting interval of igneous rock series in the presence of excess water have been determined by *Piwinskii and Wyllie* [1968, 1970], *Piwinskii* [1968], *Gibbon and Wyllie* [1969], *McDowell and Wyllie* [1971], and *Robertson and Wyllie* [1971a].

Figure 6 summarizes the results at 2-kb pressure obtained from nine igneous rocks from the Wallowa batholith, Oregon [*Piwinskii and Wyllie*, 1968, 1970]. The diagram was constructed as follows. The composition of each rock is represented by a specific value of the differentiation index (individual rocks are not shown in Figure 6). For each rock, the phase boundaries on a pressure-temperature diagram such as Figure 5a were determined experimentally. From these diagrams, the temperatures of each phase boundary at 2-kb pressure

were measured and plotted on Figure 6. The lines through these phase boundaries, as drawn in Figure 6, provide the isobaric temperature-composition section showing how the pattern of phase relationships changes as the rock composition changes from granite through granodiorite to tonalite; the results have been extended to connect with the results of *Yoder and Tilley* [1962] for a calc-alkaline basalt.

The solidus curve for the beginning of melting (see Figure 4 at 2 kb) and the curves for the upper temperature limit of quartz and potash feldspar form a rather narrow band extending from the granites to the granodiorites and tonalites of the main pluton. Basic rocks with lower differentiation indices do not contain potash feldspar. The solution of potash feldspar and quartz, together with the appropriate fraction of plagioclase, yields a liquid of granite composition within 50°C of the solidus. This represents complete fusion of the granite (number 705 in Figure 1; differentiation index of 93 in Figure 6), but for the other rocks, crystalline plagioclase, biotite, and hornblende persist to higher temperatures. *Piwinskii and Wyllie*

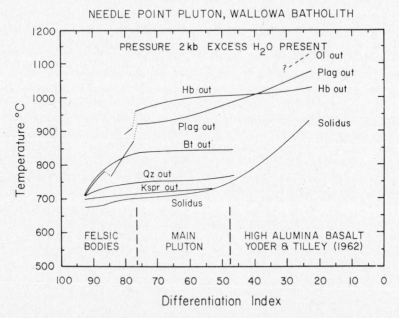

Fig. 6. Temperature-differentiation index diagram for the granitic rocks of the Needle Point pluton, Wallowa batholith, at 2-kb water pressure [from *Piwinskii and Wyllie*, 1970]. Phase boundaries for rocks on the main pluton are connected to results obtained by *Yoder and Tilley* [1962] for a high alumina basalt [*Piwinskii and Wyllie*, 1968]. Abbreviations: ol is olivine, plag is plagioclase, Kspr is alkali feldspar. For others, see Figures 5 and 2.

[1970] showed that the plagioclase composition changed only slightly through most of the melting interval, once the granite liquid had formed. The granite liquid persists with little increase in amount and little change in composition through an interval as wide as 150° to 200°C, in equilibrium with residual plagioclase and hornblende and biotite until this reacts with the liquid and disappears. Figure 6 shows that, even with excess H_2O, temperatures of the order of 900° to 1000°C at 2 kb are required to complete the fusion of rocks of intermediate composition.

MELTING IN WATER-DEFICIENT REGION OF SILICATE-WATER SYSTEMS

The effect of water content on the reactive solution temperature of a hydrous mineral in the water deficient region is illustrated by the pairs of univariant curves in simple systems for the reactions involving liquid with an aqueous vapor phase present (type IV) and with no vapor phase present (type II). Reaction pairs of this kind have been presented for the system CaO-H_2O [*Wyllie and Tuttle*, 1960c] for muscovite-quartz-H_2O [*Segnit and Kennedy*, 1961; *Lambert et al.*, 1969], for analcite in the system NaAlSiO₄-SiO₂-H_2O [*Peters et al.*, 1966; *Boettcher and Wyllie*, 1969], for phlogopite [*Yoder and Kushiro*, 1969], for phlogopite-enstatite mixtures [*Modreski and Boettcher*, 1970], and for hydrous minerals in the system CaO-Al₂O₃-SiO₂-H_2O [*Boettcher*, 1970]. In all these systems (except for CaO-H_2O at pressures less than 200 bars) the melting temperature is higher for the type II assemblage, with no vapor, than it is for the type IV assemblage, with vapor.

There are three kinds of melting behavior patterns possible for the type II assemblages, without vapor. First, the hydrous mineral melts incongruently to yield a liquid containing less dissolved water than that combined in the mineral; this is the low pressure behavior of the system CaO-H_2O, and it is doubtful whether this occurs in any process of melting within the crust and mantle. Second, the hydrous mineral melts incongruently to a liquid containing more dissolved water than that combined in the mineral; muscovite appears to be an example. Third, the hydrous mineral melts congruently to yield a liquid of its own composition. According to *Yoder and Kushiro* [1969], phlogo-

pite melts incongruently (the second behavior pattern) at low pressures, and this changes to congruency above about 2 kb.

Experimental results for solid-liquid phase relationships in water-deficient systems with water content controlled are scarce. *Yoder and Kushiro* [1969] determined the melting relationships of phlogopite in this way and applied the type II relationships to magma generation in the mantle. *Whitney and Luth* [1970] worked with granite, and *Robertson and Wyllie* [1971a] studied two syenites under these conditions. *Hill and Boettcher* [1970] and *Eggler* [1970] used H_2O-CO_2 gas mixtures with gabbro and andesite, respectively. They achieved similar conditions, with $P_{eH_2O} < P_{total}$, but with a mixed vapor phase present. Both found that the stability temperature of amphibole increased as the water content of the liquid decreased; this finding agrees with the conclusions reached from the simple systems. (For the definition of P_{eH_2O}, the equilibrium water pressure, see *Greenwood* [1961]. P_{eH_2O} is similar to the more familiar P_{H_2O}, the partial pressure of water, but is more rigorously defined.)

Burnham [1967] used type IV results obtained with excess water to deduce the phase relationships in terms of P_{eH_2O} where this varied independently of P_{total}, and he presented a detailed discussion of magma generation and uprise in the mantle and crust for water-deficient rocks in the presence of a multicomponent vapor phase.

Despite the paucity of experimental data, the positions of the phase boundaries in water-deficient silicate-water systems can be estimated by interpolating between the dry (type I) and the water-excess (type IV) experimental results.

Vapor-present and vapor-absent phase fields. Figure 7 was used by *Robertson and Wyllie* [1971b] to illustrate the distribution of vapor-present and vapor-absent fields in silicate-water systems and the melting assemblages of types I, II, III, and IV. The saturation boundary at constant pressure is defined as the line showing the amount of water required at each temperature to saturate the phase assemblage; this is the line between the shaded and unshaded fields. Equivalent points are denoted by the same symbol (a, b, c, m, et cetera). $P_{eH_2O} = P_{total}$ for all assemblages with vapor, and $P_{eH_2O} < P_{total}$ in the vapor-absent regions.

Figure 7a is an isobaric section through a

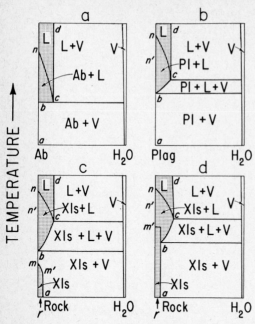

Fig. 7. Schematic isobaric temperature-composition sections for silicate-water systems [after *Robertson and Wyllie*, 1971b]. Shaded areas are vapor-absent. 7a shows a section through the system albite(Ab)-water (H_2O). Type II assemblages are absent. Compositions with less water than c are type III; those with more water than c are type IV. 7b shows isoplethal section through the ternary system albite-anorthite-water for an intermediate plagioclase (plag) composition. Note the field for plagioclase + liquid + vapor and the fact that bulk compositions between b and c will initially melt into this field; with increased temperature the vapor phase is dissolved in the liquid and the saturation boundary, b–c, is crossed. 7c shows a schematic section for a rock-water system in which a hydrous mineral breaks down below the temperature of the solidus. The rock composition is given by the point r. Note the similarity of the pattern of phase relations with those in Figure 7b. 7d shows a schematic section for a rock-water system in which the hydrous mineral dehydrates at a temperature above that of the solidus. Type II assemblages contain less water than a. Type IV assemblages contain more water than c, and type III assemblages are intermediate between a and c. For abbreviations see Figures 5 and 2.

binary silicate-water system with no hydrous minerals. The shaded vapor-absent region is delineated by the solidus b–c and the saturation boundary c–d. The point n gives the melting temperature of dry albite. This shows systems of types III and IV only.

Figure 7b is a section through the system $CaAl_2Si_2O_8$-$NaAlSi_3O_8$-H_2O for a given plagioclase composition. The phase fields intersected are ternary. The dry plagioclase feldspar melts through the temperature interval n'–n rather than at a specific temperature n. There is an additional phase field compared to Figure 7a, plagioclase + liquid + vapor. When vapor-present type III assemblages begin to melt, they pass through a melting interval with vapor and a vapor saturated liquid, until the temperature exceeds the saturation boundary b–c; then they pass into the vapor-absent region.

Figures 7c and 7d show a rock composition r. If this rock is completely dehydrated, it becomes a type I system; this rock melts in the temperature interval n'–n. If the rock is held at some temperature below the solidus in the presence of excess water vapor under pressure, additional hydrous minerals may be formed, changing the rock composition to point a. The shaded area within the saturation boundary a–b–c–d is vapor absent. Water-deficient systems of type II are shown by the shaded area for 'crystals.' Subsolidus assemblages between a and the H_2O axis include systems of types III and IV. The point m' gives the temperature at which dehydration of the hydrous mineral assemblage begins.

Figure 7c shows dissociation completed in the temperature interval m'–m, before the solidus temperature is reached, and the type II assemblages therefore do not melt; for all compositions, melting begins in the presence of excess vapor. The point c gives the amount of water required to saturate the liquid when the rock is completely melted. Therefore, for temperatures above point m, the compositions between b and c are water deficient type III, and the others are water excess, type IV.

In Figure 7d, compositions in the shaded area to the left of line a–b are water-deficient type II, compositions between points b and c are water-deficient of type III, and those on the H_2O side of point c are water-excess, type IV. The hydrous minerals are stable to temperatures higher than the solidus temperature for types III and IV; dehydration of the hydrous mineral assemblage begins at m'; a small amount of H_2O-undersaturated liquid is thus produced. The shape of the saturation boundary between b and c is specific for every rock.

Figures 7c and 7d permit comparison of the

melting relationships in systems of types I, II, III, and IV. Note the high temperature and narrow temperature interval $n'-n$ for $xls + L$ in type I. The solidus temperature for type II, if this assemblage does melt as in Figure 7d, coincides with the temperature of beginning of dehydration of the hydrous mineral assemblage and is thus independent of types I, III, and IV. The solidus temperatures for the vapor-present types III and IV are identical and much lower than for type I; the melting temperature is independent of the amount of water, and the initial liquid is saturated with water, $P_{H_2O} = P_{total}$. The melting pattern of type III assemblages is initially the same as that of type IV assemblages. Melting of type IV is completed with excess vapor present throughout, but for type III the vapor is dissolved where the temperature reaches the saturation boundary, $b-c$; further increase in temperature yields a vapor-absent assemblage; the liquid becomes progressively more under-saturated with water, and $P_{\epsilon H_2O}$ decreases. The temperature of the saturation boundary $b-c$ for type III assemblages depends sensitively on the water content and the mineralogy of the rock. The liquidus temperature in the vapor-absent region is lowered with increasing water content from type I to type IV.

The main variables in the basic pattern of Figures 7c and 7d are: the temperature of beginning of dehydration of the hydrous minerals, at m', compared with the vapor-present solidus temperature, at b; and the amount of water, at c, required to saturate the liquid (which increases with pressure), compared with the amount of water required to produce maximum subsolidus hydration of the rock, at a.

For details of the composition and amount of liquid produced during progressive fusion, we need detailed phase relationships within the melting interval in the water-deficient assemblages for specific rocks. These are considered below.

Granodiorite-water. Figure 8 is a schematic diagram corresponding to Figure 7d for a specific rock, granodiorite, with estimated field boundaries in the vapor-absent region. The phase relationships in the system granodiorite-water for type IV conditions are shown in Figure 5a, and for type I conditions they are shown in Figure 5b. Figure 8 was obtained by *Robertson and Wyllie* [1971b] by interpolation

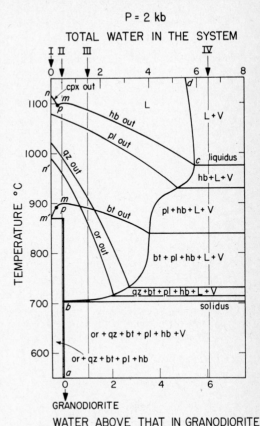

Fig. 8. Schematic isobaric T-X section for granodiorite 766-H_2O showing the estimated upper stability limits for the individual minerals in the vapor-absent region, obtained by interpolation between dry type I results and H_2O-excess type IV results. [after *Robertson and Wyllie*, 1971a, b]. There is not much variation possible in the positions of boundaries for the anhydrous minerals, but there is more uncertainty for amphibole and biotite. The congruent reactions for amphibole and biotite are modelled on results for phlogopite [*Yoder and Kushiro*, 1969]; incongruent reactions without the maximum m are alternate arrangements. The temperatures of biotite and amphibole boundaries are sensitive to the mineral compositions and to oxygen fugacity. In other rocks, and possibly in this rock, the liquidus phase in the vapor-absent region may change as a function of water content; pyroxene and plagioclase may occur on the liquidus. The temperature m' where biotite begins to dehydrate in the type II assemblages could be considerably lower in temperature than indicated, relative to the completion of dehydration at m. For abbreviations see Figures 5 and 2.

between Figures 5a and 5b at 2 kb. Isothermal boundaries for the solidus and liquidus, and the upper limit of each mineral extend from the water-excess into the water-deficient region as far as the saturation boundary, b–c. The shape of the saturation boundary was determined from estimates of the percentage of liquid at each temperature between the solidus and liquidus with excess vapor and the known solubilities of water in similar silicate liquids at each temperature. The dry rock at this pressure melts between the points n' and n, taken from solidus and liquidus in Figure 5b.

Field boundaries in the vapor-absent region were interpolated from the water-absent points between n' and n to the corresponding points on the saturation boundary between b and c. There is very little variation possible from the estimated curves shown for feldspars and quartz. More variation is possible for biotite and hornblende, because they have no stability on the water-absent axis, and their stability temperatures in the vapor-absent region have not been measured experimentally.

The estimated positions of the biotite-out and hornblende-out curves in Figure 8 are based on the review of simple systems and rock-H_2O-CO_2 systems at the beginning of this section. They increase in temperature as water content decreases within the vapor-absent region. The temperature maximums m are modelled on the known phase relationships for phlogopite [Yoder and Kushiro, 1969]. The temperature interval m'–m represents the reaction interval of biotite. An alternate arrangement is an incongruent relationship with the hydrous minerals yielding H_2O-unsaturated liquids containing more dissolved H_2O than the percentage combined in the minerals. These curves need experimental determination.

In other granodiorites, with excess water, the liquidus phase could be pyroxene rather than amphibole, and the relative stability limits of amphibole and plagioclase could be reversed. The temperatures of reactive melting of amphibole and biotite can be lowered by changing the oxygen fugacity, as mentioned above in connection with Figure 5a. The simple picture in Figure 8, with amphibole occurring as the liquidus phase until very low water contents, may thus be considerably modified. The temperature variation of the vapor-absent liquidus and the

sequence of liquidus phases, which may vary as a function of water content, merits detailed experimental study for different rock types.

Note two features in the vapor-absent region that contrast with the vapor-present fields. First, the temperature interval for the coexistence of two feldspars with liquid increases, and second, quartz is stable with liquid through a significant temperature interval. Note also that type III assemblages, with up to about 3% water in a vapor phase, exhibit only a narrow temperature interval above the solidus within which the vapor persists. Within a few degrees, all the water dissolves in the liquid, and with further increase in temperature the assemblage is vapor-absent, with H_2O-undersaturated liquid.

Figure 9 contrasts the melting interval for a rock in the earth's crust under conditions of excess water (Figure 9a) with that for vapor-absent conditions of type II (Figure 9b). Figure 9a is an extrapolation of Figure 5a for a granodiorite with excess water to 10 kb; the temperatures of the boundaries at 2 kb correspond to those indicated for composition IV in Figure 8. Figure 9b is a schematic diagram for the same granodiorite with only 0.4 weight per cent combined H_2O; the temperatures of the boundaries at 2 kb are those indicated for composition II in Figure 8. Extrapolation of the boundaries in Figure 9b to higher and lower pressures is based on Figure 9a and on the differential between temperatures for compositions IV and II in Figure 8. A vapor phase is present only in a limited low pressure region between a and d, where biotite begins to dissociate subsolidus and between d and e, where the liquid is just saturated with H_2O. At all higher pressures the liquid is unsaturated with H_2O and $P_{eH_2O} < P_{total}$.

In Figure 9b, amphibole is given as the liquidus phase above 2-kb pressure, but as discussed in connection with Figure 8, this could equally well be pyroxene or plagioclase. Figure 9b is schematic, but it illustrates significant differences for melting of type II assemblages compared with the patterns usually considered in magmatic models.

There are three significant differences between Figures 9a and 9b. If there is no vapor phase present in the subsolidus assemblage, first, the slope of the solidus boundary is reversed; it approaches parallelism with the slope of the dry rock boundary (Figure 5b). Second, the solidus temperature is greatly increased. Third, the fusion of alkali feldspar, quartz, and the appropriate fraction of plagioclase to yield a granite liquid occurs through a temperature interval of more than 100°C instead of about 30°C.

Brown and Fyfe [1970] have presented an experimentally determined solidus curve for a biotite-diorite under vapor-absent conditions of type II that confirms the arrangement illustrated in Figure 9b. They reported that liquid developed where biotite begins to dehydrate (see *m'* in Figure 8) along a linear solidus passing through the points 740°C at 2 kb and 815°C at 10 kb. For comparison, the estimated solidus in Figure 9b is about 90°C to 130°C higher in temperature. The difference in temperature could be due to several causes. *Brown and Fyfe* [1970] reported that the biotite reacted through a wide temperature interval, and possibly *m'* in Figure 8 and the solidus in Figure 9b should be located farther below the biotite-out curves. The dehydration temperature of biotite is strongly dependent on its composition and the oxygen fugacity [*Wones and Eugster*, 1965], and it is likely that both these factors differ between the experiments of *Piwinskii and Wyllie* [1968], on which Figures 8 and 9a are based, and those of *Brown and Fyfe* [1970]. The runs of Brown and Fyfe were buffered, but no experimental data were given in their preliminary report.

Gabbro-water and peridotite-water. Figure 5c compares the melting interval of dry gabbro with that in the presence of excess water and illustrates the main differences between the systems of types I and IV. Figure 10a gives the details of the phase relationships (major minerals only) within these melting intervals (I. B. Lambert and P. J. Wyllie, 1970, manuscript in preparation). From Figure 10a the general pattern of phase relationships for gabbro in the presence of a trace of water can be deduced [*Lambert and Wyllie*, 1970b; *Wyllie*, 1970], and the result is shown in Figure 10b. This isoplethal section is for gabbro-water with insufficient water present to saturate the rock in the pressure interval from about 1 to 25 kb, where the amphibole stability curve occurs above the solidus for the water-excess system. Within this pressure interval, conditions correspond to type II in Figure 7d. Below 1 kb and

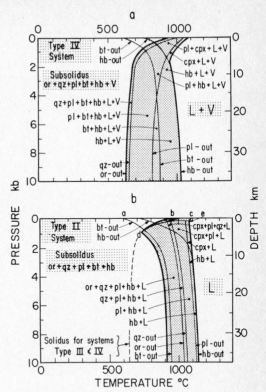

Fig. 9. Figure 9a is an extrapolation of Figure 5a to 10-kb pressure, showing the melting interval of a granodiorite with excess water. This is based on experimental measurements to 3-kb pressure under conditions specified by *Piwinskii and Wyllie* [1968]. 9b shows schematic melting relationships of the granodiorite in Figure 9a under conditions for composition II in Figure 8; no vapor phase is present except for the small low pressure region above a-d-e. The solidus at pressure greater than d is the locus of the point *m'* in Figure 8, where biotite begins to dehydrate. Estimation of the temperature of *m'*, relative to the temperature of complete reaction of biotite, suffers from the uncertainties discussed for Figure 8. The relationship of this type II solidus, involving biotite reaction, to the vapor-present solidus (types III and IV) is similar to that determined experimentally for a biotite-diorite by *Brown and Fyfe* [1970]. For abbreviations see Figures 5 and 2.

above 25 kb, conditions correspond to type III in Figure 7c. Note that this change from type II to type III occurs with no change in water content; it is caused by the changing temperature of dehydration of amphibole-bearing assemblages as a function of pressure (dehydration interval *m'-m* in Figures 7c and 7d).

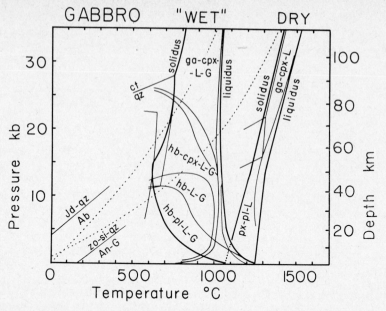

Fig. 10a Experimentally based phase relationships in gabbro [*Cohen et al.*, 1967; *Green and Ringwood*, 1967], and gabbro in the presence of excess water (I. B. Lambert and P. J. Wyllie, unpublished data; see *Lambert and Wyllie* [1968, 1970a, b]; and *Hill and Boettcher*, [1970]). The interval for the coexistence of amphibole (hb) and garnet (ga) is probably too narrow, because of the reluctance of garnet to nucleate; with seeded runs, the garnet curve would probably be determined at lower temperatures. The presence of epidote or zoisite, and kyanite or sillimanite, was suspected in high pressure runs but not proven. This shows detail for Figure 5c). Dotted lines are estimated geotherms from *Clark and Ringwood* [1964]. Abbreviations: ga is garnet, cpx is clinopyroxene, px is pyroxene, pl is plagioclase, hb is amphibole, qz is quartz, ct is coesite, Jd is jadeite, Ab is albite, zo is zoisite, si is sillimanite, An is anorthite, L is liquid, G is gas or vapor phase, aqueous (V in earlier Figures).

In Figure 10b, only a trace of silicate liquid is present within the large area between the solidus and the position of the dry solidus given in Figure 10a. Only at temperatures approaching the dry solidus does significant melting occur in the water-deficient phase diagram of Figure 10b. Thus, the effect of a trace of water on the melting of gabbro through a wide pressure range (about 1 to 25 kb) is the stabilization of amphibole and the production of incipient melting of the gabbro above temperatures where the amphibole begins to dehydrate. The main melting interval appears to be modified only slightly compared with that for the dry rock (Figure 10a).

Figure 11a shows a long extrapolation of the isopleth in Figure 10b, represented as a function of depth within the earth, for a hypothetical mantle composed of gabbro(eclogite) with 0.1% water. Figure 11b shows a similar diagram for a peridotite mantle. The construction of Figure 11 is described elsewhere [*Wyllie*, 1971a]. The subsolidus phase transformations need not concern us here. The portion of Figure 11a down to 100 kb is easily related to Figure 10b. I assume arbitrarily that the solidus with water present becomes parallel with the dry solidus at pressures corresponding to depths greater than about 200 km (see Figure 4). The melting intervals in Figures 11a and 11b are in two parts; the darkly shaded bands correspond closely to the dry melting intervals, and the lightly shaded bands show where there is a trace of silicate liquid. These latter bands show where incipient melting occurs in the presence of traces of water. On the low temperature side of the curves labelled *Hb*, all the water is stored within the amphibole (*Hb*). The position of the *Hb* curve relative to the solidus for peridotite-water is based on data by *Kushiro* [1970].

MELTING IN THE CRUST AND UPPER MANTLE

Figures 9, 10, and 11 show the conditions for melting of rock-water systems at various depths within the crust and upper mantle. We should note again that these only illustrate limiting conditions, that natural pore fluids contain components additional to water, that natural rock-fluid systems may be open with respect to volatile components, and that dynamic processes involving transportation of solid or magmatic material are involved in magma generation and differentiation. If we knew precisely the depths of magma generation, the compositions of the rocks involved, the compositions and amounts of pore fluids, and the extent of migration of pore fluids in and out of the region, we could attempt to reproduce these conditions in the laboratory and determine the precise conditions of melting. None of these factors is adequately known; therefore, the results of rock-water systems under closed conditions are used as models to show patterns of progressive fusion or crystallization. These systems probably provide quite reliable guides to natural processes, because water is generally considered to be the dominant volatile component in the crust and upper mantle; additional components in pore fluids will change the temperatures of melting and other reactions, but many sequences of reactions remain unchanged.

Rock-water systems with excess water (type IV) have provided the basis for anatectic models in the crust, and the results of dry rock studies (type I) at high pressures have been applied to mantle models of magma generation. In both crust and mantle, models for magma generation are more realistically based on water-deficient rock-water systems of types II and III.

Melting in the crust. The bitter debate about the origin of granites and batholiths subsided when *Tuttle and Bowen* [1958] presented an anatectic model based on melting experiments in type IV feldspar-quartz-water and granite-water systems as a reasonable compromise between the views of magmatists and transformists. This was extended by *Wyllie and Tuttle* [1960a] to include sedimentary rocks, metamorphosed and melted in the laboratory, and to include the effects of volatiles other than pure

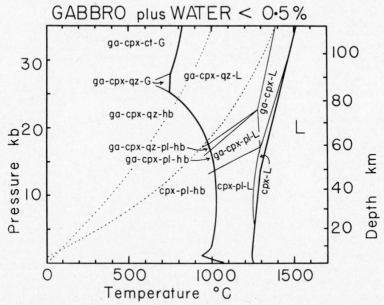

Fig. 10b. Estimated isopleth for gabbro in the presence of a small proportion of water, 0.5% by weight, based on results in Figure 10a [see *Lambert and Wyllie, 1970b; Wyllie, 1970*]. The solidus between 1 and 25 kb is the curve where amphibole in the vapor-absent rock begins to dehydrate (type II system). Notice the subsolidus transition from amphibole-gabbro to amphibole-quartz eclogite. Dotted lines are geotherms. Abbreviations as above. Compare Figure 11a.

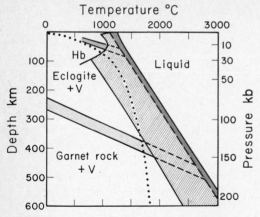

Fig. 11a. Schematic isopleth for gabbro in the presence of 0.1% water, based in part on extrapolation from figure 10b [after *Wyllie*, 1971a]. The melting interval consists of two parts: the light-shaded band represents incipient melting, and the dark-shaded band is equivalent to normal melting, as in Figure 10a. The dotted line is oceanic geotherm calculated by *Clark and Ringwood* [1964]. Hb marks the curve for dehydration of amphibole.

water. *Winkler* [1967] incorporated results from his laboratory on the melting of sediments and metamorphic rocks into the anatectic model, and he reviewed additional aspects. *Burnham* [1967] presented a detailed account of magma generation in the crust under conditions with $P_{sH_2O} < P_{total}$.

The granite-water solidus in Figure 4 is usually considered to represent the minimum temperature for magma formation in the crust and mantle, although traces of interstitial silicate liquid may be produced at lower temperatures in the presence of some pore fluids. Many different rock types in the presence of pore fluids with a wide compositional range will begin to melt within a few tens of degrees of this boundary. Figure 9a for a granodiorite serves as a model for an average crustal rock under the most favorable conditions (type IV).

The depth at which melting begins is given by the point of intersection of a geotherm with the solidus curve. *Tuttle and Brown* [1958] concluded that in orogenic environments melting begins at depths between 20 and 25 km. Interstitial granite liquids would be present in many orogenic rock types at greater depths. There appears to be no way for significant amounts of melting to occur in crustal rocks

on a regional scale at levels shallower than 20 km. This conclusion remains unchanged by more recent results on melting experiments and crustal geotherms [e.g., *Blackwell*, 1971, Figure 5].

The models reviewed above were based on experiments conducted with an excess of water (type IV systems). We have noted that if vapor is present in the crust and mantle the system is water-deficient, corresponding to type III in Figures 7c or 7d and Figure 8. *Burnham* [1967] has discussed a granodiorite anatectic model for conditions equivalent to type III. Figure 8 provides a guide to the pattern of magma generation from the dominant calc-alkaline rock types of the continental crust. The confining pressure of 2 kb corresponds to a depth of about 8 km; the requisite temperatures for fusion would not be attained at this depth, but the essential pattern remains unchanged for greater depths. Similar diagrams for the system gabbro-water at 5 or 10 kb [*Wyllie*, 1971b, Figure 18, chapter 8] could be constructed from available experimental data on dry basalts and on basalt-water

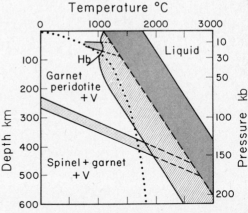

Fig. 11b. Schematic isopleth for peridotite in the presence of 0.1% water [after *Wyllie*, 1971a], based on experimental data by *Kushiro* [1970] for excess water studies, and by analogy with Figure 11a. The dotted line is oceanic geotherm calculated by *Clark and Ringwood* [1964]. Hb marks the curve for dehydration of amphibole. A similar curve for phlogopite, if present, would probably lie at somewhat higher temperatures than the Hb curve, and it would intersect the solidus for type III assemblages at a pressure corresponding to a depth on the order of 250 km, according to extrapolation of the results of *Modreski and Boettcher* [1970].

systems (Figure 10a). These show the melting behavior of gabbro, eclogite, or amphibolite in the lower crust.

The type III composition in Figure 8 represents an intermediate rock or its metamorphic equivalent with 1 weight per cent pore fluid, which is probably a generous estimate for the fluid content of a metamorphic rock in the lower crust. Melting begins with the fusion of quartz and two feldspars and the solution of vapor, producing a water-saturated liquid of granite composition. A small increase in temperature above the solidus produces much additional water-saturated liquid. The pore fluid disappears within a few degrees of the solidus. The temperature interval within which the water saturated liquid exists (the temperature interval between the solidus and the saturation boundary b–c) increases with water content, but it remains small for any reasonable amount of pore fluid.

With further increase in temperature and formation of more liquid, the liquid becomes progressively more unsaturated with water. Within the vapor-absent region, the percentage of liquid developed as a function of temperature is less than that for excess vapor. For a wide temperature interval, about 200°C in Figure 8, the undersaturated granite liquid coexists with quartz, alkali feldspar, plagioclase feldspar, and more refractory minerals until the quartz and alkali feldspar dissolve. Then through another wide temperature interval, the liquid composition becomes closer to that of a granodiorite as the granite liquid dissolves the more refractory minerals. The rock is not completely melted until a temperature of nearly 1100°C is attained.

Water saturated granite liquids produced by anatexis in the crust can exist only for a few degrees above the solidus. It appears that the normal product of anatexis is a mush composed of crystals and water undersaturated granite liquid [Piwinskii and Wyllie, 1970; Robertson and Wyllie, 1971b]. The amount of pore fluid controls the amount of liquid generated from the granitic minerals. P_{eH_2O} for a liquid within the vapor-absent region may be considerably less than the load pressure and, therefore, upward migration of such a liquid, or liquid-crystal magma, can proceed without excessive crystallization until the load pressure is decreased to a level approaching the water pressure in the undersaturated liquid [Burnham, 1967].

In deep crustal environments where a granite melt has formed and migrated upward, carrying with it the original pore fluid, the remaining rocks are type II systems [Lambert and Heier, 1967, 1968; Lambert and Wyllie, 1970a]. Subsequent anatexis can occur only at considerably higher temperature, as shown by the schematic diagrams for granodiorite in Figure 9b and for gabbro in Figure 10b. The liquid so produced is undersaturated with water, and its composition is different from that produced in type III models. There are significant differences in the products of anatexis of crustal rocks depending on whether a pore fluid is present.

At a depth of 20 km, according to Figure 9b, melting begins for a vapor-absent type II granodiorite almost 250°C higher than it does for a vapor-present type III or IV granodiorite. For the biotite-diorite studied by Brown and Fyfe [1970] with oxygen fugacity buffered, the corresponding difference is about 100°C. According to Figure 10, for a vapor-absent type II amphibole-gabbro at a depth of 35 km, melting begins at about 1000°C, a temperature more than 300°C above the solidus for the vapor-present amphibolite of type III or IV. This temperature difference can be reduced if dehydration of the amphibole occurs through a wide temperature interval, beginning at lower temperatures than estimated in Figure 10b.

Brown and Fyfe [1970] outlined a model for magma generation in the crust based on type II systems, with metamorphic rocks yielding first a liquid fraction produced by dehydration of muscovite, followed by a biotite fraction and finally an amphibole fraction. The conditions for the formation of biotite and amphibole fractions from granodiorite and basic rocks are depicted in Figures 9b and 10b.

The high temperatures required for attainment of the liquidus for a granodiorite, about 1000°C with excess water (Figure 9a) and about 1100°C for more reasonable crustal models at a few kbars pressure (types III and II in Figure 8, and type II in Figure 9b), make it unlikely that completely liquid magmas of intermediate composition can be generated by anatexis in the crust. Crustal anatexis may yield a crystal mush of over-all intermediate composition, composed of water undersaturated granite liquid and residual or accumulated crystals. This kind of intermediate magma, although appropriate for batholiths, is not a likely candidate for andesite lavas. For alternate processes, mantle material or mantle conditions must be invoked.

Melting in the mantle. Hypotheses for the derivation of andesite lavas and batholithic magmas from the mantle include: derivation from parent basaltic magma by fractional crystallization, assimilation, hybridism, or some combination of these; mantle anatexis producing primary andesitic magma; and anatexis of crustal material under mantle conditions.

There have been many recent proposals that magmas of intermediate composition are generated from crustal material carried downward into the mantle in Benioff zones [*Coats*, 1962; *Ringwood*, 1969; *Dickinson*, 1970; *Hamilton*, 1969; *Gilluly*, 1969]. For evaluation of this process, we need the equivalent of Figure 8 at a pressure of 15 or 30 kb, taking into account the significant changes in mineralogy at pressures above 15 kb [*Green and Ringwood*, 1968; *Lambert and Wyllie*, 1970a]. Magma generation in this environment could be related to rock-water systems of types II and III. Extrapolation of the results in Figures 8 and 9b to mantle pressures by analogy with data for to-nalite-water [*Lambert and Wyllie*, 1970a] suggests again that temperatures of at least 1100°C would be required to yield liquids of andesite composition. Estimates of temperature distributions within and above downgoing slabs of lithosphere are so low that it is difficult to account for melting of dry rocks except by heat generation by frictional dissipation along the slab margins [*Oxburgh and Turcotte*, 1970; *McKenzie*, 1969; *Minear and Toksöz*, 1970]. The problem requires further attention.

Discussions of melting in the mantle usually deal with the formation of basaltic liquids by partial fusion of dry mantle peridotite. Estimates of temperature in the mantle vary widely, depending on the relative roles attributed to transfer of heat by conduction, radiation, or convection. Comparison of various estimates with the peridotite solidus curve indicates that unusual temperature distributions are required for the generation of basaltic magmas [*Wyllie*, 1971a]. The estimated oceanic geotherm in

Figure 11 does not reach solidus temperatures for dry peridotites or eclogites. Models for the generation of basaltic magmas from dry mantle material have been reviewed by *D. H. Green and Ringwood* [1967], *T. H. Green and Ringwood* [1968], *Ito and Kennedy* [1967, 1968], *Cohen et al.* [1967], *O'Hara* [1965, 1968], *Gast* [1968], *Green* [1969], and *Kushiro* [1969b]. We will not discuss them further.

Figure 11 illustrates the effect of traces of water on melting in the mantle. The presence of amphibole and phlogopite in mantle-derived eclogite and peridotite nodules in kimberlites and basalts has been cited as evidence for water in the mantle [see *Wyllie*, 1970]. G. C. Kennedy maintains that this evidence has been misinterpreted, because the hydrous minerals occur only in the outer parts of nodules, as late alteration products (personal communication, 1970). Nevertheless, the effects of even traces of water in the mantle are so significant that we must evaluate them. Water was certainly present in the mantle at some earlier stage in its history, and according to the new global tectonics it is probably being carried down into the mantle beneath island arcs [*Isacks et al.*, 1968].

Figure 11 shows that for a gabbro, eclogite, or peridotite in the uppermost mantle, traces of water are likely to be combined in hydrous minerals such as amphibole; other candidates as crystallographic storehouses for mantle water include phlogopite, titanoclinohumite [*McGetchin et al.*, 1970], hydroxylated pyroxenes [*Sclar et al.*, 1967], and hydrogarnet [*Fyfe*, 1970]. The availability of water for melting in the mantle is related to the conditions for the breakdown of such hydrous minerals. Figure 11 shows the stability limits for amphibole in mantle eclogite and peridotite and the wide zone of incipient melting below the dry solidus curves, which is caused by traces of water not combined in hydrous minerals. If potassium abundances permit the presence of phlogopite, then extrapolation of the experimental data of *Modreski and Boettcher* [1970] with phlogopite-enstatite mixtures indicates that the solidus curve for a type II phlogopite-peridotite is situated at somewhat higher temperatures than the *Hb*-curve in Figure 11b and that it does not intersect the type III solidus until a pressure corresponding to a depth of the order of 250 km is reached.

The oceanic geotherm in Figure 11 passes through the region for incipient melting within the depth interval of about 80 to 300 km. *Lambert and Wyllie* [1968, 1970a, b], using the results in Figure 10b, interpreted the low velocity zone of the upper mantle in terms of incipient melting in the presence of traces of water. The production of silicate liquid by the presence of water is likely to change the position of the geotherm adopted in Figure 11, which in turn would change the depth interval within which incipient melting could be anticipated. This procedure of using extrapolated experimental data and arbitrarily selected geotherms is intended only to show patterns for the petrology of the mantle; specific temperatures and depths can be modified as required by the acquisition of extended and improved theoretical and experimental data. The available evidence indicates that if there is any water in the upper mantle, even in trace amounts, incipient melting is likely to occur (unless temperatures are very much lower than are usually assumed).

The compositions of magmas produced by the mantle melting in the presence of traces of water have been reviewed by *Kay et al.* [1970], *Green* [1971], and *Wyllie* [1971a, b].

EXTENSIVE LAYERS IN THE CRUST AND UPPER MANTLE

There appears to be no prospect of explaining the existence of an extensive layer at 10-km depth, with specific properties, in terms of melting of crustal rocks. The experimental limits in rock-water systems suggest that melting occurs only at depths greater than 20-km. Geological and petrological studies indicate that significant melting of crustal rocks occurs only in active orogenic belts; the melting is thus limited in space and time, and it is not likely to produce uniform layers with appreciable horizontal extent. *Ramberg* [1967] has demonstrated with centrifuged scaled models that partial melting in the crust is almost certainly followed by upward, diapiric motions of magmas or of mobile rock masses.

If any kind of uniformity on a regional scale is produced by melting processes in the crust, this may be in ancient basement rocks, where the rocks may represent the dessicated residues remaining after partial melting has occurred

and magmas have transported the volatile components upwards and away. The properties of rocks are very sensitive to the presence and quantity of pore fluids, and if all pore fluids have been removed, a relative degree of uniformity in physical properties can be achieved even in metamorphic rocks with different compositions. Such regions would be composed of rocks of type II (see Figure 8).

In the uppermost mantle, there is a layer identified by its physical properties, the seismic low-velocity layer, that may be produced by incipient melting caused by traces of water. This layer approaches the crust, or has possible localized connections with it, beneath oceanic ridges and via downgoing lithosphere slabs in Benioff seismic zones.

Acknowledgments. I thank I. B. Lambert and J. K. Robertson for their contributions to unpublished material included in this article.

The research has been supported by National Science Foundation grants GA-15718 and GA-10459. Apparatus and supplies were provided by Advanced Research Projects Agency grant SD-89.

REFERENCES

Blackwell, D. D., The thermal structure of the continental crust, in *The Structure and Physical Properties of the Earth's Crust, Geophys. Monogr. Ser.*, vol. 14, edited by J. G. Heacock, AGU, Washington, D. C., this volume, 1971.

Boettcher, A. L., The system $CaO-Al_2O_3-SiO_2-H_2O$ at high pressures and temperatures, *J. Petrol.*, 11, 337–379, 1970.

Boettcher, A. L., and P. J. Wyllie, Melting of granite with excess water to 30 kilobars pressure, *J. Geol.*, 76, 235–244, 1968.

Boettcher, A. L., and P. J. Wyllie, Phase relationships in the system $NaAlSiO_4-SiO_2-H_2O$ to 35 kilobars pressure, *Amer. J. Sci.*, 267, 875–909, 1969.

Bowen, N. L., and O. F. Tuttle, The system $NaAlSi_3O_8-KAlSi_3O_8-H_2O$, *J. Geol.*, 58, 489–511, 1950.

Brown, G. C., and W. S. Fyfe, The production of granitic melts during ultrametamorphism, *Contrib. Mineral. Petrol.*, 28, 310–318, 1970.

Burnham, C. W., Hydrothermal fluids at the magmatic stage, in *Geochemistry of Hydrothermal Ore Deposits*, edited by N. L. Barnes, pp. 34–76, Holt, Rinehart, and Winston, New York, 1967.

Clark, S. P., Jr., and A. E. Ringwood, Density distribution and constitution of the mantle, *Rev. Geophys. Space Phys.*, 2, 35–88, 1964.

Coats, R. R., Magma type and crustal structure in the Aleutian arc, in *The Crust of the Pacific Basin, Geophys. Monogr. Ser.*, vol. 6, edited by

G. A. MacDonald and H. Kuno, pp. 92–109, AGU, Washington, D. C., 1962.

Cohen, L. H., K. Ito, and G. C. Kennedy, Melting and phase relations in an anhydrous basalt to 40 kilobars, *Amer. J. Sci., 265,* 475–518, 1969.

Dickinson, W. R., Global tectonics, *Science, 168,* 1250–1259, 1970.

Dodge, F. C. W., and D. C. Ross, Coexisting hornblendes and biotites from granitic rocks near the San Andreas fault, California, *J. Geol., 79,* 158–172, 1971.

Eggler, D. H., Water saturated and undersaturated melting relations in two natural andesites, in *Abstracts with Programs, 2,* p. 544, Geol. Soc. Amer., Boulder, Colorado, 1970.

Fyfe, W. S., Lattice energies, phase transformations and volatiles in the mantle, *Phys. Earth Planet. Interiors, 3,* 196–200, 1970.

Gast, P. W., Trace-element fractionation and the origin of tholeiitic and alkaline magma types, *Geochim. Cosmochim. Acta., 32,* 1057–1068, 1968.

Gibbon, D. L., and P. J. Wyllie, Experimental studies of igneous rock series: The Farrington Complex, North Carolina, and the Star Mountain Rhyolite, Texas, *J. Geol., 77,* 221–239, 1969.

Gilluly, J., Oceanic sediment volumes and continental drift, *Science, 166,* 992–993, 1969.

Green, D. H., The origin of basaltic and nephelinitic magmas in the earth's mantle, *Tectonophysics, 7,* 409–422, 1969.

Green, D. H., Compositions of basaltic magmas as indicators of conditions of origin: Application to oceanic volcanism, *Phil. Trans. Roy. Soc. London A, 268,* 707–725, 1971.

Green, D. H., and A. E. Ringwood, The genesis of basaltic magmas, *Contrib. Mineral. Petrol., 15,* 103–190, 1967.

Green, T. H., and A. E. Ringwood, Genesis of the calc-alkaline igneous rock suite, *Contrib. Mineral. Petrol., 18,* 105–162, 1968.

Greenwood, H. J., The system $NaAlSi_2O_6$-H_2O-Argon: Total pressure and water pressure in metamorphism, *J. Geophys. Res., 66,* 3923–3946, 1961.

Hamilton, W. Mesozoic California and the underflow of Pacific mantle, *Geol. Soc. Amer. Bull., 80,* 2409–2430, 1969.

Hill, R. E. T., and A. L. Boettcher, Water in the earth's mantle: Melting curves of basalt-water and basalt-water-carbon dioxide, *Science, 167,* 980–981, 1970.

Holloway, J. R., and C. W. Burnham, Phase relations and compositions in basalt-H_2O-CO_2 under the Ni-NiO buffer at high temperatures and pressures, in *Abstracts with Programs for 1969,* Part 7, pp. 104–105, Geol. Soc. Amer., Boulder, Colorado, 1969.

Isacks, B., J. Oliver, and L. R. Sykes, Seismology and the new global tectonics, *J. Geophys. Res., 73,* 5855–5899, 1968.

Ito, K., and G. C. Kennedy, Melting and phase relations in a natural peridotite to 40 kilobars, *Amer. J. Sci., 265,* 519–538, 1967.

Ito, K., and G. C. Kennedy, Melting and phase relations in the plane tholeiite-lherzolite-nepheline basanite to 40 kilobars with geological implications, *Contrib. Mineral. Petrol., 19,* 177–211, 1968.

Kay, R., N. J. Hubbard, and P. W. Gast, Chemical characteristics and origin of oceanic ridge volcanic rocks, *J. Geophys. Res., 75,* 1585–1614, 1970.

Kennedy, G. C., G. J. Wasserburg, H. C. Heard, and R. C. Newton, The upper three-phase region in the system SiO_2-H_2O, *Amer. J. Sci., 260,* 501–521, 1962.

Koster Van Groos, A. F., and P. J. Wyllie, Liquid immiscibility in the join $NaAlSi_3O_8$-Na_2CO_3-H_2O and its bearing on the genesis of carbonatites, *Amer. J. Sci., 266,* 932–967, 1968a.

Koster Van Groos, A. F., and P. J. Wyllie, Melting relationships in the system $NaAlSi_3O_8$-NaF-H_2O to 4 kilobars pressure, *J. Geol., 76,* 50–70, 1968b.

Koster Van Groos, A. F., and P. J. Wyllie, Melting relationships in the system $NaAlSi_3O_8$-NaCl-H_2O at 1 kilobar pressure, *J. Geol., 77,* 581–605, 1969.

Kushiro, I., The system forsterite-diopside-silica with and without water at high pressures, *Amer. J. Sci., 267A,* 269–294, 1969a.

Kushiro, I., Discussion of the paper 'The origin of basaltic and nephelinitic magmas in the earth's mantle' by D. H. Green, *Tectonophysics, 7,* 427–436, 1969b.

Kushiro, I., Systems bearing on melting of the upper mantle under hydrous conditions, *Carnegie Inst. Wash. Yearb., 68,* 240–245, 1970.

Kushiro, I., Y. Syono, and S. Akimoto, Melting of a peridotite nodule at high pressures and high water pressures, *J. Geophys. Res., 73,* 6023–6029, 1968.

Lambert, I. B., and K. S. Heier, The vertical distribution of uranium, thorium, and potassium in the continental crust, *Geochim. Cosmochim. Acta., 31,* 377–390, 1967.

Lambert, I. B., and K. S. Heier, Geochemical investigations of deep-seated rocks in the Australian shield, *Lithos, 1,* 30–53, 1968.

Lambert, I. B., and P. J. Wyllie, Stability of hornblende and a model for the low velocity zone, *Nature, 219,* 1240–1241, 1968.

Lambert, I. B., and P. J. Wyllie, Melting in the deep crust and upper mantle and the nature of the low velocity layer, *Phys. Earth Planet. Interiors, 3,* 316–322, 1970a.

Lambert, I. B., and P. J. Wyllie, Low-velocity zone of the earth's mantle: Incipient melting caused by water, *Science, 169,* 764–766, 1970b.

Lambert, I. B., J. K. Robertson, and P. J. Wyllie, Melting reactions in the system $KAlSi_3O_8$-SiO_2-H_2O to 18.5 kilobars, *Amer. J. Sci., 267,* 609–626, 1969.

Landisman, M., S. Mueller, and B. J. Mitchell, Review of evidence for velocity inversions in the continental crust, in *The Structure and Physical Properties of the Earth's Crust, Geo-*

phys. Monogr. Ser., vol. 14, edited by J. G. Heacock, AGU, Washington, D. C., this volume, 1971.

Luth, W. C., The systems NaAlSi₃O₈-SiO₂ and KAlSi₃O₈-SiO₂ to 20 kb and the relationship between H₂O content, P_{H_2O}, and P_{total} in granitic magmas, *Amer. J. Sci., Schairer, 267-A*, 325–341, 1969.

McDowell, S. D., and P. J. Wyllie, Experimental studies of igneous rock series: The Kungnat syenite complex of southwest Greenland, *J. Geol., 79*, 173–194, 1971.

McGetchin, T. R., L. T. Silver, and A. A. Chodos, Titanoclinohumite: A possible mineralogical site for water in the upper mantle, *J. Geophys. Res., 75*, 255–259, 1970.

McKenzie, D. P., Speculations on the consequences and causes of plate motions, *Geophys. J. Roy. Astron. Soc., 18*, 1–32, 1969.

Merrill, R. B., J. K. Robertson, and P. J. Wyllie, Melting reactions in the system NaAlSi₃O₈-KAlSi₃O₈-SiO₂-H₂O to 20 kilobars compared with results for other feldspar-quartz-H₂O and rock-H₂O systems, *J. Geol., 78*, 558–569, 1970.

Millhollen, G. L., Melting of nepheline syenite with H₂O and H₂O+CO₂, and the effect of dilution of the aqueous phase on the beginning of melting, *Amer. J. Sci., 270*, 244–254, 1971.

Minear, J. W., and M. N. Toksöz, Thermal regime of a downgoing slab and new global tectonics, *J. Geophys. Res., 75*, 1397–1419, 1970.

Modreski, P. J., and A. L. Boettcher, The stability of phlogopite in the earth's mantle, in *Abstracts with Programs, 2*, 626–627, Geol. Soc. Amer., Boulder, Colorado, 1970.

O'Hara, M. J., Primary magmas and the origin of basalts, *Scot. J. Geol., 1*, 19–40, 1965.

O'Hara, M. J., The bearing of phase equilibria studies in synthetic and natural systems on the origin and evolution of basic and ultrabasic rocks, *Earth Sci. Rev., 4*, 69–133, 1968.

Oxburgh, E. R., and D. L. Turcotte, Thermal structure of island arcs, Bull. Geol. Soc. Amer. *81*, 1665–1688, 1970.

Peters, Tj., W. C. Luth, and O. F. Tuttle, The melting of analcite solid solutions in the system NaAlSiO₄-NaAlSi₃O₈-H₂O, *Amer. Mineral. 51*, 736–753, 1966.

Piwinskii, A. J., Experimental studies of igneous rock series, central Sierra Nevada Batholith, California, *J. Geol., 76*, 548–570, 1968.

Piwinskii, A. J., and P. J. Wyllie, Experimental studies of igneous rock series: A zoned pluton in the Wallowa Batholith, Oregon, *J. Geol., 76*, 205–234, 1968.

Piwinskii, A. J., and P. J. Wyllie, Experimental studies of igneous rock series: Felsic body suite from the Needle Point pluton, Wallowa batholith, Oregon, *J. Geol., 78*, 52–76, 1970.

Ramberg, H., *Gravity, Deformation, and the Earth's Crust*, 214 pp., Academic, London, 1967.

Ringwood, A. E., Composition and evolution of the upper mantle, in *The Earth's Crust and Upper Mantle, Geophys, Monogr. Ser.,* vol. 13, edited by P. J. Hart, pp. 1–17, AGU, Washington, D. C., 1969.

Ringwood, A. E., and D. H. Green, Petrological nature of the stable continental crust, in *The earth beneath the Continents, Geophys. Monogr. Ser.,* vol. 10, edited by J. S. Steinhart and T. J. Smith, pp. 611–619, AGU, Washington, D. C., 1966.

Robertson, J. K., and P. J. Wyllie, Experimental studies on rocks from the Deboullie stock, northern Maine, including melting relations in the water-deficient environment, *J. Geol.,* in press, 1971a.

Robertson, J. K., and P. J. Wyllie, Rock-water systems, with special reference to the water-deficient region, *Amer. J. Sci.,* in press, 1971b.

Ronov, A. B., and A. A. Yaroshevsky, Chemical composition of the earth's crust, in *The Earth's Crust and Upper Mantle, Geophys. Monogr. Ser.,* vol. 13, edited by P. J. Hart, pp. 37–57, AGU, Washington, D. C., 1969.

Scarfe, C. M., W. C. Luth, and O. F. Tuttle, An experimental study bearing on the absence of leucite in plutonic rocks, *Amer. Mineral., 51*, 726–735, 1966.

Sclar, C. B., L. C. Carrison, and O. M. Stewart, High pressure synthesis and stability of hydroxylated clinoenstatite in the system MgO-SiO₂-H₂O, in *Program 1967 Ann. Meet.* Geol. Soc. Amer., New Orleans, 1967.

Segnit, R. E., and G. C. Kennedy, Reactions and melting reactions in the system muscovite-quartz at high pressures, *Amer. J. Sci., 259*, 280–287, 1961.

Stewart, D. B., Four-phase curve in the system CaAl₂Si₂O₈-SiO₂-H₂O between 1 and 10 kilobars, *Schweiz. Mineral. Petrogr. Mitt., 47*, 35–59, 1967.

Tuthill, R. L., Effect of varying f_{O_2} on the hydrothermal melting and phase relations of basalt (abstract), *Eos, Trans. AGU, 50*, 355, 1969.

Tuttle, O. F., and N. L. Bowen, Origin of Granite in the light of experimental studies in the system NaAlSi₃O₈-KAlSi₃O₈-SiO₂-H₂O, *Geol. Soc. Amer., Mem. 74*, 153 pp., 1968.

von Platen, H., Experimental anatexis and genesis of migmatites, in *Controls of Metamorphism*, pp. 203–218, edited by W. S. Pitcher and G. W. Flynn, Oliver and Boyd, London, 1965.

Whitney, J. A., and W. C. Luth, Water undersaturated melting of natural granites at 2 kb (abstract), *Eos, Trans. AGU, 51*, 438, 1970.

Winkler, H. G. F., *Petrogenesis of Metamorphic Rocks*, 2nd ed., 273 pp., Springer-Verlag, New York, 1967.

Wones, D. R., and H. P. Eugster, Stability of biotite: Experiment, theory, and application, *Amer. Mineral., 50*, 1228–1272, 1965.

Wyllie, P. J., Applications of high pressure studies to the earth sciences, in *High Pressure Physics and Chemistry*, vol. 2, edited by R. S. Bradley, pp. 1–89, Academic, London, 1963.

Wyllie, P. J., Ultramafic rocks and the upper mantle, *Spec. Pap. 3, Fiftieth Anniversary Symposia*, edited by B. A. Morgan, pp. 3–32, Mineral. Soc. Amer., 1970.

Wyllie, P. J., The role of water in magma generation and initiation of diapiric uprise in the mantle, *J. Geophys. Res.*, 76, 1328–1338, 1971a.

Wyllie, P. J., *The Dynamic Earth: A Textbook in Geosciences*, John Wiley, New York, 1971b.

Wyllie, P. J., and O. F. Tuttle, Effect of carbon dioxide on the melting of granite and feldspars, *Amer. J. Sci.*, 257, 648–655, 1959.

Wyllie, P. J., and O. F. Tuttle, Melting in the earth's crust, in *Proc. 21st Internat. Geol. Congress, Copenhagen, 1960*, Part 18, 227–235, 1960a.

Wyllie, P. J., and O. F. Tuttle, Experimental investigation of silicate systems containing two volatile components, 1, Geometrical considerations, *Amer. J. Sci.*, 258, 498–517, 1960b.

Wyllie, P. J., and O. F. Tuttle, The system CaO–CO₂–H₂O and the origin of carbonatites, *J. Petrol. 1*, 1–46, 1960c.

Wyllie, P. J., and O. F. Tuttle, Hydrothermal melting of shales, *Geol. Mag.*, 98, 56–66, 1961a.

Wyllie, P. J., and O. F. Tuttle, Experimental investigation of silicate systems containing two volatile components, 2, The effects of NH_3 and HF, in addition to H_2O, *Amer. J. Sci.*, 259, 128–143, 1961b.

Wyllie, P. J., and O. F. Tuttle, Experimental investigation of silicate systems containing two volatile components, 3, The effects of SO_3, P_2O_5, HCl, and Li_2O, in addition to H_2O, on the melting temperatures of albite and granite, *Amer. J. Sci.*, 262, 930–939, 1964.

Yoder, H. S., The MgO-Al_2O_3-SiO_2-H_2O system and the related metamorphic facies, *Amer. J. Sci.*, Bowen vol., 569–627, 1952.

Yoder, H. S., and I. Kushiro, Melting of a hydrous phase: Phlogopite, *Amer. J. Sci.*, 267-A, 558–582, 1969.

Yoder, H. S., Jr., and Tilley, C. E., Origin of basalt magmas: an experimental study of natural and synthetic rock systems, *J. Petrol.*, 3, 342–532, 1962.

DISCUSSION

Kennedy: I think the absence of hornblende-eclogites is a very crucial thing. If the upper mantle were wet, we should have hornblende-eclogites as the main constituent, and you never see it in diamond pipes. This is the reason I think the upper mantle is dry.

Wyllie: The absence of hornblende-eclogites in diamond pipes does not necessarily mean that there is no water. Diamond pipes seem to have their origins at depths exceeding 100 km, whereas we have experimental data that hornblendes are not stable at depths greater than about 75 km. There are many hornblende-bearing nodules, mafic and ultramafic, believed to be derived from the upper mantle, and many investigators have cited these as evidence for water in the uppermost mantle. I know that your preferred interpretation of these is that the hornblende formed late, under crustal conditions.

Boettcher: I think you should mention that as long as you use the terms water-deficient and vapor absent interchangeably, the two are the same only as long as water is the only vapor. It can still be water deficient and still have a vapor phase as long as, for example, there is carbon dioxide as another volatile component.

Wyllie: Yes, the terms as I define them apply only to water-bearing systems. However, I was not using water-deficient and vapor-absent interchangeably. Vapor-absent and vapor-present assemblages can both occur in water-deficient compositions under appropriate conditions. With another volatile component the formal rules are modified, but I think the pattern of behavior is only slightly modified compared with the rock-water models.

H. W. Smith: What do we know about the change in electrical conductivity when rocks start to melt?

Porath: The conductivity of a synthetic basalt that I just measured increased by two orders of magnitude during the melting.

Kennedy: We measured a diopside, and we could not see the melting point on the resistivity-temperature plot.

Porath: It just depends on how the short-range order in the semiconductor changes during the melting. If this does not change you get no change in conductivity on melting. The other extreme is when you get a metallic structure after melting as happens to germanium; then you get a very large increase. Russian workers have also reported a two-orders increase with basalts on melting.

Kennedy: The melting range is about 200°C. Might not the two orders conductivity increase be continuous over this range?

Porath: That is quite a discontinuity.

Frischknecht: George Keller and I measured conductivities of 0.5 mho/m and greater over frozen lava lakes. There always seems to be a very sharp discontinuity—presumably between the solid and liquid, although it is true that there is a steep temperature gradient also.

Whitcomb: Did your phase equilibrium curves refer to assemblages with water present?

Wyllie: The only case in which water has an effect is the gabbro-eclogite transition. This can be masked if there is enough water present, owing to the formation of hornblende. No water at all is involved in the other transitions, and they are not affected.

An Experimental Study of the Basalt-Garnet Granulite-Eclogite Transition[1]

KEISUKE ITO AND GEORGE C. KENNEDY

Institute of Geophysics and Planetary Physics
University of California, Los Angeles, California   90024

Abstract. The recrystallization of a tholeiitic basalt to garnet granulite and to eclogite has been experimentally examined at temperatures ranging from 800°C up to the solidus of the rock. In addition, the solidus of tholeiitic basalt has been redetermined. The slope of the boundary separating the field of garnet granulite from that of basalt is $P_{kb} = 0.014 \ T°C - 5.4$ kb. The boundary separating the field of garnet granulite from the field of eclogite has the equation $P_{kb} = 0.020 \ T°C + 4.0$ kb. The equation of the newly determined solidus of tholeiitic basalt is $T = 6.0 \ P_{kb} + 1105°C$. Curves relating the densities of the mineral assemblages to temperatures and pressures of formation are shown. One relatively abrupt change in density occurs as basalt recrystallizes to garnet granulite, and a second relatively sharp density increase occurs as garnet granulite is recrystallized to eclogite.

The basalt-eclogite transition has attracted the attention of many geologists and geophysicists since the publication of *Fermor's* [1914] suggestion that eclogite is a high-pressure equivalent of basalt and that the volume change associated with the transition may be an energy source for crustal activity.

Eclogites are found within the crust as layers and blocks among glaucophane schists [*Coleman et al.,* 1965], among gneisses [*Eskola,* 1921], as xenoliths in kimberlite pipes [*Williams,* 1932], and in alkali basalts [*Yoder and Tilley,* 1962]. Some eclogites are believed to have been brought into the crust from the underlying mantle and others are believed to have been formed within the crust. Whether eclogites are stable within the normal range of crustal temperatures and pressures is a question that has long been debated. An excellent review is *Green and Ringwood* [1967a].

Press [1970] has recently computed density distributions within the earth that are based on free-oscillation data. His results suggest that the probable density in the earth's mantle at approximately 100 km is between 3.5 and 3.6 g/cc. Such a high upper-mantle density seems consistent only with an eclogite layer.

Laboratory experiments on eclogites should provide conclusive data on their stability field and indicate whether they should be expected to occur within the normal range of crustal temperatures and pressures. Unfortunately, the reaction of basalt to eclogite and its reverse reaction is sluggish except at very high temperatures, and the stability field of eclogite at intermediate to low temperatures is unknown. Phase equilibriums between basalts and eclogites at high temperatures have been investigated by *Yoder and Tilley* [1962], *Kushiro and Yoder* [1966], and *Green and Ringwood* [1967a], and work in our laboratory has been underway since 1964. *Cohen et al.* [1967] determined the phase relations of a typical olivine tholeiite along its melting interval. *Ito and Kennedy* [1967, 1968] investigated the effect of additional ultramafic components as well as alkaline components on the phase equilibriums of a tholeiitic basalt. *Godovikov and Kennedy* [1969] investigated the effect of additional Al_2O_3, and *Ito and Kennedy* [1970] reported density changes of the solid associated with the basalt-eclogite transition and proposed a petrological model of the upper mantle.

[1] Publication 884 of the Institute of Geophysics and Planetary Physics, University of California, Los Angeles.

303

In the present study we have redetermined the solidus of a tholeiitic basalt and have determined the phase boundaries between basalt and garnet granulite and between garnet granulite and eclogite over the temperature interval 800°–1250°C. Our results on an olivine tholeiite differ significantly from those of *Green and Ringwood* [1967a], who report on an eclogite crystallized from a quartz tholeiite.

EXPERIMENTAL TECHNIQUES

Starting material. Much experimental detail has been given in an earlier paper [*Cohen et al.*, 1967] on the preparation of our starting material. The composition of the basalt used is shown in Table 1. The basalt, NM5, is close to the average oceanic tholeiite. The major difference between the two is that our rock is slightly lower in CaO and slightly higher in FeO than average oceanic tholeiite. Glass was prepared by fusing powdered rock in an iron crucible in a stainless steel bomb. All iron in the glass was presumed to be present as FeO.

High-pressure furnace. The experiments were performed in a piston-cylinder high-pressure apparatus. The high-pressure furnace is the modified type described by *Cohen et al.* [1967]. A sleeve of 1-mm wall pyrex glass tubing was placed between the carbon tube heater and the talc pressure transmitting medium to help ensure an anhydrous atmosphere. Boron nitride, heated at 800°C for 24 hours, was used for most of the furnace parts inside the graphite heating tube. Iron capsules, machined from Armco magnetic ingot iron, were used as the

TABLE 1. Compositions of NM5 Glass and the Average Ocean Tholeiite

Composition	NM5 Glass, %	Avg. Ocean Tholeiite,* %
SiO$_2$	49.93	49.94
Al$_2$O$_3$	16.75	16.69
Fe$_2$O$_3$	...	2.01
FeO	11.40	6.90
MnO	0.18	...
MgO	7.59	7.28
CaO	9.33	11.86
Na$_2$O	2.92	2.76
K$_2$O	0.37	0.16
P$_2$O$_5$	0.19	0.16
TiO$_2$	1.34	1.51

* Values from *Engel et al.* [1965].

sample containers in most runs. Gold capsules were used in a few runs at low temperatures, when we chose to buffer the sample with a low partial pressure of water.

Pressure measurement. Nominal pressures were calculated from the ram oil pressure and the ratio of ram diameter to piston diameter. The piston friction was determined by plotting a hysteresis loop wherein nominal pressure was plotted against piston position on a compression and on a decompression cycle [*Hariya et al.*, 1969; *Akella and Kennedy*, 1971]. The value of friction as a function of temperature and pressure was somewhat variable. Representative values are shown in Table 2. They range from a low of 0.2 kb at a pressure of 5 kb and temperature of 800° up to 0.8 kb at 40 kb and 1200°. The value of friction was also estimated from measurements of the density of a sample synthesized on compression and decompression runs. These results suggested that friction ranged from 0.3 to 1.0 kb, in close agreement with values determined by observation of the hysteresis loops.

Temperature measurements. Temperatures were measured by platinum-platinum 10%Rh thermocouples. In a single experiment, the emf of the Pt-Pt 10%Rh thermocouple was intercompared with a tungsten-rhenium thermocouple at 1300°C and 24.5 kb for 3 hours. The drift between the two thermocouples was less than 3° during the experiment. Corrections for the effect of pressure on the emf of the thermocouples were made according to results of the measurements of *Getting and Kennedy* [1971]. Temperatures were controlled to within ±5° for runs shorter than 10 hours, but temperatures of some of the longer overnight subsolidus experiments fluctuated as much as ±20°.

Iron content of the samples. Inasmuch as our samples were run in iron capsules for the most part, it was important to identify the increase in FeO with time in our samples. A single sample was placed in an iron capsule and run at 1200°C for 3 hours and was then extracted from the press, crushed, and placed in a fresh capsule and rerun at the same temperature and pressure. This was repeated three times. The total FeO increased from the starting value of 11.4% to 13.8%. The iron was determined by analysis under an electron probe analyzer with a defocused electron beam. This change in total iron with time is believed to be the maximum for any of the runs.

Equilibrium. The phase boundaries were determined by an approach to equilibrium from both directions. The upper pressure limit of a phase stability field was determined by starting with a crystalline powder synthesized at a pressure just below the limit. The boundary was then determined by observation of the pressure at which a phase appeared or disappeared. The boundary was redetermined by approaching equilibrium from the high-pressure side. At temperatures of 900° and below, all reactions are extremely sluggish. Both nucleation and growth of new phases and disappearance of old phases take place at an extremely slow rate. Only minute submicroscopic crystals grow from a dry basalt glass. In order to determine the lowest pressure at which garnet appears at 800°, we fluxed our runs with 0.25% $Li_2B_4O_7$. Lithium does not enter the garnet structure, and boron probably enters only in a very minute amount. Thus, we believe the presence of a trace of lithium borate does not significantly change the stability field of garnet. These runs at 800°, fluxed with lithium borate, served to locate the boundary separating the stability field of basalt from the stability field of garnet granulite. However, even in a 125-hour run at 10 kb and 800°, the reaction did not proceed to completion, and no meaningful density measurements could be made on the sample.

Unfortunately, lithium borate cannot be used in the determination of the high-pressure limit of the stability field of plagioclase. At 800°C lithium enters the structure of both plagioclase and clinopyroxene and presumably affects their stability fields. The high-pressure limit to the stability field of plagioclase was determined by buffering the sample with partially dehydrated phlogopite powder in a gold capsule. A sample, placed in one end of a gold capsule, was in equilibrium with the vapor pressure of phlogopite in the other end of the capsule. Thus the partial pressure of water was kept sufficiently low [*Luth*, 1967] so that our rock did not melt and amphibole did not form, but the small partial pressure of water greatly aided the rate of crystallization and decomposition of plagioclase.

EXPERIMENTAL RESULTS

Results of our experiments are shown in Table 3 and 4 and are plotted in Figures 1, 2, 3, and 4. We present in these tables and figures some 33 crucial runs. However, some 120 runs were made.

Solidus temperature. Cohen et al. [1967] reported earlier on the solidus of NM5. The solidus they report represents the temperatures at which no glass could be seen among the crystalline products. Their studies indicated that NM5 recrystallized solely to garnet + clinopyroxene at high pressures. However, normative calculations show that approximately 6% quartz should be present along the solidus, but quartz was not detected among their crystalline products. Provided a small amount of quartz was present along the solidus, an addition of quartz should not have affected the solidus temperatures. Thus, we have reinvestigated the solidus of NM5 to which we have added approximately 20% quartz. The solidus we now report is shown in Figure 1. It is remarkably lower than that reported earlier, because with quartz added to our starting material we were able to detect substantial amounts of glass at temperatures we had believed earlier to be subsolidus. Thus, *Cohen et al.* [1967] presumably failed to detect quartz in NM5 because glass was always present in their runs and their temperatures were never truly subsolidus. We show in Figure 1 our currently determined solidus curve, along with prior results on the solidus and the liquidus by Cohen et al. Results by Cohen et al. lie close to the solidus of an olivine eclogite, whereas the curve we now present represents the solidus of a quartz eclogite. The slope of the new solidus is approximately 6°/kb, which is remarkably close to the 6.5°/kb value found by *Segnit and Kennedy* [1961] for the liquidus of a synthetic quaternary muscovite. This similarity suggests the possibility that all quartz-bearing igneous rocks have approximately the same solidus slope, 6.0° to 6.5° C/kb. The equation

TABLE 2. Single Value Friction

Temp., °C	Nominal Pressure, kb							
	5	10	15	20	25	30	35	40
800	0.2	0.4	0.5	0.7	0.9	...	...	...
1000	0.2	0.3	0.4	0.6	0.7	...	...	...
1200	...	0.2	0.3	0.4	0.5	0.6	0.7	0.8

TABLE 3. Runs with NM5 Glass and NM5 Recrystallized as Basalt, Garnet Granulite, and Eclogite and Quartz

P, kb	T, °C	Time, hours	Starting Material	Products
13.0	1178	4	Gl + 10%GG + 12.5%Qz	Pl, Cpx, Opx, Qz, remnant Ga
13.0	1203	3	Gl + 10%GG + 10%Qz	Pl, Cpx, Opx, Qz, trace Gl
13.7	1178	4	Ba + 5%GG	Cpx, Pl, Ga, Qz
19.6	1255	2	Gl + 10%GG + 10%Qz	Cpx, Pl, Ga, Qz, trace Gl
24.8	1256	3	Gl + 10%GG + 10%Qz	Cpx, Pl, Ga, Qz, trace Gl
26.2	1256	3	Ec + 10%Qz	Cpx, Ga, Pl, Qz
26.2	1281	2	Gl + 10%GG + 10%Qz	Cpx, Ga, Qz, trace Gl
26.9	1257	3	GG + 10%Qz	Cpx, Ga, Qz
28.7	1307	1.5	Gl + 5%GG + 15%Qz	Cpx, Ga, Qz, trace Gl
36.7	1333	1.5	Gl + 10%GG + 10%Qz	Cpx, Ga, Coes, Qz, trace Gl
39.2	1334	1.5	Gl + 10%GG + 12.5%Qz	Cpx, Ga, Coes
39.2	1359	1	Gl + 20%Qz	Cpx, Ga, Coes, trace Gl

of the solidus is $T = 6.0 P_{kb} + 1105$°C. The slope of the liquidus of our NM5 basalt is approximately 12°/kb and the interval over which this basalt melts increases from approximately 100° at 10 kb to 300° at pressures slightly over 40 kb. Quartz disappears from the solidus at pressures below 9 kb, at which anorthite and olivine become stable phases. The slope of the solidus shifts somewhat in this low-pressure region. The solidus temperature at 1 atmosphere is approximately 1080°C.

Presumably, approximately 6% of quartz ap-

TABLE 4. Runs on NM5 Glass and on NM5 Recrystallized as Basalt, Garnet Granulite, and Eclogite

P, kb	Time, hours	Starting Material	Proportion of Products			Accessory Minerals
			Cpx	Pl	Ga	
At 800°C						
5.2	72	GG*	30	70	· · ·	Ol, Opx
6.5	72	Gl* + 10% GG	29	63	8	Ol, Opx
6.5	72	GG*	29	63	8	Ol, Opx
19.7	72	Ec**	32	4	64	· · ·
21.0	146	GG**	38	· · ·	62	· · ·
At 900°C						
21.4	93	Ec*	32	7	61	Qz
22.7	72	Ec**	30	· · ·	70	Qz
At 1000°C						
9.2	7	Gl + 10% GG	26	55	19	Ol, Opx
22.4	8	Ec	37	5	58	Qz
23.4	27	GG**	35	4	61	Qz
At 1100°C						
9.3	5	GG	43	57	· · ·	Ol, Opx
10.0	5.5	GG	40	49	11	Ol, Opx
10.0	5	Ba	38	62	· · ·	Ol, Opx
10.6	7	Gl + 10% GG	39	54	7	Ol, Opx
10.6	5	GG	36	46	16	Ol, Opx
25.0	5	Ec	43	· · ·	57	Qz
At 1150°C						
10.2	6	GG + Ba	48	52	· · ·	Ol, Opx
At 1200°C						
24.7	3	Ec	39	4	57	Qz
26.5	4	Ec	38	· · ·	62	Qz
26.9	3	GG	38	4	58	Qz
28.2	3	GG	42	· · ·	58	Qz

* 0.25% $Li_2B_4O_7$ added.
** Au capsule with H_2O buffered.

pears on the solidus of NM5 when it crystallizes to an eclogite. It is remarkable that we cannot detect glass in this rock by petrographic methods even at temperatures several hundred degrees above the solidus. We suspect that the solidus temperatures of the Glenelg eclogite [*Yoder and Tilley,* 1962] were actually much lower than reported, because quartz was expected to appear on the solidus in this rock but was not detected. Our new solidus is remarkably close to that of the quartz tholeiite, andesite, and dacite studied by *Green and Ringwood* [1967a, 1968]; this relationship is to be expected, because their high-pressure mineral assemblage was also garnet + clinopyroxene + quartz.

The solidus temperature of a bi-mineral eclogite, made up solely of clinopyroxene +

garnet, is somewhat higher than the solidus of an olivine eclogite, since the addition of olivine to garnet + pyroxene lowers their solidus temperature. *O'Hara and Yoder* [1967] found the solidus of an olivine eclogite to be 50° to 100° lower than that of the bi-mineral eclogite at 30 kb. *Ito and Kennedy* [1968] found approximately 50° difference at 20 kb and 20° difference at 40 kb. We show in Figure 1 the solidus of an olivine eclogite as determined by *Ito and Kennedy* [1967]. The solidus of a bi-mineral eclogite should lie fairly close to the line shown in Figure 1 for the liquidus of NM5. If the garnets and clinopyroxene remain stoichiometric in relation to the ratio of metal to silica, the addition of even a trace amount of quartz will drop the solidus some 350° at 40 kb. Over most of this melting interval, the amount of liquid

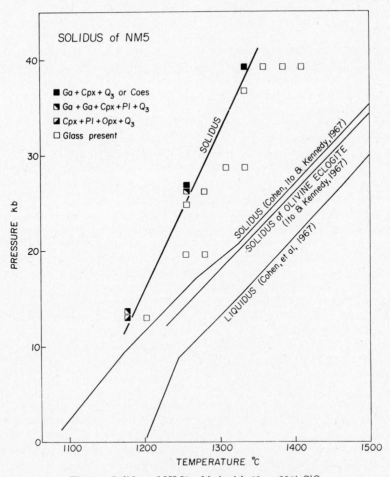

Fig. 1. Solidus of NM5 added with 10 ∼ 20% SiO_2.

present is primarily independent of temperature and is only a little more than the amount of quartz present in the eclogite. The amount of liquid begins to increase rapidly only when we reach the approximate temperatures of the solidus of a quartz-free eclogite. Thus, partial melting of a quartz eclogite whose bulk chemistry is that of oceanic tholeiite can produce a highly siliceous liquid over a wide temperature interval. This phenomenon provides us with a mechanism for obtaining a siliceous liquid fraction from material of the bulk chemistry of oceanic tholeiites. We must emphasize the fact that only a very small amount of liquid is in equilibrium with garnet-clinopyroxene assemblage over this wide temperature interval. This relationship would perhaps offer an explanation for the low-velocity zone seen in the upper mantle, if the bulk chemistry of the rocks in this region is that of oceanic tholeiites and their crystalline assemblage is garnet-clinopyroxene-quartz.

Phase boundaries. The basalt mineral assemblage, olivine + pyroxenes + plagioclase, transforms into the garnet granulite mineral assemblage, garnet + clinopyroxene + plagioclase, with increasing pressure. There is a very narrow transition zone where garnet coexists with olivine and orthopyroxene, as well as clinopyroxene and plagioclase. With increasing pressure, orthopyroxene and olivine disappear, as the garnet increases in amount. The pertinent experimental data are shown in Table 4. The results are plotted in Figure 2. This transition takes place over an interval of approximately 2.5 kb. The density of the mineral assemblages changes from approximately 3.14 to approximately 3.3 g/cc.

We have reversed the boundary separating the stability field of garnet granulite from the field of basalt. Garnet first appears at a pressure of between 9.3 and 10.6 kb at 1100°C and between 5.2 and 6.5 kb at 800°C. Permissible linear extrapolations of this boundary show that the upper temperature limit to garnet stability at atmospheric pressure is somewhere between 200° and 500°C. The equation of this boundary, which we show in Figure 2, is $P_{kb} = 0.014T°C - 5.4$ kb.

The pressure at which garnet appears in NM5 when 10 to 15 per cent quartz is added to the rock is 13.4 kb at 1175°C. This value is more than 2 kb higher than the pressure at which garnet appears in NM5. Thus, garnet becomes stable in silica undersaturated rock at lower pressures than in silica oversaturated rocks, if all other factors remain constant.

The pressure at which garnet appears may be a function of the total iron content of the rock. We noticed that the extreme peripheral parts of our specimen, next to the iron capsule, were enriched in garnet of high refractive index. This part of the sample was carefully ground off when we made density determinations. We noted, however, that the first formed garnet in the central parts of the sample is not one unusually rich in iron. Preliminary electron-probe work suggests that the iron content of our garnet remains nearly constant or that it increases slightly with increasing pressure. This finding is in contrast with observations by *Green and Ringwood* [1967a], who state that garnets become richer in magnesium with increasing pressure. *Banno* [1970] suggested that the distribution coefficient of iron and magnesium between garnet and pyroxene, i.e.,

$$\frac{(X_{Fe}/X_{Mg})^{Ga}}{(X_{Fe}/X_{Mg})^{Cpx}}$$

should increase with increasing pressure and that, therefore, the iron in the garnet may also increase with increasing pressure, in spite of the fact that the ratio of garnet to clinopyroxene increases with increasing pressure.

The amount of garnet present in our samples steadily increases through the garnet granulite zone as pressure is increased. However, the increase in the amount of garnet through the garnet granulite zone is not linear with pressure. At low pressures in the garnet granulite zone the plagioclase feldspar reacts with olivine, forming garnet and a more sodic feldspar. This first burst of garnet appears in the pressure interval 11–13 kb at 1200°. Over the pressure range 13–19 kb, only a little additional garnet appears. The stable mineral assemblage is clinopyroxene, sodic plagioclase, garnet, and quartz. As pressure is raised from 19 to 23 kb, substantial amounts of sodium in the feldspar enter the pyroxene as jadeite, releasing calcium and aluminum for a second burst of garnet. The calcium content of the garnet increases slightly. Thus, the density of the assemblage increases at a nonlinear rate as pressure is increased through the garnet granulite zone.

A small amount of highly sodic feldspar co-exists with pyroxene and garnet in the pressure interval 22–28 kb at 1200°C. Following *Kushiro* [1969], we believe that the pyroxene is jadeitic

in this pressure interval and therefore term this mineral assemblage 'plagioclase-eclogite.'

The plagioclase-eclogite assemblage transforms at the highest pressure studied to the eclogite assemblage, garnet + clinopyroxene + quartz. We have reversed the boundary separat-

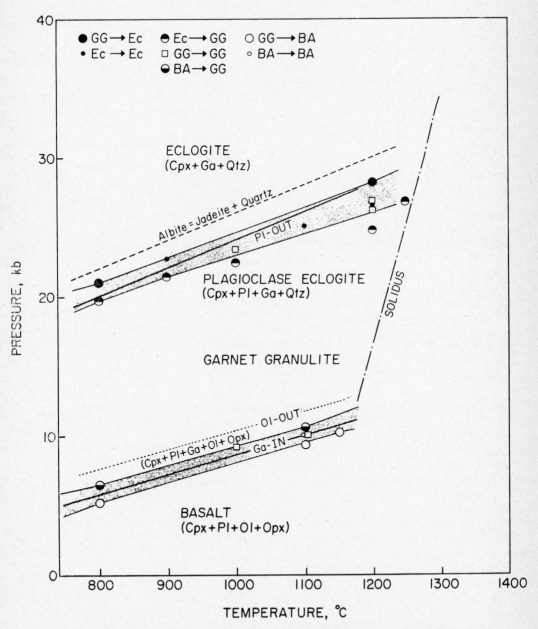

Fig. 2. Pressure-temperature diagram of NM5. Abbreviations: Ba is basalt, GG is garnet granulite, Ec is eclogite, Ga is garnet, Cpx is clinopyroxene, Opx is orthopyroxene, Pl is plagioclase, Ol is olivine, Qz is quartz, Coes is coesite, Gl is glass, and L is liquid.

ing the field of plagioclase-eclogite from that of eclogite. We find this boundary lies at approximately 27 kb at 1200°C and at approximately 20.5 kb at 800°C. This boundary is shown in Figure 2. In addition, we show the boundary for the reaction albite = jadeite + quartz, and, as is to be expected, the breakdown of sodic plagioclase to jadeitic pyroxene + quartz that we have determined in basalt is subparallel and 2 to 3 kb lower than the curve representing the reaction of pure albite to jadeite + quartz. The curve representing the equilibrium between albite and jadeite + quartz has been carefully studied by *Birch and LeComte* [1960], *Newton and Smith* [1967], *Boettcher and Wyllie* [1968], and *Newton and Kennedy* [1968]. Results of all these four investigators are almost in exact agreement. Inasmuch as the reaction albite = jadeite + quartz has been investigated to temperatures as low as 500°, we take the slope of this boundary as a guide in extrapolating the upper limit of the stability field of plagioclase-eclogite to lower pressures and temperatures. The curve representing the boundary between plagioclase-eclogite and eclogite is given by the equation $P_{kb} = 0.020T°C + 4.0$ kb.

We have attempted to estimate the composition of the clinopyroxene in our eclogite field. This estimate is based on normative calculations and partial electron-probe analyses. The results suggest that the pyroxene is 54% jadeite, 30% diopside, and 16% hedenbergite. Our eclogite boundary agrees closely with that determined by *Kushiro* [1969] from the study of the simple system $CaMgSi_2O_6 - NaAlSi_2O_6$.

Our results on the stability field of eclogite are remarkably in contrast to those published by *Green and Ringwood* [1967a]. They have extrapolated the garnet granulite-eclogite boundary determined from their experimental work and they note that the boundary extrapolates to a temperature of more than 200°C at atmospheric pressure. Thus, they deduce that, contrary to geological evidence, eclogites are the stable assemblage at all normal geological pressures and temperatures. The difference between our present results and theirs is so great that some discussion is required. They studied a quartz tholeiite that was identical in most components to our tholeiite, with the exception of a slight amount of additional quartz. This additional quartz should not have affected the

position of the stability field of eclogite; it should have changed only the ratio of garnet + clinopyroxene to quartz. Normative calculations show that their eclogite pyroxene contains 36 mole % jadeite, whereas our pyroxene contains 54% jadeite. Therefore, the boundary where plagioclase disappears is expected to be lower in pressure in Green and Ringwood's rock than in our rock. Indeed, their boundary at 1200°C, determined dry, is 4 kb lower than our boundary. However, their boundary at 1000°C is as much as 6.5 kb lower than our boundary. This great difference between the results at 1000°C is probably due to the experimental method. They made their runs at 1000° and 1100°C in unsealed platinum tubes; they allowed access of water to their samples. Their water was derived from dehydration of the talc adjacent to their heating tube during the run. They state that this caused 'a small and random amount of partial melting in some of the 1100°C runs.' We suspect that an undetectable amount of glass was present in their 1000° and 1100°C runs. We know from our own experience the difficulty of detecting small amounts of glass. Their presence would result in a lowering of the pressure at which plagioclase disappears. This factor could have drastically affected the slope they assigned to their boundary. Thus, we do not believe that their experiments can be used to derive the position of the stability field of eclogites at low temperatures and pressures, and we believe that the conclusions by *Ringwood and Green* [1966] based on this experimental data are not justified.

Density distribution. A knowledge of the change of density as one progresses from basalt through garnet granulite through plagioclase-eclogite and on to eclogite is crucial to the interpretation of these results. In an earlier paper, we reported the distribution of densities of quenched samples of NM5 along the line $P = 1070 + 0.104 T$ [*Ito and Kennedy*, 1970]. After completion of this study, we noted that the true solidus of NM5 lay below the temperatures at which we had measured density and that these samples must have contained a small but undetected amount of glass. In the present investigation, we have redetermined the density by the same methods at 1200°, 1100°, and 1000° as a function of pressure. We find that the presumed presence of a small amount of glass did not affect our earlier results in any significant way and that the conclusions in the paper [*Ito and Kennedy,* 1970] remain valid. The results of our new measurements of the density of quenched samples of NM5 are shown in Figure 3.

Significant features of Figure 3 are the two regions where density increases very rapidly as pressure is raised. Thus, two sharp changes in density are associated with the basalt-eclogite transition along any isotherm. It seems clear that the size of these density jumps is very much a function of the composition of the plagioclase. Basalts with a feldspar rich in calcium should show the biggest density jump associated with the lower pressure transition zone. Rocks with feldspar rich in sodium will show the biggest density jump associated with the higher pressure transition, which is equivalent to the reaction albite = jadeite + quartz.

The highest-density material we observed was 3.47 g/cc on a sample compressed to 35 kb. We have compared our measured density with the density calculated from X-ray data and the observed proportions of the minerals. Our calculated density is about 3% higher than the measured density. The theoretical density of this eclogite calculated from the minerals is 3.58 g/cc. The difference between the calculated and measured density is probably due to the opening up of fine cracks in our sample because of anisotropic expansion of the sample as pressure is reduced. We have assumed that this error is constant for all our measurements and thus have shown a second boundary with densities approximately 3% greater than those measured. We have extrapolated our iso-density lines on the assumption that they are subparallel to the phase boundaries; these lines are shown in Figure 4.

Interpretation of the results. The interpretation and application to the crust of the equilibrium phase relationships that we present here are difficult, and certainly there will be as many interpretations as interpreters. It is clear that, if the upper crust is not remotely of basaltic composition, then these data can have no value in deciphering seismic structure.

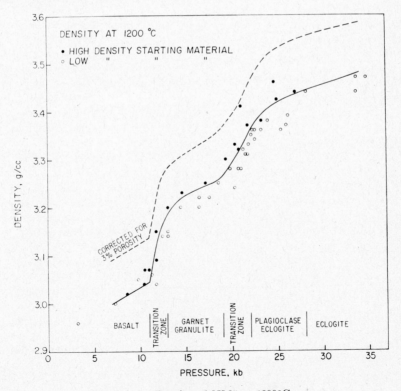

Fig. 3. Density of NM5 at 1200°C.

Further, temperature-depth profiles in the earth are of extreme uncertainty. They depend crucially on assumptions for the distribution of radio activity, contribution of heat from the upper mantle, and the values selected for thermal diffusivity. Indeed, some recent data on the pressure coefficient of thermal diffusivity [*Fujisawa et al.*, 1968] suggest that the values used for diffusivity in the deep crust may be in error by as much as a factor of 2. One also

needs to know whether equilibrium is obtained along a depth temperature profile. We have done some calculations for the temperatures at which the stable equilibrium phase will exist. We note that at 1200° in the laboratory experiment we attain equilibrium in approximately 5 minutes, whereas at 800° certainly a week or more is required for the attainment of equilibrium. From these two numbers, each of which is highly uncertain, one can calculate that, in

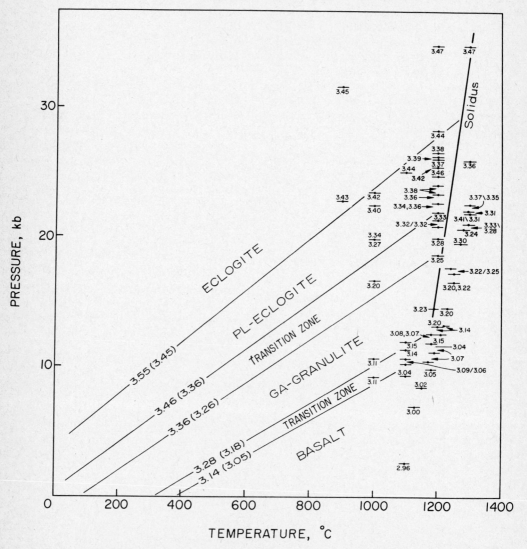

Fig. 4. Density and phase boundaries in NM5. Numbers above short bars are densities of quenched samples recrystallized from high-density starting material, and numbers beneath short bars are densities of quenched samples recrystallized from low-density starting material. Contours are phase boundaries as well as iso-density lines. On iso-density lines, the corrected densities are shown first and observed densities are in parentheses.

the geologic times of 10^6 to 10^7 years, recrystallization of basalt to garnet granulite or to eclogite should take place to a temperature as low as approximately 400°C. However, at lower temperatures basalt should remain metastable in the garnet granulite field for extremely long time periods.

In a very substantial number of places, the recent seismological data suggest a discontinuity in P-wave velocities, where velocities increase from 6.5 to 7.5 km/sec. This discontinuity occurs at shallow depths in many places in the continental crust. These seismic data have recently been reviewed by *Ito and Kennedy* [1970]. A second discontinuity appears in many places at somewhat greater depths. Here, velocities increase from 7.5 to 8.1 or 8.2 km/sec. Either one of the two discontinuities is sometimes referred to as the Mohorovicic discontinuity. Our tentative interpretation at this time is that the discontinuity of from 6.5 to 7.5 km/sec may occur along an isotherm at approximately 400°C where basalts, dacites, and andesites recrystallized to the stable assemblage of garnet granulite with more or less quartz. We also believe that the deeper discontinuity of from 7.5 to 8.2 km/sec may be the equilibrium transition of garnet granulites to eclogites. Thus, it is our interpretation that one of the discontinuities is the rate-process boundary along an isotherm and that the other discontinuity is the equilibrium boundary that occurs where the earth pressure temperature gradient crosses into the stable field of eclogitic rocks. Of course this interpretation is not unique. These two boundaries, one an equilibrium boundary and the other a rate-process boundary, will not move in harmony with each other. As temperatures increase in a crustal zone, the equilibrium boundary will move to greater depths and the rate-process boundary will move to shallower depths. The relative motions of these two discontinuities will be somewhat anharmonic, and an analysis of heat flow and the relative positions of these two boundaries could help resolve the question of whether the Mohorovicic discontinuity is a chemical change or a phase change. Our strongly prejudiced view is that in some places in the world the Mohorovicic discontinuity is marked by a sharp chemical contrast. However, we also strongly believe that, in other places in the world, particularly in the tectonic areas, the Mohorovicic discontinuity is a phase-change boundary and that it represents the local transition from basalt to garnet granulite or from garnet granulite to eclogite.

Acknowledgments. It is our pleasure to acknowledge our appreciation of the constructive criticism during the preparation of this manuscript that was provided by Drs. P. M. Bell, D. H. Green, and A. E. Ringwood.

We are grateful for grant N000-14-69-A-0200-4015 of the Office of Naval Research for partial financial support of these investigations.

REFERENCES

Akella, J., and G. C. Kennedy, Studies on anorthite + diopside$_{50}$-hedenbergite$_{50}$ at high pressures and temperatures, *Amer. J. Sci., 270,* 155–165, 1971.

Banno, S., Classification of eclogites in terms of physical conditions of their origin, *Phys. Earth Planet. Interiors, 3,* 405–421, 1970.

Birch, F., and P. LeComte, Temperature-pressure plane for albite composition, *Amer. J. Sci., 258,* 209–217, 1960.

Boettcher, A. L., and P. J. Wyllie, Jadeite stability measured in the presence of silicate liquids in the system $NaAlSiO_4$-SiO_2-H_2O, *Geochim. Cosmochim Acta., 32,* 999–1012, 1968.

Cohen, L. H., K. Ito, and G. C. Kennedy, Melting and phase relations in an anhydrous basalt to 40 kilobars, *Amer. J. Sci., 265,* 475–518, 1967.

Coleman, R. G., D. E. Lee, L. B. Beatty, and W. W. Brannock, Eclogites and eclogites: Their differences and similarities. *Geol. Soc. Amer. Bull., 76,* 483–508, 1965.

Engel, A. E. J., C. G. Engel, and R. G. Havens, Chemical characteristics of oceanic basalts and the upper mantle, *Geol. Soc. Amer. Bull., 76,* 719–734, 1965.

Eskola, P., On the eclogites of Norway, *Norsk Vidensk. Skr., Kristiania, 1, Mathematico-Naturwissenschaften* (8), 1–118, 1921.

Fermor, L. L., The relationship of isostacy, earthquakes, and vulcanicity to the earth's infra-plutonic shell, *Geol. Mag., 51,* 65–67, 1914.

Fujisawa, H., N. Fujii, H. Mizutani, H. Kanamori, and Syun-iti Akimoto, Thermal diffusivity of Mg_2SiO_4, Fe_2SiO_4, and $NaCl$ at high pressures and temperatures, *J. Geophys. Res., 73,* 4727, 1968.

Getting, I. C., and G. C. Kennedy, Effects of pressure on thermocouples to 40 kb and 1000°C, *Nat. Bur. Stand. Spec. Publ. 326,* 77–80, 1971.

Godovikov, A. A., and G. C. Kennedy, Eclogites, in *Problems of Petrology and Genetic Mineralogy,* vol. 1, (in Russian) edited by S. P. Korykobskii, pp. 48–61, Nauka, Moscow, 1969.

Green, D. H., and A. E. Ringwood, An experimental investigation of the gabbro to eclogite transformation and its petrological applications,

Geochim. Cosmochim. Acta, 31, 767–833, 1967*a*.

Green, D. H., and A. E. Ringwood, The genesis of basaltic magmas, *Contrib. Mineral. Petrol., 15,* 103–190, 1967*b*.

Green, T. H., and A. E. Ringwood, Genesis of the calc-alkaline igneous rock suite, *Contrib. Mineral. Petrol., 18,* 105–162, 1968.

Hariya, Y., W. A. Dollase, and G. C. Kennedy, An experimental investigation of the relationship of mullite to sillimanite, *Amer. Mineral., 54,* 1419–1441, 1969.

Ito, K., and G. C. Kennedy, Melting and phase relations in a natural peridotite to 40 kilobars, *Amer. J. Sci., 265,* 519–538, 1967.

Ito, K., and G. C. Kennedy, Melting and phase relations in the plane tholeiite-lherzolite-nepheline basanite to 40 kilobars with geological implications, *Contrib. Mineral. Petrol., 19,* 177–211, 1968.

Ito, K., and G. C. Kennedy, The fine structure of the basalt-eclogite transition, *Mineral. Soc. Amer. Spec. Pap., 3,* 77–83, 1970.

Kushiro, I., Clinopyroxene solid solutions formed by reactions between diopside and plagioclase at high pressure, *Mineral. Soc. Amer. Spec. Pap., 2,* 179–191, 1969.

Kushiro, I., and H. S. Yoder, Jr., Anorthite-forsterite and anorthite enstatite reactions and their bearing on the basalt-eclogite transformations, *J. Petrol., 7,* 337–362, 1966.

Luth, W. C., Studies in the system KalSiO$_4$-MgSiO$_4$-SiO$_2$-H$_2$O 1, Inferred phase relations and petrologic applications, *J. Petrol., 8,* 372–416, 1967.

Newton, M. S., and G. C. Kennedy, Jadeite, analcite, nepheline and albite at high temperatures and pressures, *Amer. J. Sci., 266,* 728–736, 1968.

Newton, R. C., and J. V. Smith, Investigations concerning the breakdown of albite at depth in the earth, *J. Geol., 75,* 268–286, 1967.

O'Hara, M. J., and H. S. Yoder, Jr., Formation and fractionation of basic magmas at high pressures, *Scot. J. Geol., 3,* 67–117, 1967.

Press, F., Earth models consistent with geophysical data, *Phys. Earth Planet. Interiors, 3,* 3–22, 1970.

Ringwood, A. E., and D. H. Green, An experimental investigation of the gabbro-eclogite transformation and some geophysical implications, *Tectonophysics, 3,* 383–427, 1966.

Segnit, R. E., and G. C. Kennedy, The reaction of muscovite and quartz to orthoclase and sillimanite at high temperatures and pressures, *Amer. J. Sci., 259,* 280–287, 1961.

Williams, A. F., *The Genesis of the Diamond,* vols. 1 and 2, Benn, London, 1932.

Yoder, H. S., Jr., and C. E. Tilley, Origin of basalt magmas: An experimental study of natural and synthetic rock systems, *J. Petrol., 3,* 342–532, 1962.

DISCUSSION

Sutton: You indicated that the depth of your double Moho might vary by as much as a factor of two. Is that so?

Kennedy: Yes, you can intersect the phase equilibrium boundary at almost any depth you like, depending on the temperature. There is a wide latitude possible in the temperature profile, depending on how the radioactivity is distributed.

Sutton: Is this true under the oceans as well as under the continents?

Kennedy: No, there is another problem here in that the rate of reaction is very slow at low temperatures, say 200°C. Basalts poured onto the ocean floor can persist almost metastably in the field where we are supposed to have eclogite. The boundary is controlled solely by reaction rate and time and not by the equilibrium situation.

Sutton: There are many places in the ocean where we have now found about 2 km of material with 7.5-km/sec velocity overlying 8.2 or 8.3 km/sec. If this is general, there are two nice breaks for you, but they come at much too low a pressure, I believe.

Kennedy: The first could be a rate-process boundary.

Landisman: Considerable support for your ideas comes from studies of the Rhine and East African rifts. Not only the seismic evidence but the rift cushion formed in the early stages seems to demand something along the lines you are proposing. However, the evidence seems to be confined to the rift systems—it is much harder to find in a normal continental area.

Kennedy: It seems to occur in the Great Plains and the Colorado Plateau.

6. WAVE GUIDE PROPAGATION

Analytical Investigations of Electromagnetic Wave Propagation in the Earth's Crust

JAMES R. WAIT

Cooperative Institute for Research in Environmental Sciences, University of Colorado and Institute for Telecommunication Sciences, Boulder, Colorado 80302

Abstract. Calculated attenuation rates for electromagnetic wave propagation in the earth-crust wave guide are discussed. Although detailed intercomparisons of published data are not feasible, it appears that there is an over-all consistency in the understanding of the important factors. It is agreed that basement rock conductivities of the order of 10^{-6} or lower are needed to provide attenuation rates of less than one db per kilometer.

Recently there seems to be a resurgence of interest in investigation of the earth crust as a low-loss propagation path for electromagnetic waves. It is the purpose of the present paper to review some of the past analytical studies that contain estimates of the expected attenuation rates of the subsurface wave guide modes. We will not consider the 'up-over-and-down' mode, since it does not deal with a horizontal transmission path directly through the earth's crust.

A number of review papers [*Ryazantsev and Shabelnikov,* 1965; *de Bettencourt,* 1966; *Levin,* 1966; *Wait,* 1966a; *Makarov and Pavlov,* 1966] have been written on the general subject of radio propagation in the lithosphere, including ground wave and ionospheric wave mechanisms for buried antennas. Therefore, we feel it is not necessary to discuss the problem in a general context, nor do we plan to consider antenna design or noise factors.

'INVERTED IONOSPHERE' CONCEPT

One of the first papers on subsurface radio propagation was by *Wheeler* [1961]. He postu-

lated the existence of a deep wave guide that has not yet been explored. He goes on to say, 'This wave guide extends under all the surface area, so it suggests the possibility of wave propagation under the ocean floor. This might enable communication from land to a submarine located on or near the ocean floor.' Wheeler envisages the wave guide as low-loss dielectric basement rock bounded above and below by conductive boundaries. The upper boundary, in this case, is formed by the geological strata located between the surface and the basement. Its conductivity is provided by electrolytic solutions in the rock pores and, to some extent, by semi-conductive minerals. On the other hand, the lower boundary is provided by high-temperature conductivity in the lower basement rocks. Wheeler refers to the lower conductive region as the 'inverted ionosphere,' since the temperature increases with depth. A more detailed picture of the electrical and thermal properties of the crust is found in papers by *Tozer* [1959], *Noritomi* [1961], *Watt et al.* [1963], and *Keller* [1963].

In Figure 1 we indicate how Wheeler en-

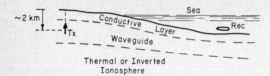

Fig. 1. The earth-crust wave guide with a buried transmitter (Tx) and submerged receiver (Rec).

visages the subsurface wave guide as a medium for communication via a shore transmitting station and an underwater receiving station. 'The latter may be a submarine on the bottom of the ocean.' The sender is supposed to launch a vertically polarized TEM (traverse-electromagnetic) wave by means of a vertical wire projecting into the basement rock. It is then intended that some power leaks out of the wave guide into the ocean; this power is sampled by the receiving antenna.

Wheeler's conjecture on how the temperature and resulting conductivity vary with depth is illustrated in Figure 2. The conductivity, in the wave guide region between depths of about 2 and 20 km, is supposed to be of the order of 10^{-6} to 10^{-11} mho/m, with a dielectric constant $K \simeq 6$. For a model of this type, Wheeler suggests (without supporting calculations) that the signal will attenuate 1 decibel for each 15 km in the wave guide at a frequency of 1.5 kHz. A practical radiator is envisaged as a vertical wire 2 km in length with a 2% radiation efficiency.

WAVE GUIDE FORMULATION

Another relatively early paper is by *Wait* [1963], who attempted to put the problem on a quantitative basis. His model is illustrated in Figure 3. Choosing a cylindrical coordinate system (ρ, ϕ, z), the earth's surface is taken to be at $z = -h$. The highly conducting surface layers are all grouped together and are represented by a homogeneous slab whose (complex) permittivity is K_g relative to free space. The plane $z = 0$ is taken to be the (sharp) boundary between the upper conducting region and the basement dielectric. The relative permittivity of the basement layer is denoted by the complex function $K(z)$, which is permitted to increase continuously with depth. In the ideal situation, $K(z)$, as $z \to 0$, would be a perfect dielectric. This initial value of $K(z)$ is denoted K_0. The form of the profile is indicated in Figure 3.

It is convenient to choose a reference plane at $z = z_d$ that can be described as the lower boundary of the earth-crust wave guide. The choice of its precise location is arbitrary, but for a number of reasons it is desirable to select z_d as the level where the conduction currents begin to dominate the displacement currents. In the earth-crust wave guide and for frequencies in the VLF band, it appears that z_d is of the order of 30 km.

The earth-crust wave guide is most conveniently excited by a vertical electric antenna that is coaxial with a vertical borehole. A second possibility is a long solenoidal coil whose axis is also coaxial with the borehole. In most applications, these two basic antenna types can be represented by Hertzian electric or magnetic dipoles, respectively. The formalism for treating these two problems is very similar [*Wait*, 1970]. By the principle of duality, the solution of one can be found from the other by a simple interchange of parameters. In this paper, only the electric dipole of moment Ids is considered.

The fields of the electric dipole at $z = z_1$ can be derived from a Hertz vector that has only a z component Π_z. The formal solution of the problem is already known. Thus, for small values of z and z_1,

$$\Pi_z = -g \int_{-\infty}^{+\infty} F(C) H_0^{(2)}(kS\rho) \frac{SdS}{C} \qquad (1)$$

where

$$F(C) = \frac{(e^{ikCz} + R_1 e^{-ikCz})(e^{ikC(z_d - z_1)} + R_2 e^{-ikC(z_d - z_1)})}{e^{ikCz_d}(1 - R_1 R_2 e^{-2ikz_d C})}$$

$$C = (1 - S^2)^{1/2} \qquad k = (\epsilon_0\mu_0)^{1/2}\omega\sqrt{K_0} \qquad (2)$$

and where S is the variable of integration.

$H_0^{(2)}(kS\rho)$ is the Hankel function of the second kind of argument, $kS\rho$:

$$g = -I \, ds/(8\pi\hat{\eta}_0)$$

where $\hat{\eta}_0 = \eta_0/\sqrt{K_0}$ and

$$R_1 = (C - Z_1/\hat{\eta}_0)/(C + Z_1/\hat{\eta}_0) \qquad (3)$$

and

$$R_2 = -e^{\alpha C} \qquad (4)$$

The quantity R_1 is a reflection coefficient at the upper boundary of the guide which is expressed in terms of the surface impedance Z_1 at $z = 0$. On the other hand, R_2 is a reflection coefficient of the thermal ionosphere that is evaluated at $z = 0$ but that is referred to the level $z = z_d$. The coefficient α in the exponent is also a function of C, but for many purposes it can be regarded as a constant.

The dimensionless factor S can be interpreted as the sine of the angle of incidence of a plane wave. In the present situation, the integration contour extends over all real values of S from $-\infty$ to $+\infty$. Thus, complex angles are included in the complete spectrum.

The poles of the function $F(C)$ correspond to the wave guide modes. This condition is written

$$1 - R_1 R_2 \exp\left(-2ikz_dC\right) = 0 \qquad (5)$$

or

$$R_1 R_2 \exp\left(-2ikz_dC\right) = \exp\left(-i2\pi n\right)$$

where n may be any integer and the solutions are designated by C_n where $n = 0, 1, 2, \cdots$. The corresponding values of S are designated by S_n.

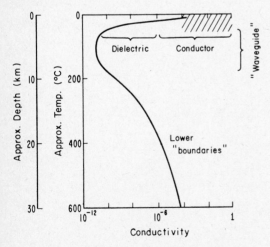

Fig. 2. Wheeler's conductivity profile.

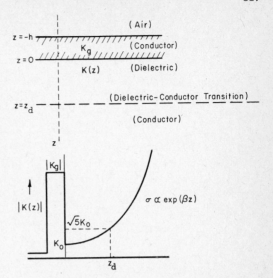

Fig. 3. Wait's model, with the appropriate conductivity profile.

The integral for Π_z can now be approximated as a sum of modes that actually are the residues of the poles at $S = S_n$. From this representation, an expression for the vertical electric field E_z can be found by differentiation. Furthermore, the Hankel function can be approximated by the first term of its asymptotic expansion, since $|kS_n\rho| \gg 1$ in cases of practical interest. The final results can be expressed in the form $E_z \cong E_0 W$, where E_0 is a reference field given by

$$E_0 = \frac{-i\mu_0\omega I \, ds}{2\pi\rho} \exp\left(-ik\rho\right) \qquad (6)$$

and W is a dimensionless factor that contains the essence of the propagation problem. It is given [*Wait*, 1963] by

$$W = \frac{(2\pi k\rho)^{1/2}}{kh} e^{ik\rho} \exp\left(-i\pi/4\right)$$

$$\cdot \sum_{n=0}^{\infty} \delta_n S_n^{3/2} e^{-ikS_n\rho} f_n(z_1) f_n(z) \qquad (7)$$

where

$$\delta_n = \left[\left(1 + i\frac{\partial R_1 R_2/\partial C}{2kz_d R_1 R_2}\right)_{C=C_n}\right]^{-1} \qquad (8)$$

The quantity E_0 can be interpreted as the field of the dipole if it is placed on the plane surface of a perfect conductor in a homogeneous medium of constant wave number k. The functions $f_n(z_1)/f_n(0)$ and $f_n(z)/f_n(0)$ can be described as 'depth-gain' functions, since they are analogous to the 'height-gain' functions in atmospheric radio propagation [Wait, 1970]. They are defined explicitly by the relation

$$2f_n(z) = [e^{-ikC_nz} + R_1e^{ikC_nz}]R_1^{-1/2} \qquad (9)$$

In most cases of practical interest, $R_1 \cong +1$, and thus

$$f_n(z) \cong \cos kC_nz \quad \text{and} \quad f_n(z_1) \cong \cos kC_nz_1$$

The form of the height-gain function given by (9) is valid only in the homogeneous region $z > 0$ at the top of the basement dielectric.

The quantity δ_n can be described as an excitation factor because it is a measure of the efficiency in which the modes can be excited. For modes of low attenuation, under the conditions $R_1 \cong 1$ and $R_2 \cong -1$, it is seen that δ_n can be replaced by unity. In the next approximation, R_2 is replaced by $-\exp(\alpha C)$, where α is regarded as a constant. Thus

$$\delta_n \cong \left(1 + \frac{i\alpha}{2kz_d}\right)^{-1} \qquad (10)$$

In the present development, earth curvature has been neglected implicitly. By analogy to VLF ionospheric propagation [Wait, 1970], this is justified when

$$\text{Re }[(ka/2)^{1/3}C_n] \gg 1 \quad \text{and} \quad |C_n|^2 \gg z_d/a$$

where a is the radius of the earth.

DISCUSSION OF THE MODE SUM

The total field E_z must be obtained by performing the summation indicated by (7). The significant terms in the sum correspond to values of n where $-\text{Im }kS_n$ is small. In fact, the attenuation of the modes is numerically equal to $-\text{Im }kS_n$ in nepers per unit distance. The phase velocity of the modes is equal to $1/\text{Re}S_n$ relative to plane waves in an unbounded medium of permittivity K_0. Numerical values of the attenuation and phase velocity, as well as the calculation of the total field, require that the solutions C_n of the modal equation (5) be obtained. In general, a rigorous solution of this equation is difficult; however, if attention is restricted to modes of low attenuation, the task may be greatly simplified.

When the upper layer is well conducting,

$$R_1 = \frac{C - Z_1/\hat{\eta}_0}{C + Z_1/\hat{\eta}_0} \cong \exp\left(\frac{-2Z_1}{\hat{\eta}_0C}\right) \qquad (11)$$

provided $|Z_1/\hat{\eta}_0C|^3 \ll 1$. Furthermore, if the thickness h of the upper conducting layer is greater than several skin depths, it is permissible to regard it as a semi-infinite medium. Thus

$$Z_1/\eta_0 \cong G^{1/2} \exp(i\pi/4)$$

where $G = [K_0/(iK_g)]$ is expressed in terms of the (complex) permittivities of the basement rock and the upper layer. For modes of low attenuation, G is a small real number, i.e.,

$$G \cong \sqrt{K_0}/(60\sigma_g\lambda_0)$$

where $\lambda_0 = 2\pi/k_0 = \lambda\sqrt{K_0}$ is the effective wavelength at top of dielectric basement.

With the approximations discussed above for R_1 and the exponential representation for R_2, it follows that (5) can be written in the form

$$C(\alpha - i2kz_d) - \frac{2G^{1/2} \exp(i\pi/4)}{C}$$
$$= -2\pi i(n - 1/2) \qquad (12)$$

This equation can be solved readily by a perturbation method. For the first approximation, $\alpha(C)$ can be regarded as a constant independent of C, and G is set equal to zero. Thus, solutions are

$$C \cong C_n^{(0)} = \frac{\pi(2n - 1)}{2kz_d + i\alpha} \qquad (13)$$

where $n = 1, 2, 3, \cdots$. To account for the finite value of G, the quantity G/C is replaced by $G/C_n^{(0)}$ in (12). Thus the corrected value of C is given by

$$C_n$$
$$\cong \frac{2\pi(n - 1/2) - 2G^{1/2} \exp(-i\pi/4)/C_n^{(0)}}{2kz_d + i\alpha}$$
$$(14)$$

Higher-order approximations can be found from further iterations. Of special interest is the function S_n as it determines the characteristics of the modes. Thus

$$S_n = (1 - C_n^2)^{1/2} \cong 1 - C_n^2/2$$

where the latter approximation is valid for low-order modes. Then, to within the same approximation,

$$kS_n \cong kS_n^{(0)} + \frac{\exp(-i\pi/4)G^{1/2}}{z_d\left(1 + \dfrac{i\alpha}{2kz_d}\right)} \qquad (15)$$

where

$$S_n^{(0)} \cong 1 - (C_n^{(0)})^2/2$$

It should be noted that in this development the mode of lowest attenuation is $n = 1$ rather than $n = 0$. The latter mode (TEM) is of very high attenuation in the present situation.

The wave number k can be regarded as slightly complex and can be written conveniently in the form

$$k = k_0(1 - i\,\delta)^{1/2} \cong k_0(1 - i\,\delta/2)$$

where k_0 is the real wave number and δ is the 'loss tangent.' Again, for modes of low attenuation, δ must be small.

We are now in the position to demonstrate the principle of 'superposition of losses.' This principle requires that the quantities $|\alpha/kz_d|$, G, and δ all be small compared with unity. For any useful modes, these conditions would be satisfied. Thus, the attenuation coefficient for mode n, in nepers per unit distance, is written

$$-\operatorname{Im} kS_n \cong A_n + B_n + D_n$$

where

$$A_n = \operatorname{Im}\left\{\frac{k_0}{2}\left[\frac{\pi(2n-1)}{2k_0z_d + i\alpha}\right]^2\right\}$$

$$\text{for} \quad n = 1, 2, 3, \cdots \qquad (16)$$

$$B_n = \operatorname{Im} \exp(i3\pi/4)G^{1/2}/z_d \cong G^{1/2}(2^{1/2}z_d) \qquad (17)$$

$$D_n = k_0\,\delta/2 \cong |60\pi\sigma_0/\sqrt{K_0}| \qquad (18)$$

The attenuation rate A_n is associated with the leakage of energy down into the thermal ionosphere. Generally, it can be seen that the attenuation increases with mode number. For the earth-crust wave guide, this can be regarded as an intrinsic attenuation, since even under ideal conditions of a dielectric basement (i.e., $\delta = 0$) and a perfectly conducting upper layer (i.e., $G = 0$) the field is attenuated from leakage. The attenuation coefficients B_n and D_n account for the finite conductivity of the upper

layer and the ohmic loss in the dielectric basement, respectively. In the first approximation, these do not depend on mode number; therefore, the subscripts on B_n and D_n can be dropped.

The calculations of the attenuation rates B and D are very straightforward. (Examples of such calculations are given as follows. For $K_0 = 9$, $\lambda_0 = 15$ km (or $\lambda = 45$ km), $\sigma_g = 10^{-2}$ mho/m, $\sigma_0 = 10^{-8}$ mho/m, and $z_d = 40$ km, one finds that $B \cong 2.8$ db/1000 km and $D \cong 5.4$ db/1000 km. If $\sigma_g = 10^{-4}$, $\sigma_0 = 10^{-7}$, and B and D are increased to 28 and 54 db/1000 km, respectively. Note that σ_0 is defined in (18).) The determination of the intrinsic attenuation rates requires a knowledge of the coefficient α. Therefore, it is necessary to understand something about propagation in an inhomogeneous stratified medium.

SPECIFIC RESULTS FOR THE EXPONENTAL PROFILE

The variation of the (complex) relative permittivity $K(z)$ with depth must now be specified. For convenience, the variation is taken to be exponential in form [Wait, 1963]. The specific form chosen is

$$K(z) = K_0[1 - i2 \exp \beta(z - z_d)] \qquad (19)$$

for $0 < z < \infty$. At $z = 0$, it is noted that

$$K(0) = K_0[1 - i2 \exp(-\beta z_d)] \cong K_0$$

provided, say, that $\beta z_d > 3$. For cases considered here, the latter restriction is always satisfied.

Defining the profile in the above fashion means that the initial permittivity of the basement rock is K_0, which, under ideal conditions, is real. As z increases, $K(z)$ becomes complex and, at the reference level $z = z_d$,

$$K(z_d) = K_0[1 - i2] \cong K_0\sqrt{5} \exp[-i1.10]$$

At very great depths

$$K(z) \cong -i|K(z)| \cong -i2K_0 \exp \beta(z - z_d)$$

An equivalent way of stating the situation is to regard the medium $z > 0$ as a variable conducting medium whose conductivity $\sigma(z)$ is defined by

$$\sigma(z) = 2\epsilon_0\omega K_0 \exp[\beta(z - z_d)]$$

where ϵ_0 is the (absolute) permittivity of free space. On the basis of published data [*Watt et al.*, 1963], it appears that β is of the order of 1 km^{-1}, where $(z - z_d)$ is also to be measured in kilometers. However, more recent estimates [e.g., *Levin*, 1966] suggest that β is more like 0.2 km^{-1}.

The plane wave reflection coefficient for an exponential layer of the form has been investigated on a number of occasions. The usual case discussed is that of the electric vector parallel to the stratification (i.e., horizontal polarization). The solution for this case was first given by *Elias* [1931]. For the case when the magnetic vector is parallel to the stratification, it is considerably more complicated. Solutions of the latter problem have been obtained by a numerical method. The details of the method are reported elsewhere [*Wait*, 1970].

The reflection coefficient R_2 for the present problem can be defined as the coefficient that connects an upgoing wave of the form exp $(ikCz)$ with a downgoing wave of the form exp $(-ikCz)$. Thus, the z dependence of the field in the region $z < 0$ has the form

$$e^{-ikCz} + R_2 e^{ikCz} e^{-i2kCz_d}$$

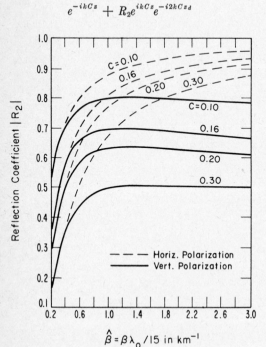

where $z = z_d$ is the reference level described above. A number of real values of C corresponding to oblique incidence are chosen; $|R_2|$ is shown plotted, in Figure 4, as a function of the parameter $\beta\lambda_0/15$ (which is also denoted $\hat\beta$). This particular parameter is chosen so that the abscissa is simply β when the wavelength λ_0 is 15 km. Curves are shown for both horizontal and vertical polarization. It is apparent that for smaller values of $\hat\beta$ the reflection coefficient may become quite small. In this instance, the medium is effectively slowly varying. It is apparent that, as $\hat\beta$ increases, the reflection coefficient becomes greater. For horizontal polarization, the reflection coefficient soon becomes quite near unity. In the case of vertical polarization, $|R_2|$ reaches a broad maximum for $\beta\lambda_0/15$; it becomes approximately equal to 1.5. For larger values of β it gradually decreases. Ultimately, it will rise again toward unity when β is very large. In the present investigation, we exploit the first broad maximum that is shown in Figure 4. We also utilize the fact that α is nearly independent of C in the complex range of physical interest. This point is discussed elsewhere in the present context [*Wait*, 1963].

The results are now readily applied to attenuation rates for the ideal earth-crust wave guide. For purposes of illustration, it is assumed that $\lambda_0 = 15$ km. It is also assumed that $G = 0$ and $\delta = 0$; therefore, the only loss is due to leakage into the deep conducting region of the basement. Thus the attenuation rate is obtained from (16) by using the complex values of α derived above. The intrinsic attenuation rates A_1 and A_2 for the modes of lowest attenuation are shown in Figures 5a and 5b, respectively. The units here are given in decibels/1000 km of path length. The abscissa is the gradient parameter β in km^{-1}. The individual curves correspond to various values of the ratio z_d/λ_0, which is a measure of the electrical thickness of the wave guide channel.

By a suitable scaling of parameters, the curves in Figures 5a and 5b can be made to correspond to other wavelengths. For example, if $\lambda_0 \neq 15$, β is replaced by $\hat\beta$ and the indicated attenuation rates are then increased by the factor $15/\lambda_0$. It is understood that z_d/λ_0 should remain constant. It is certainly apparent from the calculated results that the attenuation in such an idealized earth-crust wave guide is indeed very small.

Fig. 4. The magnitude of the reflection coefficient for the exponential profile.

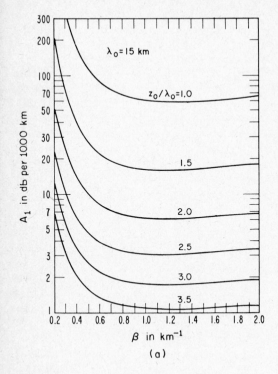

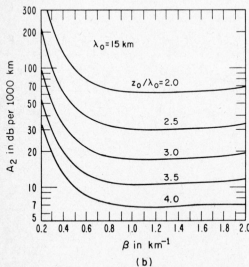

Fig. 5. Figure 5a shows the intrinsic attenuation rate for the first mode as a function of β for various values of z_d/λ_0, which is the width of the guide in wavelengths; figure 5b shows the intrinsic attenuation rate for the second mode as a function of β.

OTHER MODELS

In another approach to the earth-crust wave guide, *Brown and Gangi* [1963] considered a homogeneous slab model bounded by two half-spaces of relatively high conductivity. They used a mode series representation for the field of a vertical electric dipole within the low-loss slab. Calculations of field strength versus range were then made for several low frequencies. They compared their results with data from a scale model with a frequency scale factor of 10^5. An example is shown in Figure 6, where the scaled slab thickness h is 35 km, dielectric constant is 4, and the frequency is 8 kHz. These curves illustrate the classical type of mode interference that is reminiscent of long-distance VLF propagation in the earth-ionosphere wave guide.

A very similar slab model was considered by *Viggh* [1963]. His slab of width h with conductivity $\sigma = 10^{-6}$ mho/m and dielectric constant $K = 10$ was bounded by a half-space with conductivity $\sigma_1 = 10^{-2}$ and dielectric constant $K_1 = 10$. The modal equation for this configuration is obtained by a transverse resonance condition that is equivalent to the mode equation used in the earth-ionosphere wave guide when earth curvature is neglected. In fact, it is equivalent to equation 5 above if $R_1 = R_2 = $ Fresnel reflection coefficient. Because of the relative thinness of Viggh's slab, only the TEM mode was considered to be significant. The general nature of his results is illustrated in Figure 7, in which the interrelationship between frequency, slab thickness,

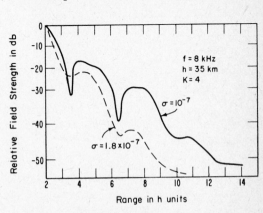

Fig. 6. Distance dependence of the field strength in the Brown-Gangi model. The scaled slab thickness is h, the dielectric constant is K, and the frequency is f.

and attenuation rate is shown. With suitable scaling of the parameters, these curves can be applied to other conductivities and dielectric constants, provided the conductivity ratio is fixed at 10^4. Clearly, other plots of this kind are needed in order to survey various situations adequately.

Low-loss geologic wave guides bounded by highly conducting mediums have also been considered by *Gabillard* [1966] and his associates in France. They were able to exploit the guiding properties of a uniform gypsum layer buried at about 80 m below the earth's surface. They used an asymptotic evaluation of the complex integral representation for the received field. For their purpose, this seemed to be more convenient than a modal or residue series evaluation. A similar approach was used by *Watt et al.* [1966].

The slab model can be generalized in a number of ways. The author had considered three- and four-layer models, with the intention of studying the relative contributions of the lateral waves that exist in addition to the guided waves [*Wait*, 1966b, c]. For propagation between deeply buried terminals, the lateral waves do not seem to be important and further discussion here is not warranted.

MORE RECENT WORK

A rather recent analytical study of the earth-crust wave guide was carried out by *Schwering et al.* [1968]. Like *Wait* [1963], they considered a special analytical form for the profile. Specifically for the dielectric wave guide region, they considered an inverse square law that is identified by the solid curve in Figure 8. Also in this model, it was assumed that the upper zone was perfectly conducting. The resultant model profile was considered to be a good approximation to an earlier profile suggested by *Levin* [1966] for the continental crust. Levin's profile is shown by a dashed curve in Figure 8. The inverse square profile permits an analytical expression for the wave functions; the resulting mode equation can be solved without too much difficulty. The results for the dominant mode (i.e., $m = 1$) are shown in Figure 9, in which the attenuation rate is plotted as a function of frequency from 0.5 to 10 kHz. To indicate how the attenuation rate varies with mode number, we list *Schwering et al.*'s [1968] attenuation rate at 1 kHz for each mode number as follows:

m	Attenuation, db/km
1	0.256
2	1.17
3	1.99
4	2.79
5	3.61
6	4.42
7	5.26
8	6.08
9	6.92
10	7.74

These results indicate that the first mode is completely dominant at distances in excess of 50 km. This behavior is not inconsistent with earlier calculations for similar models. For example, *Brown and Gangi* [1963] would predict that for a uniform slab model with $f = 3$ kHz, $h = 35$ km, $K = 10$, and $\sigma = 10^{-7}$ mho/m the attenuation rate of the least attenuated mode will be 0.052 db/km. This is to be compared with *Schwering et al.*'s [1968] value of 0.256 db/km for the same frequency and same parameters of the upper basement.

It is interesting to note that *Schwering et al.* [1968] compute the attenuation rate of 1 kHz by using *Wait*'s [1963] method and obtaining a dominant mode attenuation rate of 0.275 db/km. In view of the different approach in obtaining the formulation and obtaining the

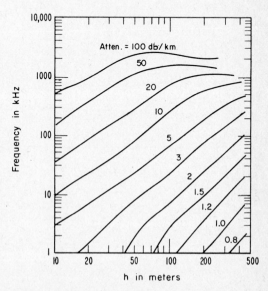

Fig. 7.   Attenuation in a thin crustal wave guide according to Viggh.

eigenvalue, this value is considered to be in reasonable agreement with the value 0.256 db/km given above.

In general, it would appear that the calculation of *Wait* [1963] for an exponential complex permittivity profile leads to lower attenuations than do the calculations of *Schwering et al.* for their power law model. However, this difference is not really surprising, since Wait used than do the calculations of Schwering et al. 1963], which suggested steeper gradients than the later more educated estimates of *Levin* [1966]. The latter should be appropriate for a crustal temperature gradient of the order of 20°C/km of depth.

FINAL REMARKS

In our discussions, we have restricted attention to uniform earth-crust wave guides whose properties vary only in the vertical direction; also, we have not considered the effects of earth curvature. Although extensions of the theory to account for nonuniformities and curvature in the direction of propagation are possible, there is some question at the moment whether additional refinements in the theory are worthwhile until we obtain reliable information on subsurface conductivity at depths of the order of 3 to 20 km, where the low-loss wave guide region will be located (if it exists at all). In any case, the modal calculations for a wide class of uniform wave guides can be used if at a later date information on the lateral changes in the basement profile is forthcoming.

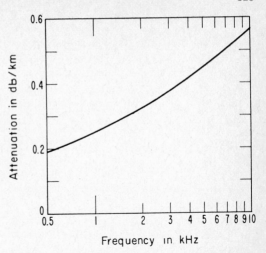

Fig. 9. Attenuation rate for the SPL model.

For example, we can exploit the similarity of mode conversion in the earth-ionosphere wave guide at a sunrise boundary [e.g., *Wait*, 1970] with the expected change in the earth-crust wave guide at the continental margins. In both cases, energy can be transferred between the propagating modes, and the relative energy loss is dependent on the nature of the transition region. Unless the nature of the transition region can be specified, at least in general terms, analytical progress will be limited.

Acknowledgments. The motivation for this review paper was the renewed interest in crustal communication stimulated by J. G. Heacock, of the Office of Naval Research. I am also grateful to Mrs. Loys Gappa, who helped prepare the manuscript.

REFERENCES

Brown, G. L., and A. F. Gangi, Electromagnetic modeling studies of lithospheric propagation, *Trans. IEEE GE-1*, 17, 1963.

de Bettencourt, J. T., Review of radio propagation below the earth's surface, paper presented at 15th General Assembly of URSI, Munich, Germany, August 1966.

Elias, G. J., Reflection of electromagnetic waves at ionized media with variable conductivity and dielectric constant, *Proc. IRE, 19,* 891, 1931.

Gabillard, R., Communications a travers le sol, in *Proc. NATO/AGARD Symp. Sub-Surface Commun.*, Paris, April 1966.

Keller, G. V., Electrical properties in the deep crust, *Trans. IEEE Antennas Propagat., 11,* 344, 1963.

Levin, S. B., Lithospheric radio propagation, *Proc. NATO/AGARD Symp. Sub-Surface Commun.*, Paris, April 1966.

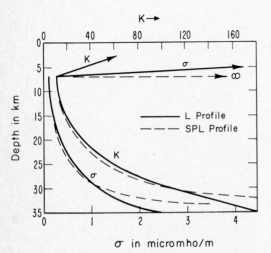

Fig. 8. Profiles studied by Levin and colleagues.

Makarov, G. I., and V. A. Pavlov, Survey of works connected with subterranean radio wave propagation, in *Problems in Wave Diffraction and Propagation, Ser.* 5, No. 4, p. 138, Leningrad Univ., 1966.

Noritomi, K., The electrical conductivity of rock and the determination of the electrical conductivity of the earth's interior, *J. Mining Coll. Akita Univ. 1,* 27, 1961.

Ryazantsev, A. M., and A. V. Shabelnikov, Propagation of radio waves through the earth's crust (a review), *Radiotekh. Elektron., 10,* 1923, 1965.

Schwering, F. K., D. W. Peterson, and S. B. Levin, A model for electromagnetic propagation in the lithosphere, *Proc. IEEE, 56,* 799, 1968.

Tozer, D. C., The electrical properties of the earth's interior, in *Physics and Chemistry of the Earth,* vol. 3, p. 414, Pergamon, New York, 1959.

Viggh, M. E., Modes in lossy stratified media with applications to underground propagation of radio waves, *Trans. IEEE Antennas Propagat., 11,* 318, 1963.

Wait, J. R., The possibility of guided electromag-

netic waves in the earth's crust, *Trans. IEEE Antennas Propagat., 11,* 330, 1963.

Wait, J. R., Some factors concerning electromagnetic wave propagation in the earth's crust, *Proc. IEEE, 54,* 1020, 1966a.

Wait, J. R., Electromagnetic propagation in an idealized earth crust waveguide, *Radio Sci., 1,* 913, 1966b.

Wait, J. R., Influence of a sub-surface insulating layer on electromagnetic ground wave propagation, *Trans. IEEE Antennas Propagat., 14,* 755, 1966c.

Wait, J. R., *Electromagnetic Waves in Stratified Media,* 2nd ed., Pergamon, Oxford, England, 1970.

Watt, A. D., F. S. Mathews, and E. L. Maxwell, Some electrical characteristics of the earth's crust, *Proc. IEEE,* 51, 897, 1963.

Watt, A. D., G. F. Leydorf, and A. N. Smith, Notes regarding possible field strength versus distance in earth crust waveguides, in *Proc. NATO/AGARD Symp. Sub-Surface Commun.,* Paris, April 1966.

Wheeler, H. A., Radio-wave propagation in the earth's crust, *Radio Sci., 65D,* 189, 1961.

DISCUSSION

Ward: Why do you like the functional form for a lower boundary rather than just choosing an n-layered medium and doing the calculations?

Wait: A partial answer is that, if you take an exponential variation of conductivity with depth, the wave functions for horizonal polarization are very simple. For vertical polarization this is not the case and a slab model is used, as discussed in the paper.

Madden: Where does the 'up-over-and-down' mode start to dominate over propagation through the crustal wave guide?

Wait: Probably the overburden should be about 0.5 km thick before you can discount the 'up-over-and-down mode,' but the best test is to look at the depth-gain functions. If you have propagation through the earth, the signal may increase with depth, whereas the 'up-over-and-down' signal will attenuate exponentially.

Phinney: Based on the calculations, for any given thickness of the high resistivity wave guide, there is an effective cut-off for the lowest mode. I just wondered if you could give us some idea of the trade-off between the lowest frequency to operate at and the thickness of the wave guide.

Wait: Actually, for this kind of propagation there is no cutoff for the zero mode. It is

a TEM (transverse electromagnetic) mode, and there will never be a cutoff.

Levin: It is not a proper cutoff, but you are losing a lot by attenuation as the effective wavelength increases.

Phinney: No, you're still gaining.

Gangi: The attenuation gets less and less.

Phinney: No. In the body it gets less and less. But now, at these wavelengths, you are seeing the highly conductive boundaries.

Gangi: A lot depends on whether you can get through, but it is essentially a trapped mode, you hope.

Phinney: Okay.

Wait: Actually there are some trade-off parameters that come into this. As you go to very low frequency, the attenuation rate is very low, but it will be very difficult to excite the modes by a physical source.

Levin: The permittivity in the crust increases with depth whereas in the ionosphere it decreases. How does this bear on the ionospheric analogy?

Wait: At the bottom of the basement, where the reflection is taking place, the conduction currents dominate the displacement currents so, as a practical matter, I think the error incurred here is very small. In any case, the ionosphere at VLF does behave as a conductor, so the analogy is still very good.

Note on Calculations of Propagation Parameters for an Idealized Earth-Crust Wave Guide

JAMES R. WAIT AND KENNETH P. SPIES

Institute for Telecommunication Sciences, Boulder, Colorado   80302

Abstract. The attenuation rates of low-order modes in an earth-crust wave guide are calculated for a uniform slab model with a finitely conducting lower boundary. The analysis indicates that the controlling factor is the conductivity of the bulk medium in the wave guide. The graphical results given here are based on assumed crustal conductivities which are considered reasonable.

In *Wait* [1971], propagation theories for the earth-crust wave guide were reviewed. Examples of calculations were given, and tentative conclusions pertaining to the limitations of radio communication in the lithosphere were presented. Although the existence of low-loss layers has not been proven, the possibility of exploiting guided modes in a subsurface wave guide should not be overlooked. With this situation in mind, we are embarking on a program to evaluate the relevant parameters associated with an idealized earth-crust wave guide. Hopefully, the results also have some general significance for guided wave propagation in inhomogeneous dissipative mediums.

We shall consider a very special case of the general stratified medium formulation given elsewhere [*Wait*, 1970]. The 'earth-crust wave guide' is assumed to be a homogeneous slab with conductivity σ and dielectric constant $K\epsilon_0$. Its thickness is h, the upper boundary is taken to be a perfect conductor (i.e., $\sigma_g = \infty$), and the lower boundary is a homogeneous half-space of conductivity σ_e and dielectric constant $K_e\epsilon_0$. The situation is depicted in Figure 1, in which the modes are excited by a vertical electric dipole located within the wave guide at depth h_1 from the top boundary.

This may seem like a highly oversimplified model of the earth's crust, and we agree that it is. Nevertheless, the essential loss mechanisms are accounted for in a reasonably satisfactory manner. In addition, it is logical to exhaust the possibilities of a simple model before elaborat-ing the formulation to account for secondary factors.

The relevant equation for the TM wave guide modes is given by

$$R_e \exp(-i2khC) = 1 \qquad (1)$$

where $ik = [i\mu_0\omega(\sigma + i\epsilon_0 K\omega)]^{1/2}$ is the intrinsic propagation constant of the bulk medium in the wave guide. R_e is a reflection coefficient associated with the bottom boundary, and C is the cosine of the (complex) angles of the waves in the guide. The quantity of principal interest is the attenuation α_n of the modes in the horizontal direction. Thus, $\alpha_n = -\operatorname{Im}(kS_n)$ nepers/m where

$$S_n = (1 - C_n^2)^{1/2}$$

and the discrete values of C_n ($n = 0, 1, 2, \cdots$) satisfy (1).

The reflection coefficient R_e is defined by

$$R_e = (C - \Delta_c)/(C + \Delta_c) \qquad (2)$$

where

$$\Delta_c = Z_c/\eta$$

$$Z_c = \frac{[k^2 S^2 - k_c^2]^{1/2}}{\sigma_c + i\epsilon_0 K_c\omega} = \frac{i[k_c^2 - k^2 S^2]^{1/2}}{\sigma_c + i\epsilon_0 K_c\omega}$$

$$\eta = [(i\mu_0\omega)/(\sigma + i\epsilon_0 K\omega)]^{1/2}$$

$$ik_c = [i\mu_0\omega(\sigma_c + i\epsilon_0 K_c\omega)]^{1/2}$$

Here, Δ_c can be interpreted as a dimensionless surface impedance, of the lower boundary,

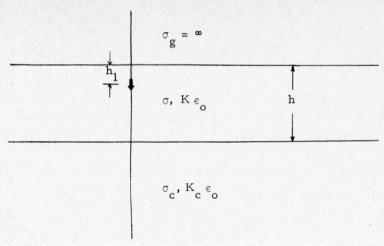

Fig. 1. Geometry.

normalized by the intrinsic wave impedance η of the bulk medium in the wave guide.

NUMERICAL EVALUATION OF C_n

The values of C_n were calculated by repeated applications of Newton's method, as outlined below. To this end, the mode equation is re-written as

$$F(C) = 0 \qquad (3)$$

where

$$F(C) = R_e(C)e^{-i2khC} - 1$$

Then, according to Newton's method, we start with an approximation $C_n^{(0)}$ to a root C_n and compute successive approximations $C_n^{(1)}$, $C_n^{(2)}$, $\cdots$, where

$$C_n^{(i)} = C_n^{(i-1)} - \frac{F(C_n^{(i)})}{F'(C_n^{(i)})} \quad (j = 1, 2, \cdots) \qquad (4)$$

Under suitable conditions, the sequence $C_n^{(0)}$, $C_n^{(1)}$, $C_n^{(2)}$, $\cdots$ converges to C_n and the iterations are terminated when

$$|F(C_n^{(i)})/F'(C_n^{(i)})|$$

becomes sufficiently small.

The key to successful application of Newton's method to mode equation 3 consists of a starting approximation $C_n^{(0)}$ that results in rapid convergence of the iterations. For the cases considered here, the following procedure was found to work well. First, for a prescribed

wave guide, construct a sequence $\sigma_e^{(1)}$, $\sigma_e^{(2)}$, $\cdots$, $\sigma_e^{(N)}$ of decreasing σ_e values, with $\sigma_e^{(j)} = \sigma_e^{(j-1)}/2$, $\sigma_e^{(1)} > 1$ mho/m, and $\sigma_e^{(N)}$ equal to the prescribed value of σ_e. By using Newton's method, roots of (3) are found for $f = 1$ kHz and $\sigma_e = \sigma_e^{(1)}$; the remaining wave guide parameters have their prescribed values. When $n = 1, 2, 3, \cdots$, the starting approximation for C_n is given by

$$C_n^{(0)} = \frac{n\pi}{kh} \qquad (5)$$

which is simply the solution of (3) for a wave guide with perfectly conducting walls. For $n = 0$, however, (5) gives $C_0^{(0)} = 0$, which will not do as a starting value in Newton's method. To handle the $n = 0$ case, we follow a procedure described elsewhere [*Wait*, 1970] to obtain the starting approximation

$$C_0^{(0)} = \left(\frac{i\Delta}{kh}\right)^{1/2} \qquad (6)$$

where

$$\Delta = \left(\frac{\sigma + iK\epsilon_0\omega}{\sigma_e^{(1)} + iK_c\epsilon_0\omega}\right)^{1/2} \qquad (7)$$

Solutions of (3) corresponding to $\sigma_e^{(2)}$, $\sigma_e^{(3)}$, $\cdots$, $\sigma_e^{(N)}$ are then found in succession; at each step the roots from the preceding step are used in a Newton's method iteration.

With the roots corresponding to $f = 1$ kHz (and prescribed wave guide parameters) on hand, we construct an increasing sequence $f^{(1)}$,

$f^{(2)}, \cdots, f^{(M)}$ of frequency values, with $f^{(1)} = 1$ kHz and $f^{(M)} = 100$ kHz. Finally, solutions corresponding to $f^{(2)}, f^{(3)}, \cdots, f^{(M)}$ are found in succession; again at each step the roots from the preceding step are used in a Newton's method iteration.

A suitable choice for the intermediate frequency values $f^{(2)}, f^{(3)}, \cdots, f^{(M-1)}$ depends on the wave guide parameters and is determined largely by trial and error. A necessary condition implied by 'suitable choice' is that all finer subdivisions of the frequency interval 1–100 kHz should result in the same root for $f = 100$ kHz. Special care must be exercised to avoid pitfalls when computations are performed in the vicinity of 'critical points,' where for cer-

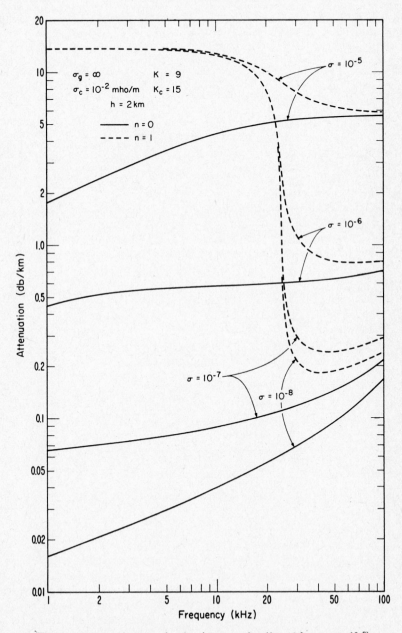

Fig. 2. Attenuation rate for dominant modes ($h = 2$ km, $\sigma_o = 10^{-2}$).

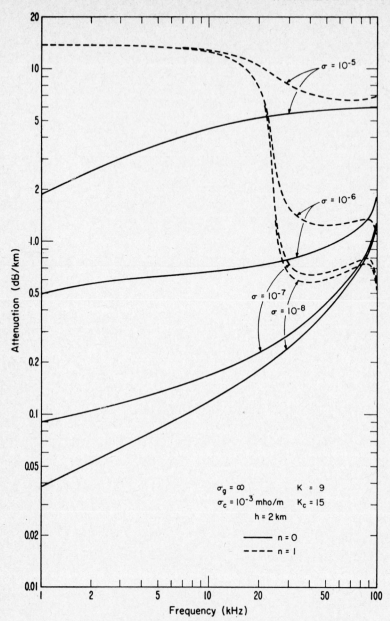

Fig. 3. Attenuation rate for dominant modes ($h = 2$ km, $\sigma_o = 10^{-3}$).

tain values of frequency and wave guide parameters, $F\ (C)$ and $F'(C)$ both vanish. Near such points, the difference $f^{(j+1)} - f^{(j)}$ must be small to ensure convergence and to avoid 'jumping' from one mode to another.

ATTENUATION RATES

As we indicated above, the attenuation of the

wave guide modes is equal to $-\text{Im } kS_n$ for a mode of order n. This quantity, expressed in decibels per kilometer, is plotted as a function of frequency in Figures 2, 3, 4, and 5, for a wave guide with bulk conductivity σ and thickness as indicated. In each case, the dielectric constants relative to free space are $K = 9$ and $K_e = 15$. Only the first two modes, designated $n = 0$ and $n = 1$, are indicated. Here, the zero-order mode is a transverse electromagnetic type (i.e., TEM), and does not suffer any cutoff.

In each of Figures 2, 3, 4, and 5, we note that the attenuation rate becomes progressively less as the conductivity σ is decreased from 10^{-5} to 10^{-8} mho/m. We also observe that the

attenuation rates in Figures 2 and 3 for $h = 2$ km are higher than those in Figures 4 and 5 for $h = 10$ km. In addition, we see that the attenuation rates shown for $\sigma_c = 10^{-2}$ mho/m in Figures 2 and 4 are lower than those for $\sigma_c = 10^{-3}$ mho/m shown in Figures 3 and 5. In general, however, the controlling factor is the conductivity σ of the bulk medium within the

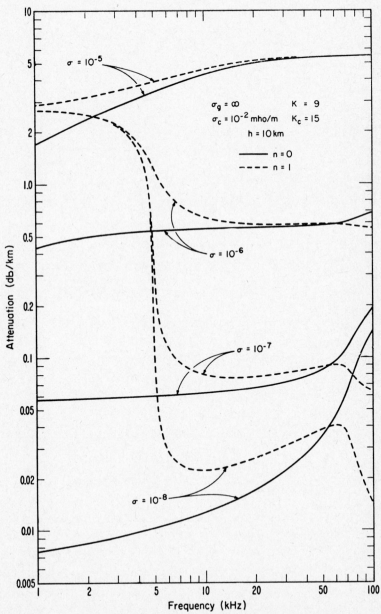

Fig. 4.   Attenuation rate for dominant modes ($h = 10$ km, $\sigma_o = 10^{-2}$).

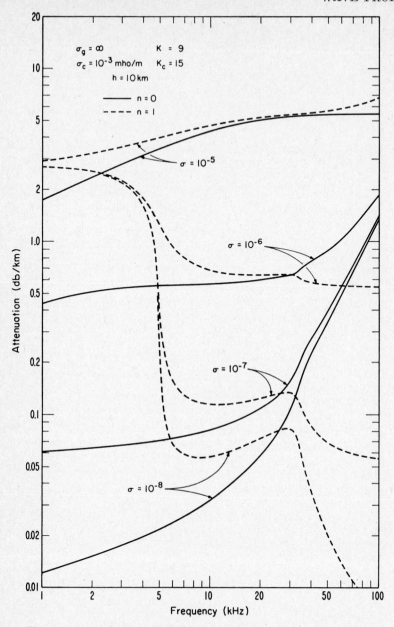

Fig. 5.   Attenuation rate for dominant modes ($h = 10$ km, $\sigma_c = 10^{-3}$).

wave guide. Only for extremely low values of σ (i.e., 10^{-7} or 10^{-8}) does the wall conductivity σ_c have any noticeable effect.

As we clearly see in the figures, the first-order mode is highly attenuated below a certain frequency. In the case of a perfectly conducting wave guide of thickness h with a bulk dielectric constant $K\epsilon_0$, we would expect the cutoff frequency for the first-order mode to correspond to the situation in which twice the wave guide thickness $2h$ is equal to the effective wavelength $\lambda_0/K^{1/2}$ in the dielectric. For $h = 2$ and 10 km, these are readily seen to be 25 and 5 kHz, respectively. These values are consistent with the detailed calculations depicted in Figures 2, 3, 4, and 5. In turn, we can expect the cutoff frequencies for the $n = 2$ mode to be 50 and 10 kHz for $h = 2$ and 10 km, respectively. Because of the various ohmic losses, however, the attenuation rates for the $n = 2$ and higher modes become, in most cases, prohibitive.

CONCLUDING REMARKS

The calculations described in this short paper are a small sample of a large amount of data obtained for this type of wave guide model. We have chosen this particular example because interpretations from certain magnetotelluric studies [*Word et al.*, 1971] have indicated that highly resistive layers may occur in the earth's crust beneath highly conductive strata. If such layers do exist, low attenuation of wave guide modes propagating in the lateral direction can be expected. However, we stress that many other considerations are needed to establish the feasibility of this mechanism of propagation for communication. For example, the excitation of the modes, nonuniformities of the wave guide, and signal-to-noise factors need to be investigated. These questions, particularly those relating to more realistic (and more complicated) propagation models, are receiving our attention at the moment.

Acknowledgment. The research described here has been supported by the Office of Naval Research, under a contract monitored by John G. Heacock, who has stimulated our interest in the subject.

REFERENCES

Wait, J. R., *Electromagnetic Waves in Stratified Media,* 2nd ed., pp. 137–143 and 551–564, Pergamon, New York, 1970.

Wait, J. R., Analytical investigation of electromagnetic wave propagation in the earth's crust, in *The Structure and Physical Properties of the Earth's Crust, Geophys. Monogr. Ser.,* vol. 14, edited by J. G. Heacock, AGU, Washington, D.C., this volume, 1971.

Word, D. R., H. W. Smith, and F. X. Bostick, Crustal investigations by the magnetotelluric tensor impedance method, in *The Structure and Physical Properties of the Earth's Crust, Geophys. Monogr. Ser.,* vol. 14, edited by J. G. Heacock, AGU, Washington, D. C., this volume, 1971.

Radio Propagation through the Crust—Retrospect and Prospect

S. BENEDICT LEVIN

Earth Satellite Corporation, Washington, D. C.   20036

Abstract. The feasibility of achieving long range communication and remote sensing by radio within the depths of the earth's crust depends on the electrical properties of the constituent rocks, the variation of those properties in the vertical (radial) direction, and their electrical homogeneity laterally (tangentially) over great distances. Laboratory measurements of dry specimens of representative crustal rocks lead to the inference that there may exist, at depths below approximately 7 km, a zone whose conductivity at very low frequencies may be less than 10^{-6} mho/m and which may therefore permit radio propagation over distances of many hundreds to a thousand kilometers. Field measurements, however, and laboratory measurements of wet rocks lead to the inference that crustal resistivities do not fall below approximately 10^{-5} mho/m and, hence, that long range radio communication is probably not feasible. The resolution of these apparently conflicting views requires further research in several interdisciplinary fields of the earth- and radio-sciences.

The successful propagation of radio signals laterally over deep crustal paths of the order of 1000 km in length would serve both practical civilian-oriented and defense-related needs; it would also contribute significant new information and an understanding of crustal structure and properties that would be applicable to other socio-technological problems. Prerequisite to such achievement is a knowledge of the spatial variation of the electrical parameters of the lithosphere sufficiently reliable for the design of feasibility experiments in radio propagation in the crust. Although lateral inhomogeneity may prove to be a major consideration, the first concern is with the vertical (radial) variation of electrical properties (the profile) that are responsive to the temperature and pressure gradients in the crust and to the progressive diminution in depth of free water in cracks and pores.

Numerous efforts directed toward estimating the electrical profile of continental crust and the probability of long-range propagation in it have been reported since 1960. Dielectric-profile models have been synthesized from laboratory measurements of the electrical properties of minerals and rocks as functions of frequency, temperature, pressure, and water content, coupled with geologically inferred crustal gradi-ents of temperature T, pressure P, and free H_2O content. Other models have been synthesized from in situ measurements of conductivity by various methods. The profiles thus derived have then been analyzed [*Wheeler*, 1961; *Harmon*, 1961; *Noritomi*, 1961; *Keller*, 1962; *Levin*, 1962; *Wait*, 1963; *Tsao*, 1963; *Watt et al.*, 1963; *Burrows*, 1963; *Brown and Gangi*, 1963; *Keller*, 1963; *Levin*, 1964; *de Bettencourt*, 1964; *Lahman et al.*, 1965; *Brace et al.*, 1965; *Levin*, 1966; *Wait*, 1966a, b; *Acker and Mueller*, 1966; *Vozoff et al.*, 1966; *Orange et al.*, 1966; *Brace and Orange*, 1968a, b; *Schwering et al.*, 1968] for signal attenuation as a function of distance, yielding results that (for 10^3 Hz) span one order of magnitude (roughly from 0.03 to 0.3 db/km) and encourage the hope that radio signal transmission over 1000-km paths may be feasible.

Direct-propagation measurements have also been made, both on scale models simulating crustal dielectric wave guides and in full scale between antennas emplaced in boreholes at different depths. Of particular interest is a propagation experiment conducted, in the period 1962–1966, with antennas in a pair of boreholes located nearly 100 km apart in the eastern United States. The holes, each roughly 3 km deep [*Acker and Mueller*, 1966], penetrated

Precambrian metavolcanics and granite-gneiss in a much-faulted geanticlinal structure. In propagation tests conducted with a 2.5-kw transmitter at a frequency of 2.2 kHz, down-hole measurements of amplitude, phase, wave form, propagation time, and noise (as functions of depth) were attempted. The results were inconclusive; path closure, i.e., receipt of transmitted signal, could not be confirmed in either direction. One critique of this experiment holds that the results cannot be meaningful because the boreholes were much too shallow and did not penetrate to a dry low-conductivity zone but, rather, involved a too-lossy water-soaked medium.

Current interpretations of laboratory measurements on dry rock samples lead to the conclusion that, between an uppermost zone which is highly conductive because it is wet and a

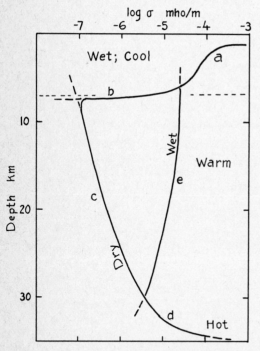

Fig. 1. Very low frequency conductivity profiles of the crust according to two hypotheses. In the figure, ab is the conductivity of superficial, fractured, ground-water saturated rocks; e is the electrolytic conductivity of wet rocks; c is the bulk conductivity of dry rocks at crustal temperatures and pressures; and d is the conductivity due to thermally activated carriers in hot rock. Profiles are abcd according to one interpretation and aed according to another.

lower zone which is highly conductive because it is hot, there may well occur a warm dry zone that, over a 10- to 20-km thickness, has low enough conductivity (σ less than 10^{-6} mho/m), permittivity (ϵ), and loss tangent (δ) to permit propagation of 1-kHz signals with less than 0.3 db/km total attenuation (α). Laboratory measurements on wet rocks, however, under varying uniaxial and hydrostatic pressures, have led to the claim [Orange et al., 1966] that water films in micro-cracks must be present at all depths in the crust and that there can be no such very low conductivity zone and, hence, no such low-loss dielectric wave guide. An opposing view [Levin, 1966] holds that micro-fractures of the type observed in the surface or near-surface rocks may not be present in dense igneous rocks, which at depth have never been mechanically unloaded from lithostatic compression. Such rocks in situ may truly be dry.

The contrasting interpretations with respect to the conductivity profile of the crust can, therefore, be summarized graphically as in Figure 1. One view [Levin, in Schwering et al., 1968] holds that the profile is composed of a wet segment ab, in which conductivity decreases with decrease in free-water content, and a dry segment cd, in which the conductivity of effectively dry crustal rock increases with temperature in keeping with the activation energies of charge carriers. In this view, the warm and dry middle zone may have a substantial thickness of low conductivity $\leq 10^{-6}$ mho/m); this medium could be reached by antennas inserted in deep (10 km or deeper) drill holes. The opposing view [Brace, 1965, 1968a, b, 1971] holds that the more probable profile is composed of wet segments a and e and a hot segment d, that electrolytic conduction controls throughout the crust down to depths where it is swamped by the greater conductivity of thermally activated carriers, and that nowhere does the conductivity fall to significantly less than about 10^{-5} mho/m.

Present uncertainty stems from both insufficient knowledge of the earth's crust and from discordant inferences that are drawn from data that are derived by different investigative techniques or that result from contradictory premises on the structure and composition of the crust. There is, therefore, a need to exploit many relevant scientific approaches in resolving current uncertainties concerning: first, the existence anywhere of a zone dielectrically favorable for telecommunication and second, the probability of the occurrence, within the crust, of such a zone laterally extensive and substantially uniform in its physicial properties.

We do not know whether crustal rocks below the 2-kb lithostatic load depth ($\sim$7km) are electrically dry or wet. We cannot now reliably assess the relative validity of in situ field data interpretations versus those based on laboratory data. We do not know in adequate detail how representative electrical profiles of continental and oceanic crusts are configured. In addition, although much excellent work has been done, we do not know whether a zone of substantial lateral homogeneity in the deep crust (below $\sim$10 km) is probable. Last, there have been no adequately geoscience- planned down-hole propagation experiments over paths longer than a few kilometers.

Some of the questions to which answers are needed are listed below.

1. *Geochemistry.* What is the H_2O content of crustal rocks in the range of P equals 2 to 10 kb and T equals 100° to 600°C? What are the states of association of that H_2O in that PT domain?
2. *Solid state physics.* What are the mechanisms of charge displacement and transport in wet and in dry crustal rock types in the P and T ranges 2 to 10 kb and 100° to 600°C respectively? What are the electrical properties (σ, ϵ, α) of crustal rocks as functions of T, P, H_2O content, frequency f, and rock type?
3. *Experimental-analytical petrology.* What modal compositions and lithologic facies will be encountered in the crust at depths below $\sim$2-kb lithostatic load? What can we reasonably expect to be the vertical and lateral variations of these facies (in continental and oceanic crust)?
4. *Seismology.* What can we reasonably infer about the vertical and lateral variations of the elastic properties or other physical properties within the oceanic and continental crust? Is there any firm evidence of zones that are laterally uniform over great distances?
5. *Geothermal geophysics.* What are the temperatures and gradients in the crust? What can we reasonably infer about the magnitude and extent of lateral variations (anomalies) at any depth levels within continental and oceanic crust?
6. *Tectonophysics and structural geology.* What can we reasonably infer about the structure of the oceanic and continental crusts in terms of vertical and lateral variations of physically identifiable entities, properties, or discontinuities?
7. *Geoelectricity and geomagnetism.* What are

the in situ conductivities and conductivity variations (as revealed by electromagnetic methods) within the crust—vertically (profile), and horizontally? What are the limitations of current methods of analysis and interpretation? Is there evidence of conductivity anomalies? How shall we account for them?
8. *Radio science.* Given various models of the structure, the lithology, the T and P distributions in the crust, and the ranges of conductivity and permittivity of the rocks, what crustal radio propagation modes are conceivable? What would be the attenuation rates, if we assume a laterally uniform propagation medium (but vertical gradation in ϵ and σ)? What would be the magnitude of scatter or multipath problems associated with regions of anomalous ϵ or σ (due to temperature or other variations) along the propagation path?
9. *Radio engineering.* What should be the design of deep (10 km) down-hole antennas (transmit and receive) and front-end systems for crustal radio communications? What should be the frequency, bandwidth, and power? What information-rate capacity would be available? Would an antenna array for beam forming be desirable? Are relay stations required? If they are required, what spacing would be optimum? How can we ensure down-hole isolation from surface (atmospheric and cultural) EM noise? What should be the design of crustal-propagation experiments for the study of signal amplitude, phase, and wave form and of propagation time as a function of depth and of frequency?
10. *Areal geology.* On the basis of desired favorable geologic setting, characterized by others, where (geographically and geologically) should we locate a pair of deep (10-km) holes for crustal research *and* radio-propagation experiments?
11. *Drilling engineering.* Can we drill, core, and log to 10-km depths in 'crystalline' basement? What will it cost? What should be the drilling strategy and plan?

Resolution of the more central of these present uncertainties would clearly constitute a major challenge to the capability for interdisciplinary research that has been developed and so effectively applied during the past three decades. It would also open the way to some technological developments of no mean significance, especially if coupled with some well planned field experiments.

REFERENCES

Acker, M., and L. J. Mueller, Some measured electrical characteristics of the earth's crust, *AGARD Conf. Proc., 20,* 211–238, 1966.
Brace, W. F., Resistivity of saturated crustal rocks to 40 km based on laboratory measurements, in *The Structure and Physical Properties of the Earth's Crust, Geophys. Monogr. Ser.,*

vol. 14, edited by J. G. Heacock, AGU, Washington, D. C., this volume, 1971.

Brace, W. F., and A. S. Orange, Electrical resistivity changes in saturated rocks during fracture and frictional sliding, *J. Geophys. Res.*, 73(4), 1433–1445, 1968a.

Brace, W. F., and A. S. Orange, Further studies of the effect of pressure on electrical resistivity of rocks, *J. Geophys. Res.*, 73(16), 5407–5420, 1968b.

Brace, W. F., A. S. Orange, and T. R. Madden, The effect of pressure on the electrical resistivity of water-saturated crystalline rocks, *J. Geophys. Res.*, 70(22), 5669, 1965.

Brown, G. L., and A. F. Gangi, Electromagnetic modeling studies of lithospheric propagation, *IEEE Trans. Geosci. Elec.*, 1, Dec., 17–23, 1963.

Burrows, C. R., Radio communication within the earth's crust, *IEEE Trans. Antennas Propagat.*, 11, May, 311–317, 1963.

de Bettencourt, J. T., Deep rock strata communication, *Elec. Progr.*, Winter, 4–10, 1964.

Harmon, G. J., Radio wave propagation through the earth's rock strata—A new medium of communication, *IRE Globecom Rec.*, 31–34, 1961.

Keller, G. V., Electrical properties of the earth's crust, Part 4 of U. S. Geological Survey report, 1962.

Keller, G. V., Electrical properties in the deep crust, *IEEE Trans. Antennas Propagat.*, 11(3), 344–357, 1963.

Lahman, H. S., A. Orange, and K. Vozoff, Deep resistivity investigations in the continental United States, *Final Rep. AF66-28*, Air Force Cambridge Research Lab., Cambridge, Mass., 1965.

Levin, S. B., Geophysical factors in electromagnetic propagation through the lithosphere, *IRE NEREM Rec.*, 38, 1962.

Levin, S. B., Geophysical factors in electromagnetic propagation through the lithosphere, *J. Geomagn., Geoelect.*, 15(4), 293, 1964.

Levin, S. B., Lithospheric radio propagation, *AGARD Conf. Proc.*, 20, 147–178, 1966.

Noritomi, K., The electrical conductivity of rock and the determination of the electrical conductivity of the earth's interior, *J. Mining Coll., Akita Univ. A*, 27–59, 1961.

Orange, A. S., W. F. Brace, and T. R. Madden, The electrical resistivity of water saturated crystalline rocks, *AGARD Conf. Proc.*, 20, 265–285, 1966.

Schwering, F. K, D. W. Peterson, and S. B. Levin, A model for electromagnetic propagation in the lithosphere, *Proc. IEEE*, 56(5), 799–804, 1968.

Tsao, C. K. H., Investigation of electrical characteristics of rock medium on Cape Cod., *Sci. Rep. 1, AF64-1*, Air Force Cambridge Research Lab., Cambridge, Mass., 1963.

Vozoff, K., T. Cantwell, H. Lahman, and A. Orange, Results of in situ rock resistivity measurements, *AGARD Conf. Proc.*, 20, 287–307, 1966.

Wait, J. R., The possibility of guided electromagnetic waves in the earth's crust, *IEEE Trans. Antennas Propagat.*, 11, May, 330–335, 1963.

Wait, J. R., Electromagnetic propagation in an idealized earth crust waveguide, *AGARD Conf. Proc.*, 20, 115–132, 1966a.

Wait, J. R., Influence of a sub-surface insulating layer on electromagnetic ground wave propagation, *AGARD Conf. Proc.*, 20, 133–146, 1966b.

Watt, A. D., F. S. Mathews, and E. L. Maxwell, Some electrical properties of the earth's crust, Proc. IEEE, 51(6), 897, 1963.

Wheeler, H. A., Radio wave propagation in the earth's crust, *Radio Sci.*, 65D(2), 189–191, 1961.

DISCUSSION

Wait: It is probable that the observed frequency dependence of the resistivity of dry granite down to 100 Hz is a spurious effect due to electrode contacts. Two electrode measurements are very susceptible to this added impedance in series, and some of the four electrode measurements do not show this effect.

Meyer: You said you measured a micro mho/m at the bottom of these holes and that there was still water there. Is that correct?

Levin: Down-hole (in situ) conductivities less than 1 micro mho/m were reported from the deeper portions of one of the holes in granite, based on 3-electrode resistivity measurements.

Laboratory measurements at 1 kHz, on a granite core sample from a 2.7 km depth, indicated conductivities of 2×10^{-7} mho/m at 25°C in the wet (as-received) state and, after drying, of less than 10^{-8} at 25°C and of less than 10^{-6} at 350°C. However, a rock which has never been unloaded to confining pressures less than 2 kb may be pretty firm, and so the laboratory measurements on core or outcrop samples may not be representative of the deep in situ conductivity, which may be significantly lower. This is an important point which I hope we may discuss later.

References are indexed for the pages on which an author's work is cited. References at the end of articles are not indexed if the author was cited in the text. Small **boldface** numbers indicate inclusive pages of articles in this monograph. *Italics* indicate 'et al.' citation.